Matthew Bolton has worked for seven years on issues of humanitarianism in conflict, as an aid worker, freelance writer and academic. His career has taken him to many countries including Iraq, Afghanistan, Bosnia and Sudan. He has a PhD in Government from the London School of Economics and is the author of *Apostle of the Poor: The Life and Work of Missionary and Humanitarian Charles D. Neff*.

INTERNATIONAL LIBRARY OF POST-WAR RECONSTRUCTION AND DEVELOPMENT

Series ISBN: 978 1 84885 216 7

See www.ibtauris.com/ILPRD for a full list of titles

Series Editor: Sultan Barakat, Director,
Post-war Reconstruction and Development Unit,
The University of York, UK

Though conflict has been the focus of much academic attention, the processes of recovering from war and conflict have been little studied. Confusion still exists as to whether post-war reconstruction is concerned with relief or development, with physical rebuilding, economic recovery, social reintegration or political reconstruction. The result is an all too frequent fragmentation and waste of effort on the ground.

An understanding of the need to plan and integrate the many different activities for reconstruction and recovery within a shared vision is therefore crucial. The *International Library of Post-war Reconstruction and Development* will set out a conceptual and strategic framework for post-war reconstruction practice, at the same time examining and illustrating specific aspects of practice for those working in the field or training to do so. The series will also act as a focus for a continuing dialogue between academics and practitioners at the forefront of developing the discipline.

Published and forthcoming titles in the series:

1. *Disaster Management and Civil Society: Earthquake Relief in Japan, Turkey and India*
Alpaslan Özerdem and
Tim Jacoby
978 1 84511 053 6

2. *Violence and Post-war Reconstruction: Managing Insecurity in the Aftermath of Peace Accords*
Christina Steenkamp
978 1 84511 769 6

3. *Post-war Recovery: Disarmament, Demobilization and Reintegration*
Alpaslan Özerdem
978 1 84511 461 9

4. *Palestinians in the Lebanon: Long-term Displacement and Refugee Coping Mechanisms*
Rebecca Roberts
978 1 84511 971 3

5. *Ethnicity and Conflict: Cultural Identity and Post-war Reconstruction*
Rob Aitken
978 1 84511 463 3

6. *Policy Transfer in Post-war Reconstruction*
Sultan Barakat and Mark Evans
978 1 84885 073 6

7. *Foreign Aid and Landmine Clearance: Governance, Politics and Security in Afghanistan, Bosnia and Sudan*
Matthew Bolton
978 1 84885 160 3

FOREIGN AID AND LANDMINE CLEARANCE
Governance, Politics and Security
in Afghanistan, Bosnia and Sudan

MATTHEW BOLTON

Published in 2010 by I.B.Tauris & Co Ltd
6 Salem Road, London W2 4BU
175 Fifth Avenue, New York NY 10010
www.ibtauris.com

Distributed in the United States and Canada Exclusively by Palgrave Macmillan
175 Fifth Avenue, New York NY 10010

Copyright © 2010 Matthew Bolton

The right of Matthew Bolton to be identified as the author of this work has been asserted by the author in accordance with the Copyright, Designs and Patent Act 1988.

All rights reserved. Except for brief quotations in a review, this book, or any part thereof, may not be reproduced, stored in or introduced into a retrieval system, or transmitted, in any form or by any means, electronic, mechanical, photocopying, recording or otherwise, without the prior written permission of the publisher.

International Library of Postwar Reconstruction and Development 7

ISBN 978 1 84885 160 3

A full CIP record for this book is available from the British Library
A full CIP record for this book is available from the Library of Congress

Library of Congress catalog card: available

Printed and bound in India by Thomson Press India Ltd
Camera-ready copy edited and supplied by the author

*To the deminers,
who risk life and limb,
so that others might live.*

'...*if any one saved a life,
it would be as if he saved
the life of the whole people...*'

– Al-Qu'ran 5:32

CONTENTS

Abbreviations	ix
Acknowledgements	xiii
Introduction	1
1. A Political History of Mine Action	12
2. The New Complexes Governing Insecurity	34
3. Donor Policymaking in the US and Norway	61
4. Implementation in Afghanistan, Bosnia and Sudan	90
5. Comparing the Performance of Tenders and Grants	117
6. Impact on Peacebuilding	146
Conclusion and Reflections	173
Photographs	188
Notes	194
Select Bibliography	236
Interviews	249
Index	257

ABBREVIATIONS

9/11	September 11, 2001
ACBL	Afghan Campaign to Ban Landmines
ADG	Afghan Demining Group
AFRICOM	US Africa Command
AFSC	American Friends Service Committee
AMAC	Area Mine Action Center (Afghanistan)
ANC	African National Congress (South Africa)
ATC	Afghan Technical Consultants
BAC	Battle Area Clearance
BHMAC	Bosnia and Herzegovina Mine Action Center
CBHA	USAID Cross-Border Humanitarian Assistance Program for Afghanistan
CCW	Convention on Conventional Weapons
CIA	Central Intelligence Agency (USA)
CMAC	Cambodian Mine Action Center
CPA	Comprehensive Peace Agreement (Sudan)
DAFA	Demining Agency for Afghanistan
DCA	DanChurchAid (Denmark)
DDG	Danish Demining Group
DMZ	Demilitarized Zone
DoD	US Department of Defense
DPKO	UN Department of Peacekeeping Operations
DSCA	Defense Security Cooperation Agency (USA)
DSL	Defense Systems Limited
EMAC	Entity Mine Action Center (Bosnia)
EOD	Explosive Ordnance Disposal
EODT	EOD Technology, Inc.

FAFO	Institute of Applied Social Science (Norway)
FBI	Federal Bureau of Investigation (USA)
FRELIMO	Mozambican Liberation Front
FSD	Swiss Foundation for Mine Action
G4S	Group 4 Securicor
GICHD	Geneva International Center for Humanitarian Demining
GNI	Gross National Income
HALO	Hazardous Area Life-support Organisation Trust
HID	Hemayatbrothers International Demining
HIV/AIDS	Human Immunodeficiency Virus/Acquired Immunodeficiency Syndrome
ICBL	International Campaign to Ban Landmines
ICG	International Crisis Group
ICRC	International Committee of the Red Cross
ICTY	International Criminal Tribunal for the Former Yugoslavia
IED	Improvised Explosive Device
IMS	Intelligent Munitions System
ITF	International Trust Fund for Demining and Mine Victim Assistance
IMAS	International Mine Action Standards
IMSMA	Information Management System for Mine Action
ISAF	International Security Assistance Force (Afghanistan)
ISI	Inter-Service Intelligence (Pakistan)
JASMAR	Sudanese Association for Combating Landmines
JIDU	Joint Integrated Demining Unit (Sudan)
JMC	Joint Military Commission (Sudan)
LIS	Landmine Impact Survey
LRA	Lord's Resistance Army (Uganda/Sudan)
LSE	London School of Economics and Political Science
MAG	Mines Advisory Group
MAPA	Mine Action Program for Afghanistan
MCC	Mennonite Central Committee
MCPA	Mine Clearance and Planning Agency (Afghanistan)
MDC	Mine Dog Center (Afghanistan)
META	Monitoring Evaluation and Training Agency (Afghanistan)
MFA	Norwegian Ministry of Foreign Affairs
MPRA	Mine Protection and Removal Agency (Bosnia)
MPRI	Military and Professional Resources, Inc.
NADR	Nonproliferation, Antiterrorism, Demining and Related Projects (USA)
NATO	North Atlantic Treaty Organization
NGO	Non-Governmental Organization
NMAA	National Mine Action Authority (Sudan)
NPA	Norwegian People's Aid
NVESD	US Army Night Vision and Electronic Sensors Directorate
OCHA	UN Office for the Coordination of Humanitarian Affairs
OECD	Organization for Economic Cooperation and Development
OHR	Office of the High Representative (Bosnia)
OLS	Operation Lifeline Sudan
OMAR	Organization for Mine Clearance and Afghan Rehabilitation

OSIL	Operation Save Innocent Lives (Sudan)
PIU	World Bank Project Implementation Unit (Bosnia)
PLO	Palestinian Liberation Organization
POW	Prisoner of War
PPE	Personal Protective Equipment
PRIO	Peace Research Institute, Oslo
PYSOPS	Psychological Operations
RONCO	RONCO Consulting Corporation
SAC	Survey Action Center
SCBL	Sudan Campaign to Ban Landmines
SFOR	NATO Stabilization Force in Bosnia
SIMAS	Sudan Integrated Mine Action Service
SLIRI	Sudan Landmine Information and Response Initiative
SLR	Sudan Landmine Response
SOP	Standing Operating Procedures
SPLA	Sudan People's Liberation Army
SSDC	South Sudan Demining Commission
UNDP	UN Development Programme
UNEP	UN Environment Programme
UNHCR	UN High Commissioner for Refugees
UNICEF	UN Children's Fund
UNMACA	UN Mine Action Center for Afghanistan
UNMAC	UN Mine Action Center (Bosnia)
UNMAO	UN Mine Action Office (Sudan)
UNMAS	UN Mine Action Service
UNMIS	UN Mission in Sudan
UNOMOZ	UN Office in Mozambique
UNOPS	UN Office for Project Services
UNOCHA	UN Office for the Coordination of Humanitarian Affairs
UNTAC	UN Transitional Administration in Cambodia
USAID	US Agency for International Development
USSR	Union of Soviet Socialist Republics
UXB	UXB International
UXO	Unexploded ordnance
VVAF	Vietnam Veterans of America Foundation
WFP	UN World Food Program
WHO	World Health Organization
WRA	US State Department Office for Weapons Removal and Abatement
WSI	Wackenhut Services Incorporated
WWI	World War I
WWII	World War II

ACKNOWLEDGEMENTS

While I must claim as my own any shadows of error in this work, any light it sheds is fuelled by others who have illuminated me or brightened my days. I wish to thank the academics whose critiques and encouragement helped shape this into a more coherent and theoretically informed work. Mary Kaldor's philosophical vision and Vesna Bojicic-Dzelilovic's cautious empiricism stretched me to think with a rigor, care and depth I would not have accomplished on my own. The comments and suggestions from David Keen, Kristian Berg Harpviken and Sultan Barakat were also erudite and crucial. I benefited immensely from the informal advice of academically-minded mine action professionals who critiqued my work, notably Simon Conway, Bob Eaton, Tim Lardner, Richard Moyes and Ted Paterson. Moreover, without the linguistic and cultural knowledge of my Bosnian research assistant, Ajla Silajdzic, I would have been lost. I also wish to express my gratitude to Joanna Godfrey, the ever-efficient editor of Tauris Academic Studies.

Of course, this work would have been profoundly lacking without the rich seam of information given to me by my interviewees and informants who were generous with their time, data and wisdom. It seems wrong to single out particular people, given that so many helped me so much, but it would be equally unjust not to mention by name those who went out of their way to meet with me multiple times, offer logistical assistance or engage with me critically. I thus want to thank personally Damir Atikovic, Per Breivik, Liza de Benedetti, John Flanagan, Zoran Grujic, Shohab Hakimi, Darvin Lisica, Fred Maio, Reuben McCarthy, Zlatan Music, Per Nergaard, Nathalie Prevost, David Rowe, Mohammed Sediq, John Stevens, Colin Wanley, Evy Van Weezendonk and Guy Willoughby.

In addition to people, I owe a great deal to several institutions. The Centre for the Study of Global Governance at the London School of Economics (LSE) was a gracious host. The research would not have been possible without grant and in-kind support from the Economic and Social Research Council, Pro Victimis Foundation, Counterpart International and Landmine Action. Moreover, while my relationship with them was at times critical and even strained, I must thank the public institutions involved in mine action in the countries I studied, including the US State Department's Office for Weapons Removal and Abatement, Norwegian Ministry of Foreign Affairs, UN Mine Action Service, UN Development Programme, UN Children's Fund, UN Mine Action Center for Afghanistan, Bosnia and Herzegovina Mine Action Center, UN Mine Action Office in Sudan, South Sudanese Demining Commission and Sudanese National Mine Action Office. They provided me with data, logistical assistance and interviews even though it was an institutional risk for them to do so.

My research did not occur in a social vacuum. I thank my friends and fellow scholars, including Alex Jeffrey, Hugh Griffiths, Phil Coticelli and my colleagues in the LSE Government Department room H421. My family also provided a great deal of unpaid editorial assistance for which I am eternally grateful. Finally, I must reserve my greatest gratitude for my wife, Emily Welty, who has supported and cared for me throughout.

INTRODUCTION

*'We need to shift more attention from
government to governance.'*
– William W. Boyer[1]

No one should have to feel their next step might be their last. Yet over 5,000 people a year fall victim to landmines and unexploded shells lying scattered across the world's current and former war zones.[2] Landmines know no ideology, no ethnicity; they know no difference between soldier and civilian; they do not respect peace agreements and can kill six decades or more after former enemies have shaken hands and laid aside their weapons. Responding to this security threat, international donors, NGOs and commercial companies have developed a new aid sector called 'mine action,' mitigating the impact of landmines and UXO through clearance, education, survivor assistance, stockpile destruction and political advocacy.[3] Since the signing of the Antipersonnel Mine Ban Treaty in 1997, governments have spent over $3 billion on mine action, which is now considered a major component of international post-conflict reconstruction efforts, contributing to the creation of a secure environment, assisting in refugee return, opening access to roads for commerce and aid, rehabilitating agricultural land and providing employment for demobilized soldiers. This book focuses in particular on foreign aid programs funding the clearance of landmines and UXO.

In his Oscar-winning existential drama, *No Man's Land*, set in the Bosnian war,[4] director Danis Tanovic depicted two opposing soldiers, trapped in a trench together. They are surrounded by minefields and threatened by a third soldier, who has woken up to find himself lying on top of a landmine. If he moves, all three of them will be killed. Into the mix, Tanovic throws Sgt. Marchand, a French soldier with the UN peacekeeping force, whose moral outrage at the war motivates him to try to save the three stranded soldiers. Though Marchand tries singlehandedly to create a moral space in the normlessness of no man's land, by the end of the film, the results of his efforts are ambiguous at best. He is stymied by an uncaring and incompetent UN bureaucracy, pestered by a fickle and sensationalist news media and confronted by the ungrateful and mutually hostile beneficiaries of his efforts. Ultimately his labors seem for naught, crushed and poisoned by the overwhelming power of the system of conflict in which he is trapped. Sgt. Marchand's dilemma is but a small indication of what has been a brutal reality of modern warfare. Long after guns fall silent, landmines, cluster bombs and unspent shells block access to farmland, prevent refugee return and maintain a constant psycho-social reminder of the violence of war.

As Sgt. Marchand's seemingly futile efforts dramatized, removing and neutralizing such dangerous devices as landmines is no easy task. Like Sgt. Marchand, deminers operate in regions where no single entity is clearly in charge. To gain access to minefields, obtain important information about patterns of mining, hire a workforce or purchase supplies, demining agencies must negotiate between a myriad of powerful actors: military factions, international organizations, shadowy underworld structures, local political machines, foreign embassies and aid providers. Many competing motivations, interests and conceptions of security are constantly interacting, competing and collaborating with each other and impacting the manner in which mine clearance work is done. Demining is not simply a technical matter, it is also political one, for mines and unexploded ordnance pose a violent threat of bodily harm – even death – to people in their vicinity. Demining is thus ultimately an act of governance – the removal of a violent threat to life. Who controls this power to defuse such a threat is a political question. Political and economic interests and the manner in which states, individuals and other powerful entities conceive security all influence the way in which mine action is organized and implemented.

Thus, while this book is about the political economy of foreign aid programs that clear landmines and other explosive remnants of war, it is ultimately about the ways in which institutions deal with the problem of security – the management, reduction, mitigation or elimination of risk, particularly the risk of violent harm. Observing demining programs allows one to study how norms, interests and the multiple shifting layers of governance can shape foreign aid and security provision in a post-conflict zone. This book contributes to the literatures on security, governance and the political economy of aid in conflict. Unlike more established foreign aid sectors, like health, food aid or community development, there is not a long pedigree of academic research into mine action. Research on demining from a social scientific, rather than a technocratic or campaigning point of view, is in its nascent stages. Part of the purpose of this study is to address this gap by exploring how the politics of the demining can be understood through the lens of social science, primarily political science. This book will draw on and contribute to academic debates on post-statist governance, realism versus idealism, peace and security, privatization and contracting, the political economy of war, and foreign aid.

Landmines, UXO and Demining

Landmines are explosive traps – devices that through mechanical, electrical or chemical fuzes are detonated when initiated by their victim. Once laid, they wait until either they claim a victim or are cleared by deminers. Military forces have generally used mines for defensive purposes, multiplying the impact of their forces by protecting a strategically important location or 'shaping terrain' by channeling enemy forces in certain directions. However, they have also been used offensively, particularly by irregular and guerilla forces, to ambush vehicles, intimidate civilians and penetrate into areas where it would be impossible to leave troops for an extended period of time.

Mines come in a variety of forms. Antipersonnel landmines are small, detonated with only a few kilograms of pressure and intended to harm individual people. They are usually designed to injure rather than kill, aiming to distract enemy resources away from fighting and towards caring for the casualty. Antipersonnel mines can cause horrific injuries. The blast can blow off a foot or hand, force dirt and debris far into the wound (causing terrible infections) and generate shockwaves in the flesh that damage other parts of the body. Fragmentation mines are a kind of antipersonnel mine intended to injure or kill other people in addition to the victims who encounter them, by spraying shrapnel or ball bearings. They are often placed above the ground on a stake, to maximize the radius of damage. Bounding fragmentation mines, like the WWII-era 'Bouncing Betties' or the Yugoslav PROM-1, have an initial detonation that lifts them into the air, before a second explosion spews fragments over 100 meters away. Directional fragmentation mines, like the US Claymore mine, spray shrapnel in a specific, predetermined direction. They are often connected to trip wires to increase the chances of detonation. Antivehicle mines, sometimes referred to as antitank mines, contain considerably more explosive material than antipersonnel mines, take more pressure to detonate and are intended to disable or destroy vehicles. They are often laid on roads to ambush convoys or prevent military traffic. If a human is unfortunate enough to initiate an antivehicle mine, it usually kills them. Some technologically sophisticated mines have the option of being 'command-detonated' by a person who lies in wait, using a fuse, switch or remote control. Such techniques have become common with the use of 'Improvised Explosive Devices' (IEDs) – homemade mines – in the Iraq and Afghanistan conflicts. Local factions in these conflicts have innovated methods of using mobile phones and walkie-talkies to remotely detonate roadside bombs. As mass-produced mines have become stigmatized in the international arena, irregular forces have increasingly turned to improvising their own devices.

Mines can be laid by hand, distributed by vehicle or scattered from the air. Regular armies, particularly when laying defensive minefields, have tended to lay mines in regular patterns, both to ensure the efficient coverage of an area, and facilitate clearance by their own troops (who know the patterns). Of course, when scattering mines from aircraft, no such precision is possible. Irregular forces have tended to use mines in a less 'linear' fashion, partly because of their lack of training, but also because they are more likely to use mines as an offensive weapon – placed in the path of military convoy or on the doorstep of 'unwanted' civilian populations.

Victim-activated landmines first emerged during the US Civil War, but came into their own as a weapon in the WWII desert tank battles in North Africa. During the Vietnam War, the US innovated methods for scattering thousands of mines from the air, techniques which were mirrored by the USSR during its war in Afghanistan. In both Vietnam and Afghanistan, local resistance fighters developed methods of using mines as offensive weapons of ambush, which were also seen in the African wars of decolonization. The 1980s and early 1990s saw unprecedentedly widespread use of mines in the Reagan-era proxy conflicts and post-Cold War 'New Wars.' This created a crisis of contamination that prompted the campaign to ban landmines, culminating in the Ottawa Convention, a total ban on the stockpiling, trade, transfer and use of antipersonnel landmines in 1997.[5]

In contaminating the land of conflict zones, mines are joined by the 30% of modern munitions that fail to detonate upon impact with their targets, leaving highly unstable explosives in the ground decades into the future. Such UXO, while sometimes easier to clear than landmines, create *de facto* minefields – as they can detonate when disturbed, tampered with or touched.[6] Cluster munitions, small bomblets that disperse over a wide area from a single bomb or rocket, are particularly prone to becoming UXO and, as a result, have also been banned by the around 100 countries that signed the Convention on Cluster Munitions in 2008.[7]

Over 75 countries are impacted by mines and UXO. A reliable quantification of contamination levels around the world is not available. However, in 2007, demining agencies cleared some 534 square kilometers of mine and UXO-contaminated land and the *Landmine Monitor* publication estimated that 'thousands' of square kilometers remained.[8] There were at least 5,426 casualties of mines and UXO (1,401 killed, 3,939 injured and 86 whose status was unknown) in 2007. Just three countries – Afghanistan, Cambodia and Colombia – accounted for 38% of the casualties.[9]

Methods of clearing landmines and UXO have changed surprisingly little since their initial development during and after WWII. The development of 'humanitarian demining' in the 1990s has made changes in the process and standards but few major differences in techniques. Humanitarian demining has different purposes and standards to military demining, also called 'counter-mining.' When faced with an enemy minefield in the heat of battle, the top priority of a military commander is to get through it as quickly as possible while minimizing casualties. In the aftermath of conflict, such emphasis on speed over safety is not acceptable when returning land to civilian use. Recognizing the need for higher standards of safety, the various nonprofits organizations, UN agencies and commercial companies that developed 'humanitarian demining' have emphasized the need for checking and clearing every square meter of suspected hazardous area and attempting to achieve near 100% clearance.

The popular media has a tendency to fixate on high-tech and bizarre demining innovations – everything from rats, bees and remote-control robots to air-balloons, radars and genetically modified mustard seeds. However, given that importing sensitive, high-tech machines and organisms into mine-affected countries is often prohibitively expensive, there are few 'quick fixes' to the world's mine problem. The vast majority of demining programs use a combination of relatively technologically simple techniques. The most common is a human deminer, armed only with a prodder and/or trowel, slowly prodding and excavating the ground along a predetermined 'lane', carefully ensuring that s/he approaches mines from an angle that will not detonate them. When s/he comes upon a mine, it is either removed, defused or destroyed *in-situ*. When deminers know that the mines used in a particular area have metal parts, and the surrounding earth is not too metallic, they may also use metal detectors to facilitate their work. Working in combination, along lanes well-spaced from each other, the deminers eventually check every single square meter of the minefield. Demining agencies also often train dogs to sniff out and locate explosives, though within the sector there is some debate about how best to use them. The dogs are rarely heavy enough to initiate mines and so can sit on or near the mine to show their handler where to excavate. Inventors have developed a variety of machines to assist in the demining process, clearing vegetation, flailing the ground to explode mines or using radar to aid detection. But, as yet, no machines have been able to match the accuracy and care of a good human deminer and so mechanical clearance methods are usually combined with other techniques.[10]

Demining and Governance Complexes

After WWII, millions of mines and tonnes of UXO contaminated continental Europe. Just as the European states had mobilized massive resources to prosecute the war, the post-war mine and UXO clearance effort was government-led and funded. Like the war, post-WWII demining was very much a nation state and military affair. In contrast, today's demining programs mirror the new forms of globalized public-private partnerships arising to deal with the rise of transnational sources of insecurity in the conflicted 'frontiers' of the international system. For example, while Bosnia does have its own military clearance program, much of its capital Sarajevo was demined by an international charity, largely funded by the Norwegian government. Likewise, in Afghanistan, the US has contracted private security companies to clear NATO military bases. Demining, like many other government services, has become globalized and privatized.

Mine clearance programs no longer mirror the Weberian hierarchical and bureaucratic structures of command and control of WWII European armies. They are made up of a myriad of competing, colluding and collaborating entities, including international and local nonprofit organizations, commercial companies, bilateral donors, local government authorities, UN agencies and military alliances. Demining programs echo security and development scholar Mark Duffield's description of 'strategic networks and complexes' that have replaced the traditional idea of 'security through government' with security through 'polyarchical, non-territorial and networked relations of governance.'[11] Duffield has a tendency to conflate all these complexes into one 'emerging system of global liberal governance.'[12] However, this book will show that there are multiple types of the networks that can operate in very different ways. The differing constituent members and institutional structures of a network shape the approach it takes to mines and UXO. Different ways of organizing networks produce different outcomes, both in terms of demining performance and the impact on the peacebuilding and reconstruction process. While there is some overlap, these 'demining complexes' can be divided into at least two broad ideal types (or perhaps two poles on a continuum):

1. **Strategic-Commercial Complexes** are shaped largely by the interests of a privileged few, in which militarized and securitized public bodies, often of the great powers, contract out significant authority to commercial companies. Within the mine action field, one finds that great power states try to limit the regulation of mines and other weapons and contract out clearance through private security companies, prioritizing military or strategic objectives. Such networks tend to produce a low cost and rapid demining process but sacrifice quality and safety. They are also more likely to compromise with the political economy of conflict and contribute to the privatization of the use of force.

2. **Human Security-Civil Society Complexes** are shaped by humanitarian norms and a more global understanding of interest, in which middle power states and multilateral agencies form partnerships with NGOs and social movements. They aim to provide protection to the general population, especially the vulnerable, through aid, advocacy, persuasion and the legal process. Within the mine action field, one finds that middle power states, in coalition with NGOs and social movements, try to heavily regulate the use of mines and prioritize humanitarian need in clearance programs. Their demining programs are often slower and more costly, but value high levels of quality and safety. They are also often

more inclusive organizations, trying to build local capacity and advocate for limits on the politics of violence.

This book will show the development, nature, organization and effects of the above two approaches in managing the threat of landmines and unexploded ordnance in conflict zones. To do this it will look at two case studies of donor countries – the US and Norway – and observe their funding of demining programs in three mine and UXO-affected countries: Afghanistan, Bosnia and Herzegovina (hereafter, Bosnia) and Sudan.

Case Studies

The USA and Norway were selected as donor country case studies because they were the top two bilateral donors in absolute terms (though Norway gives more relative to its GDP). Additional characteristics that make them attractive for comparison include the fact that they are both wealthy industrial democracies that won independence from colonization, and both have significant histories of foreign aid provision. The differences in their mine action policies reflect the findings of Jan Egeland in his 1988 comparison of US and Norwegian human rights policy.[13] Egeland found that as a superpower, US foreign policy was constricted by numerous strategic and commercial interests that hijacked its ability to shape policies according to humanistic ideals. By contrast, Norway had fewer transnational interests and more space to pursue a normative foreign policy.

Similarly, the US has resisted tight regulation of mines, cluster munitions and other explosive remnants of war. The macro structure of US funding of clearance and mitigation of explosive remnants of war was shaped largely by its strategic interests and favored a commercially-driven process. In contrast, working with NGOs, churches and other small states, Norway has been at the forefront of efforts to ban landmines and cluster munitions. Its mine action programs, implemented through international NGOs, were shaped by a more global conception of interest and normative commitments to humanitarianism, multilateralism and international law.

At the level of implementation in Afghanistan, Bosnia and Sudan, Norwegian long-term grants to international NGOs produced demining that, while sometimes more expensive and slower, was better targeted at humanitarian priorities, safer and of better quality. Such programs also attempted to build inclusive institutions and resist the politics of violence. In contrast, US efforts, shaped by strategic concerns and often tendered out to commercial companies, were frequently cheaper and faster but also less safe and of lower quality. These companies were also embedded in the political economy of war and may have contributed to the fragmentation of the public monopoly on force.

The three implementation countries – Afghanistan, Bosnia and Sudan – were chosen carefully to try to avoid case selection bias, while ensuring that the cases would have enough material to make for an interesting study. The countries had to have large mine action programs and be among the top recipients of US and Norwegian mine action funding (to ensure there would be enough activities to study and that the donor would have a coherent funding policy). To guarantee some variation between them and see if similar trends could be spotted in diverse circumstances, they had to be located in very different regions, with different political contexts, economic situations, climate and soil conditions (important in demining performance). To facilitate historical comparisons, the countries selected had mine action programs that started in different time periods (Afghanistan is among the oldest programs, Sudan among the newest). Finally, a variety of logistical and security constraints ruled out a variety of cases (like Angola, Iraq and Lebanon) that would have satisfied the above criteria.

Afghanistan was chosen as a case study because it is one of the oldest and largest mine action programs and has one of the highest levels of mine and UXO contamination in the world. Researching Afghan mine action enabled the exploration of the genesis of the sector and tracing the roots of trends that appeared in other countries. Afghanistan also offered potential for tracing the varying impact of humanitarian strategic interest. Afghan demining began at the end of the Cold War, in the context of the massive humanitarian and covert efforts in the Soviet-Afghan War. Three different models of demining emerged at that time: a) commercialized and securitized demining in support of US-backed paramilitary efforts, b) international NGO demining claiming humanitarian neutrality and c) UN coordinated mine action, implemented by local NGOs and also claiming neutrality. As Afghanistan dropped off the global radar from 1992 to 2001, the UN-led model was ascendant and was supported by both the US and Norway. However, US re-engagement in Afghan politics after 9/11 saw it return to a 'securitized' and commercialized model of demining. In contrast, objecting to the massive top-heavy growth of the UN program, Norway switched to funding an Afghan-led international NGO, the HALO Trust, in order to support an Afghan-led civil-society actor espousing traditional humanitarian values.

Following the 1995 endgame of Bosnia's war, the frontlines dividing Bosnian Government territory from its separatist statelets – the 'Republika Srpska', 'Herzeg-Bosna' and the 'Bihac pocket' – were littered with extremely high levels of mine and UXO contamination. Bosnia makes a good case study because it has a very diverse mine action sector, with many different actors, allowing for some interesting intra-country comparisons. It is also a 'middle-aged' mine action program, its genesis lying in growing trends of international intervention in the 1990s rather than the Cold War roots of the Afghan program. Finally, like Afghanistan, Bosnia has been of variable strategic importance over time. From 1996 to 2000, Bosnia received considerable international attention and significant numbers of US and Norwegian troops deployed as part of the NATO stabilization mission. However, since then, especially after 9/11 when attention shifted to Afghanistan and Iraq, the US and Norway have reduced their involvement in

Bosnia. US support of Bosnian demining took two forms: a) support to the local militaries, as part of a larger effort to transform them from a threat to European security into future NATO members, and b) a commercial tendering system that for several years was captured by a criminalized ethno-nationalist elite. This may have functioned as part of a broader US 'passive policy' toward such elites, aimed at getting their buy-in to the peace and reconstruction process. In contrast, while Norway gave some token assistance to military deminers, the vast majority of its assistance was channeled through an international NGO, Norwegian People's Aid (NPA), that employed a multiethnic staff and supported efforts to build a cosmopolitan polity in Bosnia.

At war since 1983, Sudan has an as yet undefined landmine and UXO contamination problem, caused largely by fighting between the Northern government and Southern rebels. Sudan was primarily chosen to act as a potential 'spoiler case.' Following field research in Afghanistan and Bosnia, the author felt it was necessary to look at a third case that would challenge and test the predictive powers of his tentative hypotheses. Because the demining program in Sudan is quite new, beginning in earnest only in 2004, there has been little written about it. Sudan also differed from the other two cases in terms of strategic interest. While both Norway and the US have paid significant attention to the situation in Sudan, neither have found it of enough importance to make major commitments of troops to the country. For all these reasons, the Sudan case had the potential to prove the author's ideas wrong, or at least force them to become more sophisticated. US support for demining between the 2002 Nuba Mountains ceasefire and the 2005 Comprehensive Peace Agreement (CPA) followed the commercialized and securitized patterns seen in Afghanistan and Bosnia. In 2007, a USAID demining contract continued in this vein. However, following the CPA, the US State Department concentrated on funding international NGOs and local government capacity building. Indeed, its funding of the Scandinavian NGOs Norwegian People's Aid and DanChurchAid (DCA) overlapped with Norway's choice of implementing partners. This contrasted with the UN model of commercializing and integrating mine action into the politico-military objectives of the UN peacekeeping mission, which actually resembled the US strategic-commercial approach in Afghanistan and Bosnia. This shows that the impulse to control and securitize the process when there are more interests at stake may not be limited to the US. When strategic interest is lower, donors seem more willing to grant their demining funding to NGOs. This reflects a tendency among Western countries to frame conflicts of low strategic importance as zones for 'humanitarian' rather than politico-military intervention.

A Brief Note on Methodology

This project would be impossible without immersion into mine-affected countries, the offices of mine action agencies and visits to the actual minefields themselves. Therefore, while an anthropologist would never allow this to be called an ethnography (as the author's immersion in each of the case study countries was not lengthy enough and the focus of study was political economy rather than culture) the author tried to use ethnographic techniques such as participant-observation, interviewing key informants and 'thick description.'[14] The author thus uses the "loose definition" of ethnography, described by political scientists Lorraine Bayard de Volo and Edward Schatz as 'those methods that seek to uncover emic (insider) perspectives on political and social life and/or ground-level processes....'[15] Diverging from traditional ethnography, the data gathered is placed into the framework of comparative politics – comparing across places and times to seek potentially generalizable information and trends. Rather than seeing 'the field' in the traditional sense of a bounded *geographic* area, the field of study was conceived of as the bounded *sector* of mine action in a variety of locations. The point was less to learn the culture and politics of a particular place, but rather to be immersed in the culture and politics of the demining sector generally.[16] Therefore, this book aims to trace the politics of mine action from the top to the bottom, from the global level to the level of bilateral donors, and on down to the level of implementation. This approach to defining the field is necessary in a time of globalization, where few phenomena can be easily bounded to a particular location. Therefore, unlike traditional field research, this required a global methodology; the author's research took him to London, Washington DC, New York, Geneva, Oslo, Kabul, Sarajevo, Juba and Khartoum.

Overview

Chapter One gives a historical overview of the global politics of mine action, illustrating the ways in which relations in the international arena have shaped the response to the problems of mine and UXO contamination. After a description of post-WWII state-centric demining efforts, it traces the development of alternative models of mine action, through the Indochinese Wars to contemporary 'New' and 'Post-Modern' Wars.

Chapter Two, though not specifically focused on mine action, lays the theoretical foundations of this study, creating a typology of responses to insecurity, since the threat of mines and explosive remnants of war is essentially a threat of physical violence. It shows that traditional state-centric responses – realist 'National Security' and idealist 'Collective Security' – are no longer appropriate in responding to 'New Wars' and other transnational threats. In their place, two new forms of 'post-statist' networked governance, comprising both public and private actors, have arisen: Strategic-Commercial Complexes and Human Security-Civil Society Complexes. The rest of the book uses this typology as its theoretical framework, with which to understand and compare different ways of structuring mine action programs and the implications for the outcome of mine clearance as an element of post war reconstruction. It traces the development, operation and impact of these two models in the mine action sector, from the macro-level of global politics down to the micro-level of implementation in affected countries.

Taking a closer look at the internal workings of these complexes, Chapter Three argues that the mine action policies of the US and Norway can be useful as rough proxies for comparing the Strategic-Commercial and Human Security-Civil Society approaches to mine action. Borrowing Jan Egeland's argument that the US is more constrained by strategic and commercial interest than Norway, the chapter shows that the US has, with only a few exceptions, consistently tried to block tight regulations on mines and cluster munitions, while Norway has championed them. Likewise, US aid for demining is influenced heavily by military and security concerns and much of it is contracted out to commercial companies. In contrast, Norwegian demining aid is rooted in humanitarian concerns and is largely implemented by international NGOs.

The next three chapters focus down on implementation in three mine and UXO affected countries: Afghanistan, Bosnia and Sudan. Chapter Four provides background on demining efforts in each country, focusing on US and Norwegian supported programs. It then shows that, in general, when a donor's strategic and commercial interests were higher, they tended to opt for a commercial tendering model. When they had less strategic interests at stake, they were able to act in a more humanitarian fashion and give long-term grants to international NGOs. The performance of these two models of funding – commercial tendering and grants to NGOs – is then compared in Chapter Five. Basic statistical analysis shows that while it tends to be slower and more expensive, the granting model tends to concentrate on more difficult demining tasks and conduct the process to a higher standard of quality and safety. Chapter Six then looks at the wider impact of the two 'Demining Complexes' on the broader socio-political context of transition from war to peace. It finds that while implementing agencies operating in a Strategic-Commercial mode may contribute to strengthening state security organs, they are also more likely to strengthen the fragmentation and privatization of security. In contrast, the Human Security-Civil Society Complex's greater freedom from the constraints of expediency enables it to resist the politics of violence, advocate for limits on the technologies of war and set up systems that distribute protection according to need. The book concludes by offering a summary of key findings, policy recommendations for the mine action sector and, in closing, final reflections on security in a post-statist world.

1

A POLITICAL HISTORY OF MINE ACTION

*The history of the landmine mirrors a
century of social and military history....'*
– Lydia Monin & Andrew Gallimore[1]

Landmines continue to kill and maim years after armed factions have shaken hands and sat down to negotiate. Moreover, up to 30% of modern munitions fail to detonate upon impact with their targets, leaving highly unstable explosives in the ground decades into the future. These weapons project the impact of war onto future generations. Throughout the 'century of war' beginning in 1914, people have attempted to mitigate this threat to civilian life. Initiatives have come from states and from non-states, military and civilian entities. Most dramatically, in the late 1990s a combination of NGOs and medium-sized states were able to institute what is now almost a universal norm prohibiting the use of anti-personnel landmines. The struggle to restrict the threat of remnants of war continues as the movement to ban cluster munitions aims to universalize the cluster munitions ban treaty, signed in December 2008.

This chapter will show that the evolution of attempts to control, mitigate and clean up the contamination left by landmines and unexploded ordnance (UXO) has been intimately linked to the changing nature of the state, warfare and the international system since 1945. Since the end of the Cold War, the regime governing explosive remnants of war has undergone a remarkable transformation, from individual sovereign states managing the problem internally to collective management through global systems that include, in addition to states, multilateral institutions, NGOs and commercial companies. This chapter tracks the shift in the governance of the remnants of war from state to non-state actors, superpowers to small states, military organs to civilian agencies, national governments to institutions of global governance.

The chapter is thus framed as a political history. It examines the rising prominence of norm-based, non-state and commercial actors in the international system by showing how global politics has written its biography across the world's minefields. Imitating Olivier Razac's history of barbed wire[2] and John Ellis' *Social History of the Machine Gun*,[3] this chapter is an 'archeological' investigation of the response to the global mine and UXO problem, through an examination of four cross-sections of its history: the aftermath of WWII, the wars in Indochina, the 'New Wars' and the development of 'Post-Modern Warfare' into the Global War on Terror. While other wars (such as those of decolonization in Africa or the Arab-Israeli conflicts) contributed to the development of mine and UXO governance regimes, in the interests of brevity, the author believes focusing on the four periods chosen is sufficient to make the key theoretical points. The chapter will close with a brief description of how, at the beginning of the 21st Century, two alternative models of organizing demining had emerged: one rooted in strategic politics of great powers and contracted to commercial companies and the other based on coalitions between middle power states and NGOs.

World War II

Victim-activated landmines probably first appeared in the American Civil War, though their predecessors – booby traps and metal spikes laid to disrupt horses – go back to the Roman era.[4] However, mines came into their own in the era of mechanized warfare. While their initial innovation as a means to disable the newly invented tank saw some use in WWI,[5] it was the WWII tank battles of North Africa and constantly moving frontlines in Eastern Europe and Russia that bred the first crisis of landmine contamination.[6] By WWII's end, there were an estimated 300 million landmines strewn across Europe and North Africa. Reflecting the technological and sociological character of WWII's industrial and modernistic mode of 'total war', these mines were laid mostly by state militaries in massive, patterned constellations. The German army even designed precise algorithms to guide mine-laying, aimed at covering maximum area for minimum cost.[7] They represented an example of how states had the power to regulate and reshape territorial space to an unprecedented degree.

In addition to the mine crisis, the aftermath of two mechanized World Wars left large swaths of Europe contaminated with millions of tons of unexploded ordnance. Artillery and strategic aerial bombing dumped unprecedented quantities of munitions in civilian areas, such as farms and urban settlements. So much of this was left behind that European countries continue to dispose of WWI and WWII mines and UXO even today.[8]

Unlike the imperial 'Great Game' in Central Asia or the proxy wars of the Cold War, the two World Wars directly pitted the great powers against each other. The battle for state survival was fought right at the very core of the international system, with limited skirmishes at the peripheries. Even the running battles in North Africa or East Asia, while outside the traditional center of global power, were fought largely between the principal great power antagonists. As a result, the majority of mine and UXO contamination in the aftermath of WWII was located at the very heart of the global system, in Europe and the European part of the USSR.

The aftermath of WWII saw what is still the largest ever landmine and UXO clearance campaign. While the technical side of demining and ordnance disposal has changed little since WWII, the political organization of the process was far more state-centric than today. It reflected the nature of the state and the international system, in which WWII saw the peak of government and military control over European politics. Firstly, as much of the contamination was at the very center of the international system – in Europe – there was political and economic pressure to complete demining and explosive ordnance disposal quickly – especially in major cities. In more peripheral areas, such as North Africa, clearance was longer in coming. Secondly, the military, which had been at the center of all social activity in WWII Europe – with entire societies mobilized in total warfare – took control of clearance.[9] This, in effect, reinforced the notion that mine and UXO contamination was the purview solely of state security organs.

Therefore, clearance mirrored the unchallenged territoriality and sovereignty of the state. Unlike post-Cold War demining, funded by international donors and implemented by NGOs and companies, clearance was funded and handled internally by the state whose territory was mined. Even today, the ongoing clearance efforts in Europe are handled largely by government agencies.[10] Moreover, the demining programs were state interest – rather than human – centered. They displayed a certain callousness to life (reflective of the prosecution of the war) with a high toleration for casualties, acceptance of a less than 100% clearance standard and even press-ganging of prisoners of war (POWs) to work as deminers.[11] As such, demining was conducted in the same statist manner as war.

The Hague Regulations at the turn of the Twentieth Century had seen several humanitarian attempts to control 'inhumane weapons' such as 'dum-dum' bullets[12] and poisoned gases[13] and the aftermath of WWI led to the Geneva Protocol banning gas and bacterial warfare.[14] However, the widespread landmine contamination after WWII provoked no such proscriptions in international law. Mines were seen as perfectly legitimate weapons in the nation-state's arsenal. The 1949 Geneva Conventions made no attempt to regulate them. The only regulation specifically concerning landmines was a ban on further use of POWs as deminers.[15] Of course, mine warfare was to be subject to same the principles of discrimination and proportionality, which were primarily aimed at constraining the use of airpower that had rained down ordnance on European and Japanese cities. However, as will be seen in the next sections, these proscriptions had limited impact on the massive and indiscriminate use of both mines and aerial bombing in Korea, Indochina, Afghanistan and many other places. In 1956 the International Committee of the Red Cross (ICRC) raised concerns about the post-conflict impact of landmines, calling for a new legal instrument requiring the mapping of minefields, but this proposal met with disinterest from the world's states.[16]

War in Indochina

The years after WWII saw 'a terrific growth in the killing power of antipersonnel weapons.'[17] By the late 1960s, ordnance manufacturers had developed new innovations in fragmentation (such as the US Claymore mine that was more effective at wounding personnel than traditional blast mines) [18] and cluster munitions (in which an artillery or airdropped missile releases many smaller bomblets or landmines over a large area).[19]

As the US commitment to counterinsurgency in South Vietnam deepened from 1965 to 1975, the US struggled to target an elusive and hidden enemy. The Viet Cong was difficult to identify, hard to find and supplied by a complex network of trails through Cambodia and Laos, nicknamed the 'Ho Chi Minh Trail.' 'The solution,' opined one commanding US general was 'more bombs, more shells, more napalm ... till the other side cracks and gives up.'[20] US warfare in Indochina (Vietnam, Cambodia and Laos) was thus the logical extension of its position as a super-modern industrial power, pitting the might of technology and heavy manufacturing against a poor, lightly-armed guerilla force. While there was an 'Other War' – covert operations such which attempted to single out and eliminate individual Viet Cong cadres – much of the US effort consisted of blanket targeting of whole areas. By the war's end, the US had expended 14.3 million tons of ordnance in Indochina. This was twice the total amount expended in all WWII theatres together.[21] In Laos alone, between 1964 and 1973, the US dropped an average of one 'planeload of bombs...every eight minutes.'[22]

While there were important similarities to the strategic carpet bombing of WWII – targeting a wide area, hoping to hit something valuable – there were key differences. While a WWII bomb killed only those in the vicinity of its explosion, cluster munitions covered an area with a more fine-grained distribution – pixilation in computer parlance – of ordnance. Likewise, aerially dispersed mines and delayed detonation cluster munitions distributed detonations over time – projecting the deadly power of ordnance into the future, beyond the moment of the munitions' impact with the ground.[23] Thus the US could continue 'bombing' Indochina even when its planes were not above its skies.[24] These developments caused enormous civilian suffering. For instance, a US Information Service survey found that 80% of the victims of the bombing in the Laotian Xieng Khouang province were civilians.²⁵ The legacy of the bombing continues to kill and wound civilians; in 2006 there were 92 new UXO casualties in Vietnam, 58 in Laos and 259 in Cambodia.²⁶

The Vietnamese Communist forces also laid mines, but followed a very different strategy than the US. Unlike the large defensive constellations of minefields in WWII, the Viet Cong used mines as offensive weapons, setting booby traps, mining trails and laying them in non-Euclidian patterns. As such, their mine warfare reflected their guerilla strategy in which forces avoided concentration and relied on ambush and surprise. Therefore, while they used far fewer mines than WWII armies, the mines were more difficult to detect and in low densities, spread out over large areas. ²⁷ Indochina continues to be heavily contaminated with mines and UXO. Over 21% of Vietnam's territory, mostly in the former DMZ, is considered hazardous, and as much as 600,000 tons of ordnance remain scattered throughout the country.²⁸ In Laos, the UN estimated that about 500,000 tons of UXO remained in the ground in 1996.²⁹

Despite the massive contamination problems during this era, attempts at clearing mines and UXO for civilian purposes was done only on a small and ad hoc basis at the fringes of other programs. While there are accounts of internal military and governmental organs of North and South Vietnam and Laos conducting some clearance, the extent of the problem completely overwhelmed this indigenous capacity.³⁰ In many cases it was local farmers and informal 'village deminers' that conducted what little clearance occurred. There was certainly no replication of the massive clean-up seen following WWII, nor the international financing of clearance seen in the post-Cold War era. This was probably because although Indochina represented a key Cold War battleground, it was at the periphery of the international system and did not have the resources or state strength to induce or coerce clearance. Moreover, in the Cold War context, such a massive and expensive effort would have required the financial and technical backing of one of the superpowers, which were reluctant to accept responsibility for the problem.

That said, both superpowers and other agencies did sponsor some small clearance efforts. While they never had much impact, in these programs one can detect the nascent development of two broad political models of mine and UXO clearance for civilian purposes: one rooted in state strategic interest and the other initiated by humanitarian organizations. Both the US and USSR developed small clearance programs which benefited civilians, but were designed for military or Cold War political advantage. In response to Viet Cong mining and booby-trapping, the US Army and Marine Corps developed their capacity for mine detection and destruction and began training mine dogs. One major effort was keeping roads open, which the Viet Cong re-mined repeatedly.³¹ While this clearance was for military purposes, it also benefited the civilian population. In the early 1970s, US Air Force Explosive Ordnance Disposal (EOD) teams working in Thailand were occasionally sent to Laos to clear UXO ostensibly for 'humanitarian' purposes but operating in the broader context of substantial US covert and economic aid used to prosecute the 'secret war' in Laos.[32] One of these EOD operators said he 'worked in civilian clothes with the CIA, picking up unexploded ordnance so the roads were clear and the farmers could work their paddies.'[33] The teams operated at least until mid-1974, when it appears they were withdrawn to comply with the 1973 Vientiane Peace Agreement.[34] For its part, in what appears to be a unique case, the Soviet Union sent 12 experts to Xieng Khouang, Laos, in 1979 to initiate an UXO clearance project at a state farm. This was part of a wider package of support to the revolutionary regime. For 18 months, these experts trained 120 Laotians to use jeep-mounted metal detectors, clearing some 5,000 hectares of 12,700 UXO. After the Soviet advisors left, the trainees were spread out over a wider area, but lacking funding, expertise and spare parts, the program slowly fizzled.[35]

While these programs were novel for the time, they were pittance in comparison with the massive contamination of aerially dispersed mines, cluster munitions and other UXO littering Vietnam, Cambodia and Laos. The US has never accepted full responsibility for the problem and did not begin funding clearance programs in these countries until after the Cold War ended. Since the US was to blame for the majority of contamination, it had little incentive to publicize this problem by acknowledging responsibility for clean-up. Likewise, the USSR seemed uninterested in making much of an effort to assist its client states clean up, despite having extensive experience of landmine clearance on its own territory. One might have expected the USSR to see the propaganda value in clearing the mess caused by the other superpower. However, there was an institutional bias within the national security establishments of both superpowers to maintain strong control of all security issues. Acknowledging mines and remnants of war as a humanitarian issue would disadvantage the Soviet Union in other potential conflicts, especially since mine warfare and air power were so integral to the USSR's military doctrine. The USSR may not have wanted to frame mines and UXO clearance as a humanitarian issue because to do so would delegitimize using them in their own future conflicts. One can detect the hints of a superpower collusion game (which was more visible in the Conventional Weapons Convention negotiations described in the next subsection), where superpowers had an implicit arrangement maintaining military hegemony over perceived security issues.

Surprisingly absent too, was significant response to the landmine and UXO issue from the UN system, despite its ostensibly pivotal role in both global disarmament and humanitarianism. Other than helping to organize the Convention on Conventional Weapons (see below), a UNEP study on explosive remnants of war[36] and WHO research on the impact of antipersonnel weapons, the UN was rather inactive on the issue. This (lack of) UN response was symptomatic of the paralysis that affected the UN during the Cold War over any issue that impinged a) on the national sovereignty of its member states and b) the interests of the superpowers.

Though also small and ad hoc, an alternative model of clearance and mitigation, shaped by humanitarianism, religious imperative and solidarity, developed among small international NGOs working at the level of the population (rather than the state) in Indochina. Driven by their opposition to the war, the American Friends Service Committee (AFSC), a humanitarian arm of the American Quakers, and the Mennonite Central Committee (MCC) did relief work in both South and North Vietnam (this was illegal under US law) and began to see the impact unexploded munitions and mines were having on the civilian population. In Quang Ngai, AFSC supported a project rehabilitating war victims and producing prosthetics.[37] MCC actually experimented with clearance efforts in rural areas of the same province, finding 'plowing such ordnance-littered fields with an armored tractor – dangerous as that was – was safer than tilling by unprotected farmers.'[38]

After the 1975 communist revolution in Laos, the AFSC and MCC were the only Western relief agencies allowed to stay in the country. Both organizations, which coordinated their efforts closely, again experimented with ways to address the ordnance problem, funded mostly by church-affiliated agencies. While they experimented with metal detectors and armored tractors fitted with chain flails, it was only after the end of the Cold War that MCC was able to develop UXO clearance into a viable program.[39] AFSC and MCC had more success with projects aiming to mitigate the impact of ordnance. For instance, the AFSC worked in partnership with a new French NGO, Handicap International, providing prosthetics and UXO survivor assistance. Handicap, which later became a prominent voice in the International Campaign to Ban Landmines (see next section), was founded in 1982 to provide orthopedic services to war amputees in the refugee camps of Indochina.[40] By the late 1980s, Handicap had an office of its own in Vientiane and scaled up its operations, supporting provincial prosthetics workshops around the country.

The AFSC and MCC also discovered that many UXO accidents were caused by farmers using traditional Laotian hoes to till the soil. The hoe, lifted above the head and then swung down hard, had a tendency to detonate cluster bomblets and strike them in the direction of the farmers' own body. AFSC and MCC found that American-style shovels, which entered the ground more gently, were less likely to initiate the explosive. As a result, between 1977 and 1991, the AFSC and MCC distributed 30,000 shovels to farmers and helped a Laotian manufacturer start domestic production. This project attracted many more funders than any other mine or UXO mitigation project of the time, but these were mainly private donors, NGOs (such as Oxfam America) or Quaker affiliated groups. The only countries that provided substantive bilateral assistance were middle powers, including Norway.[41] While the shovels project was less dramatic that the clearance efforts, it probably had a much greater impact on mitigating the impact of UXO.[42]

Compared with the clean up after WWII and in the post-Cold War era, however, efforts to mitigate and neutralize the threat of landmines in Indochina were minimal. Affected states, in the periphery, lacked the resources to conduct clearance themselves, and superpowers were loath to take responsibility for the problem. Though a few humanitarian NGOs did what they could, their possibilities were limited. States were not willing to part with information on mine laying, bombing patterns or even counter-mining techniques, which they considered military secrets. Without the support of large bilateral donors, NGOs were also unable to mobilize the necessary resources to make a real difference.

Unsurprisingly, the US actively prevented information on ordnance contamination in Indochina reaching the rest of the world. Indeed the bombing of Laos was kept secret until 1969. While prompted by North Vietnam, it was non-state actors – radical journalists and peace activists – who first sounded the alarm.[43] Likewise, the 1967 International War Crimes Tribunal, an inquiry organized by philosophers Bertrand Russell and Jean-Paul Sartre, sent investigators to North Vietnam to examine whether US use of cluster munitions constituted a war crime.[44] The most comprehensive account of the impact of the bombing of Laos was by a former Laos International Voluntary Services volunteer Fred Branfman who interviewed hundreds of people displaced from Xieng Khouang province and in 1972 published their stories in a book called *Voices from the Plain of Jars*.[45]

As a result of their work in Laos, MCC and AFSC managed to get coverage of the UXO contamination in major news outlets. Both NGOs saw 'public advocacy and awareness-raising [as] a main part of their response – an effort which overall has been considerably more successful than actual UXO clearance.'[46] The AFSC, in a coalition of Minnesotan and national anti-war groups, participated in a particularly high profile campaign exposing the impact Honeywell, a key manufacturer of parts of cluster munitions, was having on Indochinese civilians.[47] After it was discovered in 1977 that Honeywell had collaborated with the FBI in illegal surveillance of the anti-war movement, peace groups filed for damages. When the federal government and Honeywell finally settled, the campaign garnered much publicity and donated some of the settlement to the Laos shovel project.[48]

However, discussion of the effects of explosive remnants of war on Indochina remained largely at the margins of Western, especially US, society. The mainstream American press was reluctant to publicize an issue that would attract accusations of sympathy with the communists. A few congresspersons asked questions about the humanitarian impact of cluster munitions and aerially dispersed mines, but these were rare and met with euphemism and denial from the Pentagon.[49] North Vietnam too, while it had fed information to anti-war activists on the effects of cluster munitions and had partially funded the International War Crimes Tribunal for propaganda purposes, was not interested in strict international regulations on weaponry. Rather than calling for a ban or restrictions on mines and cluster munitions, it felt an 'imperialist weapon' became 'a sacred tool' when borne by a 'liberation fighter.'[50]

While the US and North Vietnamese governments were unwilling to question the unrestricted right of states to use mines and cluster munitions, some middle powers, particularly Scandinavian and Non-Aligned countries, began to suggest new legal frameworks to regulate them. Reacting to widespread domestic discontent with the Vietnam War, Sweden convened a committee of military and medical experts to look at the effects of antipersonnel weapons. The resultant report, released in 1973, floated the idea of a ban on certain kinds of weaponry.[51] Sweden found an ally in the International Committee of the Red Cross (ICRC), whose unique status as both the mandated guardian of international humanitarian law and an NGO, has put it on the forefront of pressing for new norms on mines and cluster munitions. In addition to its trusted role in developing humanitarian regulation of war, the ICRC had operational experience in the field, where it was able to observe the actual impact of such weapons.[52] Therefore, Sweden, along with 18 other countries, asked the ICRC to convene a Conference of Government Experts on Weapons that May Cause Unnecessary Suffering or Have Indiscriminate Effects to be convened by the ICRC in Lucerne, Switzerland in 1974.

At the conference, Sweden, Norway, Mexico, Egypt, Switzerland, Sudan and Yugoslavia presented a surprisingly direct and bluntly worded proposal to heavily restrict the use of antipersonnel weapons, including a ban on fragmentation cluster bomblets and aerially dispersed mines. The proposal was met with disdain and disinformation from the US and NATO countries, which cast aspersions on the information gathered by the Swedes and misled conference participants on the human impact of these weapons. As a result, participants could only agree to meet again in Lugano, Switzerland two years later. In the intervening years, the Non-Aligned Movement and the UN General Assembly passed resolutions calling for a response to the pernicious effects of explosive remnants of war, but concrete results were also lacking from the 1976 Lugano Conference. NATO countries continued to denigrate the data gathered by Sweden and Switzerland on the effects of antipersonnel weapons.[53] The 1976 and 1977 conferences preparing the Additional Protocols to the Geneva Conventions again came up with nothing more concrete than vague prohibitions on weapons that cause 'superfluous injury' or 'unnecessary suffering'[54] and a resolution recommending another conference on the issue of specific weapons.

While the Stockholm International Peace Research Institute (SIPRI) and the Friends World Committee for Consultation had observer status in the Lucerne and Lugano meetings, and the ICRC was deeply involved in calling for tighter restrictions on weapons, there was little civil society involvement in discussions and little press coverage. Many countries where such weapons were used were afraid to speak up – 'one does not want to tell one's adversary which blows are landing hardest'[55] – and two of the most affected countries, Laos and Cambodia failed to show up. This left considerations of weapons' impact on ordinary civilians in the hands of the military and security establishment. This meant norm entrepreneurs wishing to replace *raison d'état* with *raison de l'humanité* were constrained to the margins of the regimes governing international security.

Moreover, there was a struggle over which model of international law should regulate antipersonnel weapons like mines and cluster munitions. So-called 'prohibitionists' – middle powers and the ICRC – wanted the strongest proscriptions possible and for antipersonnel weapons treaties to be rooted in the humanitarian law tradition, which prioritized protection of war victims over military concerns.[56] In contrast, the 'realist' great powers argued that humanitarian law should not cover specific weapons. If there were to be any regulations at all – one US negotiator admitted the US 'was not particularly desirous of concluding a weapons agreement' – they preferred the model of 'arms control', which was more about preserving stability and balance of power than the possible humanitarian impact of weapons.[57] As former Swedish Minister for Disarmament Alva Myrdal has argued, the 'arms control' model was often a superpower collusion game, enabling the US and USSR to maintain and legitimize their military hegemony.[58]

The UN finally convened the Conference on Prohibitions or Restrictions of Use of Certain Conventional Weapons which May be Deemed Excessively Injurious or to Have Indiscriminate Effects in Geneva in 1979.[59] The result, commonly known as the 1980 Convention on Conventional Weapons (CCW), placed weak and loophole-filled restrictions on mines and failed to regulate cluster munitions at all. 'Indiscriminate' use of mines was prohibited, though the definition of this was left rather vague.[60] The use of scatterable mines (dispersed by air, machine or artillery) had to be preceded by a warning to civilians, 'unless circumstances do not permit,' and unless they had self-destruct mechanisms.[61] The location of both hand-laid and air-scattered minefields was to be recorded (a nearly impossible task from an airplane).[62]

While some commentators felt this was an important step forward, at least in establishing a new norm of special restrictions on mines, in reality the CCW was wholly inadequate in protecting civilians from the indiscriminate effects of landmines and UXO. The CCW did practically nothing to prevent or restrict further use of mines and cluster munitions, or to spur an effort to clear them up and mitigate their effects. Reflecting the 'arms control' rather than humanitarian agenda of the great powers, the CCW's restrictions on mines gave 'the impression of having been written to satisfy the needs of military forces, which may later have to occupy a mined area, rather than to protect civilians.'[63] As will be discussed in the next section, in the late 1980s and 1990s the CCW was actually followed by the most widespread crisis of mine and UXO contamination the world had ever known. The fact that the Soviet Union, a signatory to the CCW, used mines so widely and indiscriminately in the Afghan War (see next section) is a testament to the CCW's inability to change states' behavior.

The conferences between 1974 and 1980 failed to develop substantive restrictions on mines and cluster munitions because they were conducted in the great-power-centric context of the Cold War. Thus while humanitarian concern about the effects of mines and cluster bomblets were raised by norm entrepreneurs (such as activists, NGOs, the ICRC and middle powers) their energy was dissipated, depoliticized and stifled in the highly technical discourse and proceedings of Cold War disarmament meetings. While radical suggestions were put on the table by Scandinavian and Non-Aligned countries at the Lucerne conference, the superpowers and great powers domesticated them into weak, unenforceable provisions. The Cold War reification of 'national security' placed concerns about weapons firmly in the hands of the military, arms control negotiators and arms 'experts' rather than affected or concerned civilians.[64] As one Norwegian diplomat explained, 'Defense had a primacy at the time...it was possible for either Washington or Moscow to call the rest of their respective kinds to order in a way that you couldn't do now.'[65]

The Post-Cold War Era and the New Wars

The 1980s and 1990s saw an 'epidemic' of mine contamination in developing countries torn apart by proxy wars and the New Wars. Fortuitously, this coincided with the end of the Cold War and the opening of space for 'norm entrepreneurs', such as international NGOs and middle powers, to link up and raise their concerns in the international arena. As a result, there was a rapid expansion of NGO demining programs and a coalition of middle powers and NGOs succeeded in achieving an Antipersonnel Landmine Ban Treaty in 1997. However, a variety of 'realist' states have been unwilling to sign the treaty, which was also unable to stem the use of similar weapons such as cluster munitions and improvised explosive devices (IEDs).

During the Cold War, nuclear deterrence and fear of total war prevented any direct military confrontation between the East and Western blocs. However, disturbed by what it perceived as the growth of Soviet power in the Third World in the 1980s, the Reagan administration broke with the 'Containment' consensus and adopted an aggressive 'forward policy,' aimed at 'rolling back' Soviet expansionism. The resulting policy, often called the 'Reagan Doctrine,' supported anti-government insurgencies in Third World socialist countries (Afghanistan, Angola, Cambodia and Nicaragua) through covert action. A keystone of this strategy was the creation of havens – safe and secure rear areas – for proxy forces in neighboring non-Socialist countries (Pakistan, Zaire, Thailand and Honduras, respectively). Rebels set up bases on the safe side of the border, often under the cover of refugee camps, and used these as staging grounds for cross-border raiding and infiltration. [66] In each case, the Communist forces quickly recognized the strategic importance of these cross-border infiltration routes and tried to cut them off by deploying extra troops to seal off the border and laying massive constellations of minefields – both by hand and by air. While much of the mine-laying in these wars was usually done by the Communist forces, the US proxies also engaged in mine warfare to protect their 'liberated areas' and bases vulnerable to cross-border raiding. Both sides often mined areas used primarily by civilians – in contravention of international law – and with little regard for proper marking, recording or regular patterns.[67] By the end of the 1980s, all of these proxy war countries had significant landmine problems – three of them, Afghanistan, Angola and Cambodia, are among the ten most mine contaminated countries in the world.[68]

The targeting of civilians in the proxy wars presaged the use of mines in the so-called 'New Wars' of the 1990s. These wars were fought in the periphery of the international system, in developing and post-communist states and sometimes morphed out of the proxy wars.[69] Mine laying in the New Wars reflected the non-linearity of its strategy, with mines often strewn in unrecorded and non-Euclidean patterns and used to target and/or contain civilian populations.

The majority of mine and UXO contamination from the proxy and New Wars was in territory of weak states, which were unable to marshal the fiscal and technical resources needed for clearance. Thus demining and mitigation efforts largely came at the initiative of outsiders – aid donors, the UN and international NGOs. As will be discussed in the next chapter, the early US response to landmine contamination caused by the proxy wars was rather small and rooted in the logic of the Cold War, often assisting the logistical supply chains of their proxies. But by the early 1990s, the increasing tendency of external actors to intervene in the politics of conflict in the periphery spurred a massive increase in humanitarian demining programs. For instance, in 1989, the UN launched its first demining program as a part of Operation Salam in Afghanistan, one of the first massive UN 'post-conflict' operations. Further large scale UN interventions in Cambodia, Mozambique, Bosnia and Kosovo included mine action programs. In 1993, the General Assembly adopted a resolution to establish UN mine clearance program, which eventually became the UN Mine Action Service (UNMAS), located in the Department of Peacekeeping Operations.

Indicative of the increasing role of non-state actors in international politics, international NGOs and commercial companies played a leading role in the new clearance efforts. Several existing development NGOs, such as Norwegian People's Aid, DanChurchAid, HELP and Vietnam Veterans of America Foundation saw the terrible destruction caused by mines in the countries where they worked, or were asked by their donors to start up programs and expanded into mine clearance operations. At the same time, a whole new set of NGOs specializing in mine action, such as Danish Demining Group, Mines Advisory Group, the HALO Trust, Landmine Survivors Network (now Survivor Corps), Menschen gegen Minen and the Swiss Foundation for Mine Action, started up in the late 1980s and early 1990s.

Non-state demining operations were suddenly made possible because donors wanted to show their electorates that they shared sentiments with the mine ban campaign. A good example of the huge growth in interest in clearance and mitigation efforts following the Cold War is the UXO clearance programs in Laos. As discussed in the previous section, MCC's UXO clearance efforts during the 1970s and 1980s were rather sparse, underfunded and unguided. However, by 2006, the Lao UXO clearance programs had an annual budget of over $13 million.

On the supply side, the demobilization of Cold War and Apartheid era armies enabled NGOs to hire military experts in mine warfare, explosive disposal and countermining. Prior to the end of the Cold War, only states were able to mobilize this kind of expertise. Now it was available to NGOs and commercial companies on the open labor market. Demobilization and democratization also led in some places to greater openness with military secrets, such as the location of minefields and bombing targets. The handover of Soviet minefield maps to HALO Trust by the communist regime in Afghanistan in the late 1980s was a watershed moment. It was the first time that superpower mine data – sensitive national security information – was handed to a civilian non-state actor. This represented a sea change toward a partial humanitarianization of national security information. In the 1970s and 1980s, the AFSC and MCC had toiled to get the US Department of Defense to release bombing data for Laos.[70] But in the aftermath of the Kosovo, Afghanistan and Iraq bombing campaigns, US bombing data was handed over almost voluntarily, with little fuss. A significant shift in the normative framing of mine and UXO issues had occurred. Mines and UXO were no longer simply a national security issue, with secret locations, hidden from the civilian population. They were a humanitarian, human security issue, whose locations should be revealed to the public and which should be dug up and destroyed.

The humanitarian rather than strategic spirit of the new NGO clearance organizations was embodied in the 'Bad Honnef Framework', a 1999 statement developed by the main mine action NGOs that laid out guidelines for running mine action programs 'from a development-oriented point of view.' They called for mine action to be guided by the 'needs and aspirations of people affected by mines', rather than strategic interests or commercial profit. When compared to the mine clearance programs after WWII, this was a radically different framework.[71]

The early 1990s also saw a revolution in advocacy on mines. Following the 1991 publication of a Human Rights Watch report on the impact of landmines on Cambodia,[72] several NGOs involved in mine action and conflicted regions began discussing ways to campaign for the bold goal of banning antipersonnel landmines. In 1993, they formed the International Campaign to Ban Landmines (ICBL), which expanded rapidly to include hundreds of NGOs, religious groups and citizen's associations from around the world. Based on their experience working with the many civilians who had fallen victim to mines and UXO, the core members of the ICBL were convinced that antipersonnel mines were a particularly egregious violation of the principles of proportionality and discrimination enshrined in international humanitarian law. However, they felt that with the lack of specificity in the Geneva Conventions and CCW, such principles meant nothing on the battlefield. Indeed, the USSR and the US-backed proxy armies had paid no attention to the regulations on mine use in the CCW. ICBL campaigners felt not only the use of mines had to be proscribed, but also the production, sale and transfer.

The ICBL found common cause with the ICRC that, in an unprecedented move, launched its first ever public advocacy campaign, calling for a ban on antipersonnel mines. ICRC medics in field hospitals in conflict zones were exhausted by the number of amputations they had to perform on mine survivors and pressured their headquarters to seek new restrictions on landmines.[73] In 1996, the ICRC published a crucial study, endorsed by over 50 senior military officials from around the world, arguing that the negative humanitarian impact outweighed the military utility of landmines.[74] This reportedly persuaded many political and military leaders to back or at least not object to a ban.[75]

While the ICBL drew leaders and lessons from the Cold War disarmament and peace campaigns (its executive director Jody Williams had risen to prominence raising awareness of US abuses in Central America in the 1980s), it also represented something new. Firstly, though largely rooted in rich countries, it was able to reach out on a more global scale than previous disarmament campaigns. Operating in countries affected by mines, constituent NGOs could gather field data to contradict misinformation of the militaries that had defused dissent in during the Cold War. Secondly, the ICBL was highly decentralized, allowing local campaigns to tailor messages to particular political contexts. Thirdly, using new communications technology they shared information and strategic planning rapidly and effectively. The campaign also effectively used the newly globalized news media to raise public consciousness, through dramatic images of mine victims and street theater (such as the 'mountain of shoes' representing amputees outside negotiation conferences).

Perhaps most importantly, the ICBL drew upon a much wider base than traditional arms control groups (limited largely to a specialized policy community), the ICRC or the peace movement (limited to the political left), which had sought regulations on mines and UXO during the Cold War. It was rooted more in the human rights and aid communities, which had broad appeal across the political center. Moreover, endorsements from celebrities, a tactic used effectively by NGOs during the Ethiopian famine, drew considerable attention. Nelson Mandela and Princess Diana were crucial in raising public awareness.[76] Diana endorsed the mine ban campaign at the height of her fame (to the consternation of the UK Tory government that was against a ban) and according to interviewees, her death just a day before mine ban negotiations began in Oslo made many diplomats feel pressure to pass the ban as a way to honor her life.[77] Even today the landmine issue is largely associated with Princess Diana in the UK and US general public.

The NGOs found sympathetic partners in a core group of middle powers – states such as Canada, Austria, the Scandinavian countries, South Africa and Australia. Indeed, 82% of states with a population under 50 million supported the Mine Ban Treaty in 2007, compared with 56% of countries with more than 100 million. Likewise, 84% of countries with smaller military budgets (ranked 26 to 168 in the world rankings) signed the treaty, compared with 60% of the 25 top military spenders.[78] As noted previously, middle powers found more space to assert themselves in the international arena following the Cold War. In international NGOs, middle powers sought humanitarian legitimacy and information gathering power, while the middle powers gave NGOs state advocates to represent them in international conferences. The campaign to ban landmines probably became the most successful example of this alliance.[79] Together, the ICBL, ICRC and middle powers were able to frame the landmines issues as a part of what some commentators were optimistically calling a 'revolution of moral concern' – a surge in public feeling that the end of the Cold War should also lead to a new and more norm-based international political system.[80] By reframing landmines as a humanitarian or human rights issue, rather than a security problem, they were able counter the arguments of military necessity and the monopoly of military personnel and arms control specialists over weapons issues.[81] This use of moral discourse by a middle power-NGO coalition has been described as 'a new model of diplomacy' rooted in negotiation of norms, rather than competition and collusion around narrow strategic interests.[82]

Under pressure from the ICBL, the UN decided to convene a review conference of the CCW in 1995. However, all NGOs (except the ICRC) were excluded from the meetings leading up to the conference. It was also determined that decisions in preliminary meetings would be made by consensus, effectively giving veto power to the most conservative states.[83] As a result, the CCW was unable to come to an agreement on whether to place further meaningful restrictions on mines, since all the great powers, including the US, were against doing so. While President Clinton had initially favored a ban, he became wary of a clash with the Pentagon, and US policy remained in a realist Cold War framework. The US was willing to see some restrictions on mines, but not on technologically advanced ones (which they argued caused fewer civilian casualties) and with an exception for UN-mandated operations (i.e. the US-maintained minebelt in Korea).[84]

Rather than playing into a traditional great power-dominated arms control game, where decisions were always made by consensus and the lowest common denominator,[85] the ICBL-middle power coalition decided to create a completely new game with a different set of incentives.[86] In October 1996, Canada unilaterally announced that it would sponsor negotiations for a complete ban on antipersonnel landmines outside the CCW process, a move that was quickly backed by the ICBL and several other middle powers, including Norway. By demanding a maximalist position and daring other countries to join, Canada, the ICBL and the other middle powers created a game with strong incentives for countries to accept the treaty to avoid international stigmatization and domestic public opinion backlash. Unlike the CCW, the 'Ottawa Process', as the Canadian-led negotiations came to be known, encouraged the involvement of NGOs and the ICRC at each stage of the process. The ICBL and ICRC, the main drafters of the treaty, were represented in the negotiating chamber and many countries (especially middle powers) incorporated mine ban campaign NGOs into their national delegations. This was a radically different way of doing business than the CCW, as it prevented the veto-induced 'race to the bottom' and isolated and embarrassed potential 'spoilers' to the process.

In December 1997, 122 countries signed the finalized Antipersonnel Mine Ban Treaty, also known as the Ottawa Convention, which came into force in March 1999. To date, the treaty had been signed by 155 countries.[87] In contrast to the vast majority of international treaties, which exclude non-state military actors, the campaign even persuaded several significant rebel movements, including the SPLA, PROSARIO, PKK and many Somali factions, to sign an equivalent document called the Geneva Call Deed of Commitment.[88] The Ottawa Treaty banned the production, stockpiling, transfer, sale or use of antipersonnel landmines. It also required countries to destroy their existing stockpiles and clear mines from their territories, conduct mine awareness and assist mine survivors.[89] It was the first disarmament treaty with provisions for those impacted and victimized by the weapon. Non-affected countries were called upon to provide assistance for mine action programs, which many aid donor countries did. In the five years after the treaty was signed, donors gave some $1.3 billion to mine action.[90]

Though several great powers (like the US, Russia and China), regional powers (such as India, Pakistan, Saudi Arabia and Israel) and a few 'rogue' states (such as Cuba, Burma and Iran) refused to sign, the treaty has succeeded in creating a broadly recognized norm against the use of mines. While there have been a few isolated incidents, the production, trade and use of antipersonnel mines has largely come to a halt. Even the US, which maintains stockpiles and has been developing new technologically sophisticated mines, has not used them in Kosovo, Afghanistan or Iraq.

The ICBL and mine ban treaty have not been without critics, aside from those who feel mines are legitimate weapons of war. While the ICBL did include members from all over the world, its most dominant campaigners came from rich Northern countries, which some have argued confirms the continuing power of metropoles to determine the agenda. Moreover, the fact that the landmines issue gained more publicity than less glamorous but more deadly problems such as malaria or diarrhea shows how public opinion of what is an important 'humanitarian' issue is shaped by the nature and public image of a campaign. This has been accompanied by concerns that the ICBL vastly inflated the extent of landmine contamination.[91] Others have criticized the narrow focus, leaving states and non-state actors with the ability to continue deploying anti-vehicle mines (including ones with anti-handling devices that in practice turn them into antipersonnel mines) and cluster munitions. It has also not stemmed the widespread use of Improvised Explosive Devices (IEDs) – homemade mines – by the insurgencies in Afghanistan and Iraq. Thus some have argued that the treaty is merely part of a performative Western discourse 'civilizing' warfare, legitimizing it and making it an acceptable political practice by 'humanizing' it through accepting restrictions on marginal weapons like mines.[92]

Nevertheless, the mine ban treaty was a remarkable achievement. In the space of five years, a coalition of NGOs and middle powers was able to achieve a much tighter and more effective treaty than the international conferences that had met since 1974. It also operated outside of the traditional security structures of international politics. The treaty was not negotiated in the Conference on Disarmament, the Security Council or the CCW – and it was led by actors that until 1989 had been rather marginal. Moreover, implementation of the treaty has been closely monitored not by an international public organization, but by the ICBL itself, in its annual *Landmine Monitor* publications[93] that are widely accepted by states parties as the definitive source of information on compliance. Together, these characteristics represented a truly new way of dealing with international security issues. The ICBL-middle power coalition had taken a security problem out of the hands of the great powers, the military and, to a certain extent, even states. This would have been unimaginable in the aftermath of WWII and practically impossible during the Cold War. In recognition of their contribution to international peace and security, the ICBL and its director Jody Williams were jointly awarded the Nobel Peace Prize in 1997.

Post-Modern Warfare and the War on Terror

Developing concurrently with the New Wars, and sometimes interacting with them, were so-called 'Post-Modern Wars,'[94] where great or regional powers intervened in regional conflicts. Examples include the Falklands/Malvinas War, Gulf War, NATO intervention in Bosnia and Kosovo; the British intervention in Sierra Leone; the US in Afghanistan and Iraq; Russia's role in Chechnya and Georgia; and Israel's recent war with Hezbollah in Lebanon. These interventions were characterized by enormous asymmetries between local factions and the intervening power. The intervening great power had access to the massive use of firepower and airpower; super-sophisticated technology in guidance systems, communications and processing intelligence; and mastery of media management and broadcast propaganda. It was also characterized by close partnerships between government and private industry, which were increasingly involved in actual security operations.[95]

While local factions often engaged in widespread mining (such as Iraqi mining of Kuwait or Serb-laid minefields in Kosovo) or the use of IEDs, in these post-modern wars, the employment of staggering amounts of air-dropped or artillery munitions by the intervening great power left considerable UXO contamination, especially of cluster bomblets (such as in Chechnya,[96] Kosovo,[97] Afghanistan,[98] Iraq,[99] Lebanon[100] and Georgia[101]). For instance, the US fired some 23 million air-dropped and rocket dispersed submunitions in the first Gulf War.[102] As a result, some 30,000 tons of unexploded ordnance and millions of mines laid by Saddam Hussein's forces contaminated Kuwait at the end of the war.[103] In the first six months of its recent Iraq war, the US and its allies unleashed some 2.2 million submunitions on Iraq.[104] Human Rights Watch has documented the extensive harm to civilians caused by these cluster bomblets.[105] The current Iraqi insurgency has killed over 1,500 US soldiers with IEDs (some of which are often basically homemade mines).[106]

While the 1990s saw the rise of NGO mine and UXO clearance, they also saw the concurrent development of demining programs run by commercial companies contracted either to governments or other companies.[107] This model of clearance began in the late 1980s and early 1990s with RONCO's contract to USAID in Afghanistan and the South African mine warfare research company MECHEM winning contracts in Angola and Mozambique.[108] The development of this model, which like NGO demining is an advent of the post-Cold War world, is similarly linked to the rise of privatizing and contracting out public services since the 1980s. It was also helped by the supply of demobilized soldiers from First World and South African armies that brought expertise in explosives, mine warfare and counter-mining techniques into the private sector.[109] The commercial model came into its own in the aftermath of the first Gulf War, when Kuwait let massive contracts worth $800 million to a variety of companies, notably Royal Ordnance, CMS and Sofremi, most of which were actually weapons developers and had produced both mines and cluster munitions. It was a poorly run program, with 284 demining accident casualties.[110]

While commercial demining has occurred in many mine-affected countries, it is particularly prevalent in places where the US has been involved in the conflict. The massive boom in the private military industry provoked by the Global War on Terror has strengthened the commercial demining sector, especially in Iraq and Afghanistan, where the US has let contracts worth millions of dollars for demining, UXO clearance and destruction of abandoned caches. This has contributed to the increasingly blurred lines between the commercial security and demining industries, with the involvement of companies like ArmorGroup and DynCorp spanning both sectors. While commercial demining companies have been employed to do a wide range of tasks, they have been much more likely than NGOs to take contracts from militaries and commercial clients. One major demining company, MECHEM, is owned by an arms manufacturer, DENEL, that until recently continued to produce cluster munitions, even while MECHEM cleared them in the field.[111]

The UN has also moved toward private contracting of demining, especially since its operations in Kosovo began in 2000. Power over demining has shifted from OCHA and UNDP (rooted in the humanitarian and development communities) to UNMAS in the more militaristic Department for Peacekeeping Operations (DPKO). Whereas OCHA and UNDP had set up local NGOs to do demining in Afghanistan and Mozambique respectively, UNMAS operations have been managed by the UN Office for Project Services (UNOPS), which prefers a commercial tendering model of service provision.

While the Mine Ban Treaty has largely stemmed the use of antipersonnel landmines, the 'Post-Modern Wars' have demonstrated its limits. Indeed the War on Terror seems to represent the possibility of a return to great power strategic politics and diminishing space for norm entrepreneurs on the international stage. Insurgents in Iraq and Afghanistan have shown that despite the end of trade in mass-produced mines, they have been able to improvise sophisticated IEDs. Moreover, advocacy aimed at persuading the great and regional powers to join the Ottawa Process has met little success. While the US has not used mines, the Bush Administration was less receptive to landmine campaigners than its predecessor and developed several new landmine systems. In February 2004, it issued a statement declaring that 'Landmines still have a valid and essential role protecting United States forces in military operations.'[112] Similarly, in December 2006, Pakistan threatened to lay mines along the Afghan border.[113] The great and regional powers have continued to block or stall efforts to fill the loopholes in the CCW and the Ottawa Convention. Despite repeated efforts, there has been no success in creating a legal instrument to regulate the use of anti-vehicle mines. Moreover, the CCW Protocol on Explosive Remnants of War, added in 2003, aimed to 'minimize the risks and effects of explosive remnants of war' other than mines, but was full of escape clauses like 'as far as practicable', 'subject to these parties' legitimate security interests' and 'where feasible.'[114]

However, the middle power-NGO partnership has pushed back with a new campaign to ban cluster munitions – the Cluster Munition Coalition – modeled explicitly on the ICBL. While some people who had been involved in the anti-cluster bombs campaigns in the 1970s continued to write about the problem in the 1990s (NATO's use of submunitions in Kosovo piqued some interest) they tended to take a back seat to the mine ban campaigners. The success of the mine ban allowed those who had been involved in the ICBL to turn their attention to the continued use of cluster bombs in Chechnya, Afghanistan and Iraq.

Cluster bomb campaigners were suddenly thrust into the public eye following the 2006 Israel-Hezbollah war, when Israeli forces fired thousands of cluster rockets at Lebanon, leaving behind some 1 million unexploded bomblets scattered largely in civilian areas.[115] According to a UN mine action official in Lebanon, 'The scope was extensive and unprecedented in any modern use of these types of cluster weapons' – some 26 percent of arable land in Southern Lebanon was contaminated.[116] Such a flagrant disregard for the laws of war dramatically demonstrated the indiscriminate effects these weapons could have on noncombatants, even long after the war had finished. This put cluster bombs back on the agenda of world leaders and campaigners saw the opportunity to push for a ban.

While the middle powers tried to push for an amendment to the CCW regulating or banning cluster munitions, they met considerable resistance.[117] This is unsurprising, given that cluster munitions, unlike landmines, were still considered integral and important weapons in the arsenal of modern militaries. Therefore, campaigners began calling for a maximalist voluntary treaty similar to the Ottawa Convention. Again, they found common cause with middle powers. In November

2006, the Norwegian government declared that it would host negotiations for a cluster bomb ban outside the CCW, in what came to be known as the 'Oslo Process.'[118] In December 2008, some 100 countries signed the resultant Oslo Convention, which, mirroring the mine ban treaty, banned the production, stockpiling, transfer and use of cluster munitions.[119] Given that cluster munitions were probably more integral to modern military planning than landmines, the Oslo Convention is an indicator of how powerful and effective the middle power-NGO coalition could be.

Nevertheless, displaying a similar attitude shown to the mine ban treaty, the great powers such as the US, Russia and China and regional powers such as Israel, Iran, India and Pakistan rebuffed the Oslo Process, arguing that restrictions on cluster bombs should be dealt with in the CCW, where they have effective veto power,[120] if at all. Analogous to mine ban convention, the Cluster Munition Convention is more likely to be supported by smaller states with lower military expenditure. Fifty-six percent of states with up to 50 million people signed the cluster bomb ban, compared with 36% of states with over 100 million people. Similarly, 62% of lower military spenders (ranked from 26 to 168 in the world) supported the treaty, compared with 48% of the top ranked 25.[121]

Two Emerging Political Complexes of Mine Action

Toward the end of the first decade of the 21st Century, it appeared that the statist model of mine action seen at the end of WWII was a thing of the past. Very few demining programs were operated wholly by states marshaling resources from taxation in their own territory. Even a superpower like the US had lost the ability to conduct mine clearance itself. When US Army occupied mine and UXO contaminated bases in Afghanistan in 2001 it found it no longer had the expertise to conduct demining itself. Rather, it had to rely on RONCO, a private contractor, to make the land safe for their troops.[122] This demonstrates a major fragmentation of the state's monopoly on managing violent threats, even to its own soldiers. Mine action, reflecting trends toward multi-level and multi-actor models of governance, is now conducted with funding from a wide range of international donors and implemented by partnerships between public and private for-profit and non-profit actors. Mine action has become a form of networked governance, or in the words of Mark Duffield, an 'Emerging Political Complex.'[123]

However, these complexes have been organized in a variety of ways, incorporating different actors with different sets of motivations. There are obviously infinite ways these networks can be organized, but abstracting out from the preceding historical description, one can discern the two competing versions of the political complexes that will be described in greater detail in the next chapter.

The first complex is that of a great or regional power that reserves the right to create a secure enclave for itself by creating rings of minefields and cluster munitions strikes around its borderlands and strategic nodes. For instance, one can see similarities between the wall Israel has built to separate itself from the West Bank and its use of cluster munitions to create, in effect, a wall of depopulated territory in southern Lebanon, discouraging Hezbollah infiltration.[124] This divides the world into safe 'green zones' that are mine and UXO free and dangerous 'red zones' where the use of mines and cluster munitions is acceptable. The danger of explosive remnants of war in the red zones is managed through commercial contractors, who focus on clearing areas of strategic significance so that they can be turned into new green zones or beachheads to control the red zones (such as new US bases). This complex is made of actors that are primarily motivated by state strategic interest or commercial profit. As a result, they will be described as 'Strategic-Commercial Complexes.'

The second complex of actors is typified by the middle power-NGO coalition that has pushed for the bans on mine and cluster munitions. They argue that strategic 'realist' models of international politics result in a proliferation of minefields and cluster munition strikes, actually reducing international security. Focusing more on the security of human populations, especially civilians, rather than the security of states (or enclaves) they see the abolition and clearance of all mines and UXO as a way to make the world a safer place. The distribution of clearance is also guided by this principle, with international NGOs and humanitarian UN agencies allocating their demining services according to need, rather than strategic interest or ability to pay. In short, the middle power-NGO coalition is primarily shaped by a humanitarian vision, in which the whole of the collective is made into a unified green zone, or at least a 'greener zone.' As such this type of institution will be described as a 'Human Security-Civil Society Complex.'

Conclusion

The way that mines and other explosive remnants of war have been used and, in turn, managed through a governance regime of policy, mitigation and clearance, has been shaped throughout history by the dominant political constellations of the age. In the aftermath of WWII, demining was conducted in a manner very similar to the way the war itself was fought – through mass mobilization of the military. Likewise, the developing international law of war placed few restrictions on states' abilities to deploy mines to guard their national security. During the Cold War, strategic bipolar politics exported insecurity, including remnants of war, out to the peripheries. Therefore, the superpowers had little motivation to engage in mine and UXO clearance, except in the occasional circumstances when it benefited them strategically. Likewise, the superpowers were loath to place any significant restrictions on their arsenals and ability to wage war. That said, this period also saw the beginnings of popular social movements seeking to transform the militarism of the era. Rising out of the peace movement, one sees the beginnings of attempts to restrict the use of mines and cluster munitions and small NGO programs to mitigate and neutralize the threat to civilians.

The end of the Cold War suddenly opened up space for normative actors, such as middle powers and international NGOs, to link up and push for a universalistic agenda of peace, human rights and humanitarianism. Through mass mobilization and tactics employing new media and communications technology, they were able to create a new political constituency dedicated to the abolition of mines and cluster munitions and raise millions of dollars to clear them, raise awareness and assist survivors. At the same time, however, great and regional powers were less receptive to this agenda. In the massive asymmetries of post-modern warfare, they felt free to use cluster munitions with few restrictions. The resurgence of militarism with the advent of the Global War on Terror led to the explosive growth of the private security industry that increasingly consolidated with the commercial demining sector. Thus there appeared to be two 'Emerging Political Complexes' of mine action: the Human Security-Civil Society Complex – a coalition of middle powers, UN humanitarian agencies and international NGOs – and the Strategic Commercial Complex, a network of a great power, military organizations (including sometimes the UN DPKO) and commercial companies. The next chapter looks at these two types of complexes in more theoretical depth. It takes a step back from the particularities of mine action and explores how these complexes of mine action are rooted in broader political trends in the international governance of insecure regions.

2

THE NEW COMPLEXES GOVERNING INSECURITY

'Frontiers are indeed the razor's edge on which hang suspended the modern issues of war or peace, of life or death to nations.'
— Lord Curzon.[1]

To understand both mine contamination and mine action complexes, one must understand the political systems that are emerging in what Mark Duffield has called the post-modern 'Global Borderlands' – areas of lawlessness and conflict beyond the reach of effective government.[2] In fact, mines and cluster munitions are often used to demarcate the frontiers of these 'zones of war.' To comprehend what is happening in these regions of the world, it is important to understand that traditional 'state-centric' conceptions of global politics (in which the primary sources of insecurity were other states or internal subversion) no longer hold true.

In his perceptive and provocative study of the architecture of the Israeli occupation of Palestine, Eyal Weizman describes the development of a West Bank settlement called Migron.[3] Responding to Israeli settlers' complaints of poor mobile phone reception in the West Bank, the telecommunications company Orange erected an antenna on a hill. Because secure communications are considered a security issue by the Israeli military, Orange did not have to seek permission from the Palestinian owners of the land. Supposedly to support the construction work, electric and water companies connected the hill to utility grids and a private security guard was hired to protect the site. Slowly, with encouragement and protection from the Israeli government, an outpost settlement developed around the antenna. Weizman claimed the Migron settlement was a form of political organization and action that

> ...cannot simply be understood as the preserve of the Israeli government executive power alone, but rather one diffused among a multiplicity of – often non-state – actors ... [including] young settlers, the Israeli military, the cellular network provider and other capitalist corporations....[4]

Meanwhile, an equally complex network has engaged in a variety of activities to publicize, condemn and counteract the development of settlements and the 'Separation Barrier.' This 'diffused global campaign' was 'waged via the UN, the Israel High Court of Justice, local and international NGOs, the International Court of Justice, the media and scores of foreign governments acting along visible or backstairs diplomatic channels....'[5]

Both these complexes represent efforts to provide security in a manner quite unlike the traditional centralized bureaucratic state security apparatus. They are, however, not unique to Israel and Palestine. Countries like Afghanistan, Bosnia and Sudan are governed by an array of UN agencies, NGOs, bilateral donors and multinational companies all interacting with local public and private powerholders. Faced by new transnational and non-state forms of insecurity, states and other powerful actors now try to penetrate and manage the territories and populations of the world's insecure 'borderlands' through complex multilevel networks of public and private actors. Duffield has argued that this is a form of 'neo-medievalism', in which political authority is diffused over multiple levels and overlaps many different actors.[6] The 'borderlands' have thus seen a shift from government to networks of governance. Duffield has a tendency to conflate these networks into one 'emerging system of global liberal governance.'[7] However, this can actually hinder understanding. For this system includes multiple types of actors, who negotiate relationships between themselves in different ways. Indeed, there are at least two ideal types (or poles on a continuum) of these complexes, as described in the previous chapter:

1. **Strategic-Commercial Complexes** are networks of great power states, military alliances and commercial companies. They are shaped largely by the interests of a privileged few, in which militarized and securitized public bodies contract out significant authority to commercial contractors. Security is derived from the construction of 'externally alienated and internally homogenous ethno-national enclaves.'[8]
2. **Human Security-Civil Society Complexes** incorporate middle power states, multilateral agencies, NGOs and social movements. They are shaped by humanitarian norms and a more global understanding of interest, in which public and multilateral agencies

form partnerships with NGOs and social movements. They aim to provide protection to the general population, especially the vulnerable, through aid, advocacy, persuasion and the legal process.

The rest of the book will map how these complexes govern the threat of mines and UXO, and the different outcomes they produce: the former, preventing legal restrictions on weapons and prioritizing militarized, commercialized, low-cost and low-quality demining; the latter, strong legal restrictions on weapons and prioritizing humanitarian, high-cost and premium quality demining.

In order to place the new systems of governance in context, this chapter will start by reviewing the traditional state-centric understanding of the world system and how it conceived of insecurity in terms of threats to the state. It will then show, however, that not all states responded to this threat in the same way. Some took a 'realist' or 'national security' position, arguing that the state's security dilemma was such that, in the anarchy of the international arena, a state's interests were constantly threatened by other powerful states. Thus realists argued states should build up a strong military apparatus to deter potential challengers and to secure national strategic and economic interests. Other states took an 'idealist' or 'collective security' position. They sought to eliminate the security dilemma by trying to entrench supranational legal norms and institutions. They thus conceived of strategic interests in a more international sense of the global public good.

The chapter will then show how both the traditional understanding of government and the classical understanding of war has been challenged by the rise of the New Wars, the growing power of non-state actors and the tendency to view insecurity as a threat to a population. In the international arena, states are now embedded in diffuse networks of public and private actors. Again, however, not all these networks react to the insecurity of the 'New Wars' in the same way. Some take a kind of 'post-statist realist' position and organize themselves into Strategic-Commercial Complexes. Others take a 'post-statist idealist' position and organize into Human Security-Civil Society Complexes. Finally, the chapter shows how relationships between organizations within these two complexes are governed by differing approaches to the 'principal-agent problem'. The Strategic-Commercial Complex is held together by contracts, in which principal organizations engage the profit motives of their agents. In contrast, the Human Security-Civil Society Complex is held together by a sense of trust and shared commitment to similar values.

Statism: Old War, Realism and Idealism

The modern international system – with its zenith from approximately 1860 to 1970 – was structured around twin pillars of 'territoriality' and sovereignty.[9] Having replaced multilayered, overlapping forms of medieval authority, the age of 'Statism' nationalized and centralized political power around sovereign governments claiming sole legitimate monopoly of force. Government was characterized by hierarchical and bureaucratic relationships of command and control, with authority extending throughout state territory and ending at its boundaries. To reduce contesting territorial claims, areas considered 'ungoverned' – for instance, the American and Russian frontiers and continent of Africa – were colonized and/or annexed, incorporating them into the statist system.[10] International politics were marked by mutual recognition that the sovereign of each state had the right to determine the nature of government, content of policy and even religious and ideological affiliations of its citizens.

As with many political institutions, the threat to the statist international order came from agents that were unwilling to play by its basic rules – in this case, mutual recognition of territoriality and sovereignty. Insecurity arose when one state refused to recognize the sovereign and territorial right of another or when local rebels contested the right of the government to rule and sought to replace it with an alternative sovereign. Since there was no supranational authority to guarantee that other states would abide by the rules, international relations were characterized by what scholars have called a 'Security Dilemma' – in which distrustful states were drawn into an unstable arms race to protect themselves that actually made all states more insecure.[11] When this broke into outright conflict – either nation-state aggression or internal subversion – war was conceived of in a Statist framework. Hostilities were conducted by uniformed agents of the state (or the rebels who wanted to be the state), regulated by formal codified laws of war and occurring in confined space and time.

This classical conception of warfare has been described by some as 'Old Wars.'[12] In this framework, mines were laid by regular armed forces (or an armed rebel movement), either as an offensive weapon targeted at an enemy state or to defend defined areas of strategic national importance such as frontlines and bases. Given the regulated nature of Old War, mines tended to be laid in accordance with the laws of war on discrimination, proportionality and the protection of noncombatants. For example, minefields in WWII were largely placed between lines on battlefields.

It would be inaccurate, however, to think that the Statist system was a monolith. Governments reacted to the basic threat of Old Wars in different ways, generally divided into two broad camps: realism and idealism.

Realist[13] conceptions of security were derived from a 'pessimistic' understanding of the security dilemma. Realists believed that the international arena, lacking a supra-national sovereign, was tragically characterized by a Hobbesian 'anarchy', a struggle for state survival amidst ruthless competition for strategic and economic interest. Stability, if possible, could only come through a balance of power – 'mutually assured destruction' – or through a *Pax Romana*, imposition by a superpowerful hegemon. Security for a state, then, could only be achieved through strength and ability to defend itself from other states. Drawing on the tradition of Sun Tzu[14] and Machiavelli,[15] many realists argued that moral reasoning could not provide adequate guidance for a leader wishing to protect the territorial integrity and sovereignty of their state. As Clausewitz wrote, 'in such dangerous things as war, the errors which proceed from a spirit of benevolence are the worst.'[16] Force was necessary to protect 'benevolent' politics – democracy, non-violence, rule of law – from external invasion or internal subversion. Likewise, international law, such as it existed, was seen as taking a back-seat to the necessity of state survival. For instance, Condoleezza Rice has cautioned against 'pursuing symbolic agreements of questionable value' that might limit ability to secure US vital interests.[17]

Realism therefore privileged military, police and intelligence agencies – the institutions of 'National Security' that fortified the state against insecurity. For instance, Rice has argued that the top priority of US foreign policy should be ensuring 'that America's military can deter war, project power, and fight in defense of its interests if deterrence fails.'[18] As a foreign policy doctrine, realism has guided many of the 'great powers'– larger states (whether economically or geographically) that have the resources and military power to maintain international dominance.[19]

Within a realist framework, mines and cluster munitions were seen as an integral part of the state's arsenal. As with other forms of arms control, great powers engaged in negotiations over how to control the use of mines and cluster munitions largely in order to maintain their freedom to deploy their own kinds of weapons, while banning the kinds used by other states. For example, the US has often argued for banning the technologically simplistic mines used by other countries, while allowing loopholes for its own sophisticated hi-tech ones. Since knowledge of explosives were seen as state secrets, closely guarded and monopolized by the military, the realist state saw demining as a task for the state. Therefore, demining priorities were determined according to their strategic importance to the security of the state and actual clearance was planned, managed and conducted by the armed forces or police. Clearance of civilian areas was conducted for strategic advantage (e.g. winning hearts and minds) or concentrated within the boundaries of the state at the end of the conflict. For example, in the aftermath of World War II, demining and clearance of unexploded ordnance was organized and funded internally by each state, which generally used its military to manage clearance.

The traditional challenge to realism came from the diverse set of thinkers collectively known as 'idealists.' They argued that while the threat to the state may be very real, realism was a recipe for instability and injustice by legitimizing the use of violence for narrow self-interest. They conceived of the security dilemma in less tragic terms, believing that by creating norms, legal rules and multilateral institutions, states could reduce the potential for mistrust and miscommunication.[20] By doing so, they believed they could eliminate, or at least reduce, the development of a security dilemma. For idealists, security was collective and came from formalizing the rules guaranteeing territoriality and sovereignty and building international mediating structures.[21] Modern idealists[22] have thus looked to intergovernmental institutions, such as the United Nations, as guardians of global security. Idealism was also rooted in a long philosophical tradition arguing that a stable and just society must be rooted in moral reasoning.[23] For instance, Kant believed that 'all politics must bend its knee before morality'[24] and that all people, even foreigners, must be treated 'as an end and never merely as a means to an end.'[25] This idea has influenced the development of human rights and humanitarian norms.

Unsurprisingly, the concept of 'Collective Security' has been more popular among smaller powers, such as the Scandinavian states, which have seen international law and institutions as a means to blunt the aggressiveness of great powers.[26] Cynics have suggested that collective security was a way for little powers to tie down superpower Gullivers with treaties and international regulation. For this reason, some have argued that 'idealism' was thus really only a 'realism of the weak.' For instance, some realists, notably the 'English School', believed that, driven by interests, states came together to form a stable 'international society', including norms and institutions, to reduce unpredictable behaviour.[27] Collective security was thus still self-interested, only a broader and more global understanding of state interest that was willing to sacrifice the immediate gain of narrow national self-interest for long-term security.

Like realism, idealism was still deeply rooted in the classical Old War conception of conflict, with nation-state aggression seen as the primary source of insecurity. Traditional UN peacekeeping, carried out by uniformed military personnel, respected the sovereignty of antagonistic states and typically deployed along the former frontlines. Likewise, traditional humanitarian groups like the Red Cross tried to stay neutral and apolitical, as war was seen as the preserve of states.

Predictably, traditional idealists wanted mines and cluster munitions regulated by international humanitarian law. They still saw demining as a military and state-led activity but put it under the control of UN peacekeeping operations. Humanitarian or potential peacebuilding impact, rather than national interest, were to guide demining priorities. For example, in the early 1990s, the UN took a leading role in mine clearance in regions such as Afghanistan, Angola, Mozambique and Cambodia.

Post-Statism: Global Governance and Privatization

Globalization – the increasing breadth and depth of transnational 'interconnectedness'[28] – has led to a hollowing out of the state, as governments have privatized key industries and services and liberalized regulations on trade and finance.[29] While governments are still among the most powerful actors on the globe, the primacy of the statist model of world politics is being challenged. Non-state actors grow more prominent on the world stage. Many multinational companies now command greater resources than small countries and NGO movements can instigate new international treaties. Moreover, states are beginning to form strategic partnerships with these non-governmental agents in complex networks of public and private actors.

Political scientists have described this as a shift from government to governance.[30] Political power can no longer simply be traced upward through hierarchical bureaucratic relationships. Rather, understanding politics now requires mapping the distribution among and exercise of power by nodes embedded in networks.[31] Echoing Hedley Bull, Mark Duffield describes these complexes as 'neo-medieval' or 'neo-feudal' as they hearken back to pre-Westphalian Europe, constituting networks of public and private actors with overlapping sources of authority and power.[32] Moreover, according to Duffield, insecurity is perceived in biopolitical terms – a threat to a population, whether a specific few or as a whole, rather than the state.[33] An international system comprised of global governance networks implies both a very different perception of and response to insecurity than the classical Old Wars.

Both the traditional realist and idealist understandings of security have been challenged in the post-Cold War era by the growing realization that states, even rebels with statist ambitions, are not the only potential sources of disorder. On one hand, humanity faces existential threats from natural phenomena such as HIV/AIDS and natural disasters like the Indian Ocean Tsunami or Hurricane Katrina. On the other hand, globalization has contributed to the growth and interconnectedness of transnational non-state networks incorporating organized crime, warlords, profiteers and extremist movements.[34] These networks are at the center of the contemporary conflicts and, argues Mark Duffield, represent a new form of social organization; parodying the characterization of conflicts as 'complex political emergencies' he calls them 'emerging political complexes.'[35] As a result, in many parts of the world, a violent threat to a person's life is now more likely to come from a warlord's private militia than regular government forces. Some have argued that these deregulated, privatized and globalized conflicts are a type of 'New War', characterized by targeting of civilians, prolonged, hostilities and exclusivistic ethnic, religious and sectarian ideologies.[36]

Mine-laying in these New Wars, was significantly different from the Clausewitzian patterns of WWII. Arising out of the strategic logic of the New Wars, minefields were used to target or control the migrations of civilians. Displaying a lack of coherence, discipline and commitment to the laws of war, troops in these conflicts often laid mines in non-linear patterns, failing to map and mark areas of contamination.

Traditional understandings of national and collective security are unprepared to deal with the threats from New War Complexes. Armed forces are less well adapted to managing insecurities like climate change, international terrorism or organized crime. Moreover, contrary to traditional realist thought, overcoming such global problems requires states to transcend narrow self-interest. Traditional idealist answers also fall short. International law is only signed by governments; it is difficult to persuade a mafioso or warlord to abide by it. In addition, vast international bureaucracies, relying on state contributions of funds, troops and materiel, lack the political will and flexibility to respond quickly to massive disorder in places like Rwanda and Bosnia. Therefore, the security dilemma has become far more complex than in the statist era. It has more layers, occurring at the sub- and supra- state level, with companies, ethnic groups and individuals also facing their own security dilemmas. It also involves more actors, as states face threats from non-state agents that cannot be deterred by traditional military means, nor engaged with in international treaties and institutions.

Just as the imperial metropolitan states annexed 'ungoverned' frontiers in the statist era, the New Wars have prompted the powerful actors (both state and non-state) to innovate new technologies of power for governing the new zones of global insecurity, which Duffield calls the 'global borderlands.'[37] It is in these very frontiers of the international system that one finds the concentration of mine and UXO contamination.

Duffield argues that while globalization and privatization are seen primarily as the retrenchment of the state, they have also enabled states to project their power in new ways. He argues that the powerful 'metropolitan states' at the center of the global system attempt to regulate insecurity in the 'periphery' through trade, diplomacy, humanitarian assistance, development aid and military intervention. Contracting these activities out to private actors enables states to penetrate farther into the 'borderlands', while avoiding the political cost of direct annexation or occupation.[38] Instead of operating solely at the inter-state level, contractors are able to operate 'biopolitically' at the level of the population.[39] Duffield sees the resulting 'polyarchical, non-territorial and networked relations' between metropolitan governments and their non-state partners as an 'emerging system of global liberal governance.'[40]

While this is an important theoretical starting point, Duffield's 'emerging system of global liberal governance' is a conflation of many different complex systems. The 'metropolitan states' and the 'international non-state actors' that he refers to are not monoliths. Great powers may not act in the same manner as middle powers. International NGOs may not act in the same way as commercial companies. In order to understand the emerging complexes of liberal governance, one has to understand the different constituent actors, their motivations and the institutions that structure relationships between them. Different actors and institutional structures may produce different outcomes, as this case study of mine action will illustrate.

Indeed, just as traditional statist understandings of international relations maintained that there were two basic rationales for international action (realist and idealist), there are at least two ideal types (or poles on a continuum) of these new complexes of global governance. Both have moved away from state-centrism to develop institutional complexes incorporating both public and private actors. However, one type, called here a Strategic-Commercial Complex, is 'post-statist realist', as it maintains a pessimism about the security dilemma, believing that in the new insecurity, one is tragically forced to privilege the security of one's 'own people' and possessions over that of 'the other.' The other type, called here a Human Security-Civil Society Complex, is 'post-statist idealist', in that it is shaped by a commitment to multilateralism and a perception of the primacy of global public interest over narrow private or parochial interest. These two developing forms of governance are explored in detail below.

Post-Statist Realism: Strategic-Commercial Complexes

The actors within the Strategic-Commercial Complex inherit from traditional realism a pessimism about the security dilemma and the tragic necessity of forceful protection. However, the post-statist realist conception of the state moves away from the centralized Weberian state bureaucracy toward the out-sourced neo-liberal 'hollow state.'[41] Therefore, post-statist realism attempts to govern insecurity through creating enclaves of security for a subpopulation (rather than the state) and managing the insecure 'borderlands' through a complex of commercial security companies in contractual partnership with public military, intelligence and police agencies. Non-military aspects of foreign policy, such as humanitarian and development aid, are both 'securitized' (made to serve security objectives) and 'marketized' (outsourced to competing private actors). Other commentators have described this phenomenon as a 'strategic complex,'[42] 'new security-industrial complex'[43] or 'disaster capitalism complex.'[44] However, in this study, this institutional system will be called a 'Strategic-Commercial Complex' to indicate both its realist understanding of strategic issues and extensive involvement of for-profit actors.

The evolution of this system is explained below by four interrelated factors: 1) a growing distrust of the public sphere and enclavization, 2) the neo-liberal critique of the state, 3) the growth of the private security sector and 4) the utility of contractors in dispersing political accountability.

The first factor that distinguishes post-statist realism from its traditional cousin is a growing distrust in government and fear of insecurity in public spaces, even inside states' borders. The ancient Greeks celebrated the public sphere; their word for 'private' – *idios* – is the root of the English word 'idiot.' However, the word 'public' is now often associated with poor quality – such as public works, public education and public health care.[45] Public space also has connotations of danger; discourses about 'public toilets' or the 'the streets' portray them as places of violence, crime and predation.

These fears of the 'public' have occurred in tandem with increasing doubts about the ability of the public sector to provide security, indeed suspicion that the government itself may be a source of insecurity. In the US, the Vietnam War, CIA 'dirty tricks', Watergate scandal, Iran-Contra and 1993 Waco siege, have all contributed to public suspicion of the government.[46] This has developed alongside growing fears of violent crime and international terrorism. In much of the developing world, citizens fear authoritarian, corrupt or weak public security forces and are forced to look for private solutions to their security problems, such as militias, vigilantism or private guards.

This suspicion of public security provision has contributed to a trend of 'enclavization' or 'forting up'[47], in which individuals, communities, corporations and even government agencies seek 'protection through separation,'[48] by barricading private space behind rings of protective walls, barbed wire, armed guards and electronic surveillance systems.[49] In many countries, oil companies and other extractive industries no longer rely on state security organs to protect their business, preferring private security companies that guard highly secure enclaves for their expatriate workers.[50] Likewise, in South Africa,[51] the US[52] and elsewhere, there has been a precipitous growth in gated communities. Researchers have found that, driven by a fear of crime and dissatisfaction with public services, residents of such suburban fortifications seek to privatize public spaces, such as roads, playgrounds and parks, and exclude access to those people perceived as security threats. In many of these communities, control of public services, such as protection, waste management, street maintenance and even governance of the community itself are contracted to private companies.[53] The walled enclave took on a new meaning in the war in Iraq, with a significant portion of the US command and control holed up in the so-called Baghdad 'Green Zone.' Journalist Rajiv Chandrasekaran has described the Green Zone as a cloistered 'Little America' where 'Iraqi customs and laws didn't apply' and 'Whatever could be outsourced was.'[54]

Gated communities form at the sub-state level, and actually hearken back to the medieval practice of fortification, in which the sovereignty of the state overlapped with the privatized protection of the feudal elite. Enclavization and fortification also occurs at levels beyond the nation-state. In the case of Israel's 'separation barrier' and settlements, fortification may envelop space and 'extraterritorial islands'[55] that are outside the official boundaries of the state. Likewise, the 'Green Zones' in Iraq and Afghanistan exist in one state, but are controlled by the US and its allies, similar to overseas imperial fortifications in strategic ports or crusader citadels in the 'Holy Land.' They are, in effect, aiming to protect specific and privileged subpopulations rather than the borders and institutions of the state.[56]

While landmines are now used less now than 20 years ago, they are a weapons system designed for protecting borders and enclaves. For example, in the Bosnian war, factions used landmines to demarcate and protect 'ethnically-clean' enclaves. Several great and regional powers, such as the US, Russia, China, India, Pakistan and Israel, continue to assert a right to create secure enclaves for themselves by creating rings of minefields and cluster munitions strikes around their borderlands and strategic locations.

A related factor behind the development of Strategic-Commercial Complexes is the neo-liberal belief that states are inefficient, create market distortions and encourage rent-seeking behavior. Based largely on the 'Chicago School' of economic theory, represented particularly by Friedrich Hayek,[57] Milton Friedman,[58] and George Stigler,[59] neo-liberals called for a 'rolling back' of the state through deregulation, liberalization and privatization. They believed markets, when freed from public interference, would result in optimality and equilibrium. Privatizing public services would spur innovation, increase quality and save costs.[60]

The resultant system, in which the government acts basically as a contract manager, outsourcing much of its activities to the private sector, has been described as the 'Hollow State.'[61] For instance, many mine and cluster munition weapon systems are developed for militaries by private industry. The new US landmine systems, for example, are being developed by defense contractor Alliant Techsystems.[62] Contracting out has not only been limited to the domestic sphere. Several countries have begun to contract significant elements of their foreign policy to commercial companies. For instance, US foreign aid programs have increasingly been implemented through private for-profit development contractors, with NGOs also receiving significant funds.[63] USAID especially has pressured NGOs to act in a more commercial manner, making them bid for contracts and grants, putting them in competition with each other and causing them to 'behave like for-profit organizations.'[64]

Mine action has been particularly affected by the neo-liberal trend of contracting-out. One commentator observed that it might be the most commercialized sector of humanitarian aid.[65] This is because many mine action policymakers and researchers believe the rigors of competition lead to better demining performance. Fitz-Gerald and Neal, for instance, argued in favor of using commercial companies, saying they were efficient.[66] Likewise, Eddie Banks has called for 'a more business-like response to mine action.' Using quantitative analysis, he demonstrated that tendering, rather than granting, demining funding led to higher productivity and lower costs in Bosnia. Subtly criticizing the good intentions of international NGOs, Banks wrote that if demining wished to be 'truly humanitarian' it must adopt a contracting model.[67]

While there was initially some reluctance among governments to contract out activities related to the use of force – once considered the very core of state competency – this is dissipating in many countries. Private security guards have long guarded buildings or valuable objects. However, in the 1990s the industry shifted from guarding 'things' to guarding people, as companies became involved in transporting prisoners, running prisons and managing asylum seeker detention centers.[68] Concurrently, the US began to contract elements of peacekeeping operations, the 'War on Drugs' and management of military bases to private security companies, notably DynCorp and Kellogg, Brown and Root (until recently a subsidiary of Halliburton).[69]

In addition to growing demand for private protection, there were also supply side factors. The military draw-down at the end of the Cold War and collapse of apartheid in South Africa released into the private sector significant amounts of military personnel and materiel. With many of the former Soviet arms stocks up for the highest bidder, and a large unemployed workforce of people trained in military operations, conditions were ideal for 'military entrepreneurs' to capitalize on the growing insecurities of the New Wars. The 1990s saw a steady growth of the private military sector, with firms like Executive Outcomes, Sandline International and Military and Professional Resources, Inc. (MPRI) in the headlines.[70] The rise of the private soldier, like the return of the fortress, echoes back to older, medieval phenomena, such as the Italian *condottieri* (literally 'contractors', private soldiers employed by the Italian city states in the medieval and renaissance eras) and the Vatican's Swiss guards.[71] However, modern private military and security companies are distinguished from their feudal predecessors by their corporate and globalized nature.[72]

The enthusiasm for private security is due partially to the way in which it can reduce political liability for a state's actions, making the chains of authority and accountability more diffuse and networked. Contractors give the state 'plausible deniability' and lower possibility of exposure. If seen as abhorrent by the electorate, a contractor's actions can be dismissed as those of a rogue company rather than the fault of the state. This allows states room to 'innovate' around the edges of legal and moral norms when faced with security threats, while minimizing blame. Likewise, electorates seem less concerned when a contractor is killed than a regular soldier. By outsourcing to the private sector, state leaders can reduce the risk that soldiers will be killed and thus reduce oversight from public and legislative bodies.[73]

The 'War on Terror' has seen a massive expansion in the willingness of countries like the US and Britain to contract out military services once seen as the reserve of the state, including managing supply chains, interrogating prisoners, guarding military bases, destroying abandoned ordnance, training military and police forces and analyzing secret intelligence. This has created explosive growth in the private security market, with older companies like ArmorGroup and DynCorp expanding rapidly, and newer start-ups, like Blackwater, Erinys and Triple Canopy enjoying a meteoritic rise. According to conservative estimates, there are now some 21,000 commercial security personnel in Iraq.[74] This was partly because the retrenchment of state security organs following the Cold War meant that the post-9/11 rapid expansion of security capabilities required hiring many of these personnel back, through the private sector.[75] However, there is also an enduring belief that the private sector is able to do many activities more efficiently than the state, and that its flexibility enables it to more effectively combat the diffuse and *ad hoc* 'New War Complexes.'[76] For example, one commentator has referred to the private military company Blackwater as 'Anti-Qaeda' or 'Al Qaeda for the good guys', arguing that its diffuse, networked and private nature makes it more attuned to fighting terrorist networks than the US government.[77]

The commercial demining market has risen in tandem with the private security market and there are many linkages and overlaps between the two industries. There are many companies provided both services since employees are often recruited from similar backgrounds and operations occur in the same conflicted locations.[78]

The result of the above trends is a Strategic-Commercial system in which the world is pockmarked with privileged enclaves for specific subpopulations with high levels of protection, insurance and services. These enclaves are protected by rings of fortification, both real (such as barbed wire, minefields and concrete bollards) and 'virtual' (such as CCTV), and access to both public and private security forces. Those outside the ring, in the 'Red Zone', are surveilled, managed, combated and governed by complexes of public agencies and contractors. This system is suggestive of Garrett Hardin's 'Lifeboat Ethics.' Hardin argued that the world's environment and resources tragically can only support a small portion of the population; those privileged to live in the 'lifeboat' of the developed world must unfortunately prevent 'boarding parties' – i.e. immigrants – who might threaten to overwhelm the boat.[79]

Figure 1: Enclavization in a Strategic-Commercial Complex

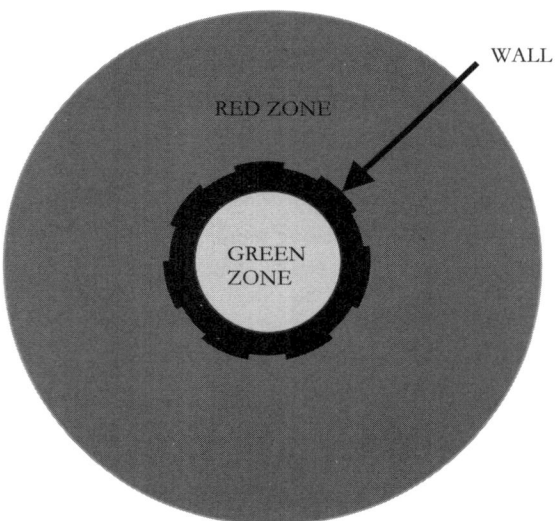

The Strategic-Commercial Complex is thus a segregative system that allocates physical and social protection to populations according to mixed criteria of the market (ability to pay for protection) and nation/ethnicity (chance of birth in a well-protected country or community). Though some have argued that private security and services relieve the burden on central public services, Blakely and Snyder found that gated communities actually resulted in 'a two-tiered system of security: more for those who pay to supplement police with private security, and less for those who cannot or do not do so.'[80]

Thus post-statist realism maintains realism's sense of a 'great divide' between the privileged inside the ring of protection and 'the others' outside. However, unlike traditional realism, these rings are not necessarily constructed at the borders of the nation's territory. Instead the divide can exist below or beyond the state boundaries. Similarly, post-statist realism maintains the interest-driven nature of realism. However, the Strategic-Commercial Complex is not a unitary actor, but rather a network of actors, each with their own particular interest-driven agendas. While the state has always consisted of competing agencies and interest groups, it is the complexity of this public-private network that creates especially diffuse and complicated systems of decision-making and authority.

Figure 2: A Strategic-Commercial Complex

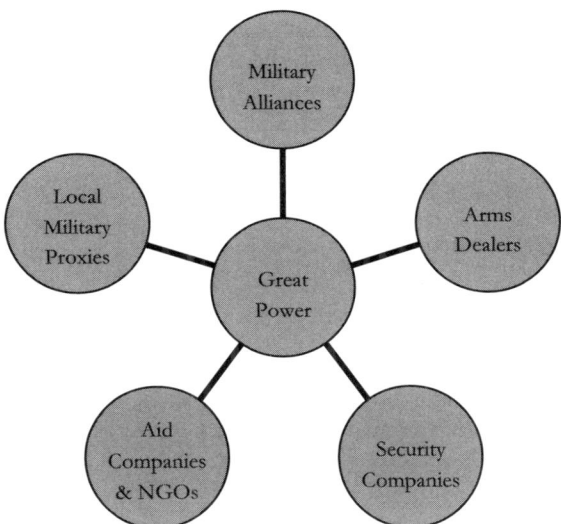

In a post-statist realist understanding, the Strategic-Commercial Complex develops landmine and cluster munition systems through public-private partnerships and reserves the right to deploy them to protect the privileged. Likewise, mine action activities are conducted either to secure key 'green zones' or for other strategic reasons, through contractual relationships with commercial demining companies or local security forces.

It is important to understand how all these disparate networks hold together and coordinate action. While states were never completely unitary actors, the Strategic-Commercial Complex, incorporating both public and private actors, faces many more difficulties than governments in coordinating movement in one particular direction. Indeed, many studies of privatization have shown that contracting relationships inevitably run into 'principal-agent' problems. That is, why should one organization do what another organization tells it to do? What is to stop it hijacking or subverting the overall agenda or following its own altogether? As one study put it, 'The chance that agents do not share the same interests and utility choices as their principals is substantial.'[81] The literature on contracting and management has paid a great deal of attention to this issue and proposes a whole host of methods, incentives and institutional structures to overcome it.

One such model is that of a 'principal-agent' relationship, in which the principal (in this case a state agency) pays an agent (such as a business) to do the work for them. It thus relies on the agent's pursuit of self-interest – profit – to ensure the principal's objectives are met.[82] To avoid being overcharged for the service, and to choose a contractor best suited for the job, the principal will often hold a competitive tendering process. Once the contract is awarded, the principal faces several problems. Since the agents are the ones actually doing the work, they often have more information than the principal about the context in which they are working, the quality of the job they are doing and ways to bend the contract to their own advantage. These information asymmetries mean that principals risk being deceived about the actual progress of the work they wish to see done. If the agent fails to meet standards, the principal has fewer institutional levers to control the process than if it had conducted the work itself. Therefore, contracting procedures often require very specific language as to the work needed, thorough systems of audit, monitoring and evaluation and close supervision.[83] An agent has little incentive to question the overall appropriateness of the work assigned, question the political agenda of the principal or suggest changes in the principal's behavior in any way other than to make the terms of the contract more favorable.

As will be shown in greater detail in later chapters, the principal-agent model governs many of the relationships between the public and private actors within Strategic-Commercial Complexes doing demining. It seems that when strategic or commercial interests are at stake and the principal wants to control the process very tightly, they opt for a principal-agent model of contracting. This is probably because short-term commercial contracts give the principal much greater political control over the activities of the agent.

In the case of demining, a donor will put out competitive bids for specific mine clearance tasks and then award them to commercial demining companies. To avoid vagueness that could be exploited by the agent, contracts usually specify the precise geographical area to be cleared and impose penalties for shoddy or inefficient work. As a result, the demining agency has little room to determine the nature of the project, no incentive to go outside the terms of their contract and, when there is significant competition, little room to negotiate for better conditions for their workers. Finally, in this relationship, the only accountability is that of the agent to the principal. If the agent has ideological objections to the principal's activities, they have little incentive to raise them. Thus commercial demining companies have not joined the campaigns to ban landmines or cluster munitions. Moreover, since the agent is only accountable to the principal (and its shareholders), there may be little 'downward' responsibility to the community in which they work. They may feel little responsibility for the welfare of the deminers, the appropriateness of their activity or the broader impact of their demining operations on the peace and reconstruction processes.

Figure 3: The Principal-Agent Relationship

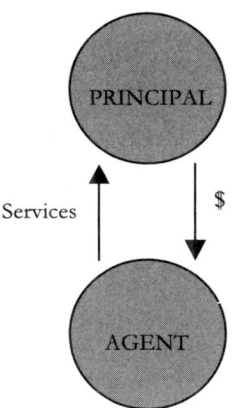

There are three main criticisms of the post-statist realist position. Firstly, some have normative concerns about a system that allocates security and services to the privileged. Such a system is seen as inherently unjust, reinforcing patterns of social exclusion through a form of segregation that can only be sustained through violence. Some argue that the very process of segregating into 'green' and 'red' zones' generates discontent and resentment, leading to disorder and insecurity.[84]

Secondly, neo-liberalism, privatization and contracting-out have not always led to the expected gains in quality and cost-savings in social service provision.[85] This is partly due to prosaic technocratic factors such as poorly written contracts, lack of sufficient competition, high cost of contract oversight and misplaced faith in commercial companies' ability to deliver. For instance, in the demining sector, the analysis of commercialization advocate Eddie Banks focused only on the impact of tendering on speed and price,[86] while this author has shown that commercial demining often produces poorer quality and safety practices.[87]

More fundamentally, some say the diffuse nature of authority in the Strategic-Commercial Complex and the profit-motive of commercial contractors is prone to abuse. With information asymmetries and low standards of rule of law in conflict zones, it is difficult to impose discipline on companies, as one can see in the stunning levels of corruption and overcharging in Iraq reconstruction.[88] In his study of privatization in Bosnia, Timothy Donais argued that

> Attempts to push through liberalization and privatization in the absence of a viable and enforceable legal framework can be expected to almost invariably produce unanticipated and distorted outcomes.[89]

Within the demining sector, the World Bank,[90] Geneva International Center for Humanitarian Demining (GICHD)[91] and Bolton and Griffiths[92] have all warned that tendering systems can be corrupted in post-conflict countries (where mine action inevitably takes place). Likewise, private security (and demining) companies often hire personnel that have been rejected from the state sector, either because they were somehow unfit for their job or involved in more nefarious activities (e.g. apartheid or Pinochet-era soldiers).[93] The seeming impunity for private security companies involved in human rights violations in Iraq and Afghanistan is particularly worrying.[94]

Finally, mirroring President Eisenhower's concerns about the military-industrial complex,[95] some commentators worry that an expanding private military sector creates a 'war lobby' – vested interests in favor of an aggressive and combative foreign policy. For example, military contracting companies lobbied in favor of a US intervention in Iraq.[96] The presence of such a sector may also make public policymakers feel that war-making is easier, due to the presence of extra troops, less subject to public scrutiny. Certainly private military companies are less likely to lobby against starting new wars. Lastly, there is the danger that, like the Italian *condottieri*, privateers may use their training, resources and materiel to carry out their own agenda. The involvement of several former private military personnel in the 2004 attempted coup in Equatorial Guinea displayed the danger that security contracting has the potential to actually be destabilizing and create new insecurity.[97]

Post-Statist Idealism: Human Security-Civil Society Complexes

While some who object to the Strategic-Commercial Complex wish to return to either of the statist conceptions of security, several states and organizations have developed a post-statist idealism, which shares idealism's optimism that creating 'international society'[98] – international norms, law and institutions – will abolish security dilemmas, through regulating, governing and demobilizing sources of insecurity. However, post-statist idealists direct their efforts not only at relations between states, but also at diffusing security – in the form of aid, institutions and norms – out into the 'global borderlands.' The Human Security-Civil Society Complex thus operates 'bio-politically' at the level of human populations, rather than states. In order to achieve this penetration into the 'borderlands', it operates through complex networks of partnerships between public, multilateral and private organizations.

Unlike the Strategic-Commercial Complex, the private actors that hold prominent roles in this mode of foreign policy tend to be NGOs and social movements. Moreover, while the Strategic-Commercial Complex tends to be favored by great or regional powers (like the US, Britain and Israel), this model is supported by smaller states, particularly internationally engaged middle powers, like the Scandinavian countries and Canada. This model has been variously called the 'Middle Power-NGO Coalition,'[99] the 'Norwegian Model' of foreign policy[100] or, by Mary Kaldor in parody of her earlier work, 'new peace.'[101] However, in this study it will be called a 'Human Security-Civil Society Complex.' As will be described in more detail below, this brings together two concepts. 'Human Security' is an attempt to focus security on the human being, rather than on the nation-state.[102] The term 'Civil Society' refers both to voluntary organizations like NGOs and social movements, but also to a more normative 'notion of minimizing violence in social relations ... [and] the public use of reason as a way of managing human affairs....'[103]

The evolution of this type of system can be explained by three interrelated factors: 1) the increased activism of middle powers following the end of the Cold War, 2) growing norms of humanitarian intervention, human rights and human security and 3) the rise of global civil society and 'socially responsible' business.

International relations scholars have found that 'middle powers'[104] often play the role of innovating norms, providing third party mediation, advocating multilateralism and championing generous foreign aid appropriations.[105] While a broader and more global understanding of self-interest than great powers, such action is not purely self-sacrificial on the part of the middle powers. They have an interest maintaining a high profile in the international system, so they are not overlooked and ignored by the great powers. They have an interest in a law-governed world, to reduce great power bullying as well as the chance of insecurity spreading to their own territory in the form of organized crime or terrorism.

During the Cold War, middle powers did not have tools to effectively pursue these interests. They had neither the military clout, economic power nor Security Council vetoes necessary to coerce other nations into following their agenda. Indeed, 'most middle powers would be expected to toe the line behind one or the other superpower....'[106] The end of the Cold War has given middle powers more space to raise these concerns in the international arena and they have been able shift the agenda through the use of 'soft power.'[107] By linking with international NGOs and multilateral agencies, they have contributed to the development of new international norms constraining the actions of the great powers or appealing directly to the electorates within them.[108]

As Duffield would indicate, this is not simply a hollowing out of the middle power state; rather it is simultaneously an augmentation of it, for the middle power is able to assert itself and project its power into the international arena and penetrate the global peripheries in ways it was never able to do before. Through these partnerships, middle powers and NGOs have been able to carry out democratization, humanitarian and development programs in the developing world and campaign for the International Criminal Court, nuclear nonproliferation, disability rights, small arms control and a ban on child soldiers. The campaigns against mines and cluster munitions are particularly indicative of the Middle Power-NGO partnership.

The post-statist idealist position has evolved in part from critiques of the state from the left (rather than the neo-liberal critiques from the right). Throughout the 1960s, 70s and 80s, the peace movement and human rights community began to see the state as a potential source of insecurity, rather than protection. They argued that state structures, especially those of the executive branch, have authoritarian tendencies toward violent repression and excessive surveillance. As a result, they believed the state needed to be held in check by society, both domestically and internationally.[109] Occurring concurrently with the ongoing expansion of human rights and humanitarian norms, this has contributed to developing ideas of conditional sovereignty and humanitarian intervention.[110] Unlike idealists of the past, who saw collective security as a means to protect sovereignty and humanitarianism as a neutral activity, today's idealists often believe in the right of the international community to intervene at the level of the population in countries that fail to protect human rights. As then UN Secretary General Boutros Ghali declared in 1992 'The time of absolute and exclusive sovereignty...has passed.'[111] In its place, the UN General Assembly has declared that the international community had the 'Responsibility to protect populations from genocide, war crimes, ethnic cleansing and crimes against humanity.'[112] The manner in which populations in the metropolitan core of international politics were sensitized and became politically mobilized around the 'humanitarian' issues of landmines and cluster munitions in the global periphery illustrates this diffusion of humanistic and interventionist norms.

These related strands of thought have coalesced into the concept of 'Human Security', which adherents offer as a kind of neo-Kantian critique of national security. According to this view, the objective of security must be to manage and reduce threats to a human being's right to life and dignity, rather than to the strategic interests of the state.[113] As the Human Security Centre explained, 'While national security focuses on the defence of the state from external attack, human security is about protecting individuals and communities from any form of political violence.'[114] This calls on policymakers to consider ways to protect populations of human beings from a wide range of threats generally not considered matters of 'national security', such as domestic violence and crime. Moreover, it calls on governments to see security through a universal lens. Therefore, the security of a person from another nation is just as important as that of one's own.[115] It is, then, a broader and more global understanding of interest than the national interest of realists.

The development of Human Security-Civil Society Complexes owes much to the rise of global civil society – the 'emergence of horizontal transnational global networks' demanding 'global rule of law, global justice and global empowerment.'[116] Such networks include the increasing numbers of international NGOs and grassroots organizations, as well as transnational social movements such as the anti-globalization movement, the International Campaign to Ban Landmines and the campaign for the International Criminal Court. The increased prominence of NGOs can in part be explained by the distrust of the state from both the left and the right, and trends toward privatization of government services. The current aid orthodoxy believes that NGOs are often more effective at delivering social services than government or multilateral agencies because they are considered less bureaucratic and closer to the communities in which they work. By the end of the 1990s, NGOs were distributing more humanitarian and development aid than the UN.[117] While some scholars believe the growth and effect of these movements has been exaggerated,[118] there is no doubt that increasing amounts of international aid are dispersed through NGOs and that such groups have more influence in the shaping of international opinion that 50 years ago.[119] For example, within the demining community, several donors and advocates have argued in favor of using NGOs, saying their humanitarian motives put them on the leading edge of advocacy efforts, developing standards and putting the community before contract.[120]

Moreover, there has been a growing interest in corporate social responsibility. This represents an attempt by companies to do business in a way that is conducive to and works in partnership with civil society. Therefore, ethical commercial phenomena like Code Red, the Fair Trade label or the Cooperative Bank show that the Human Security-Civil Society Complex can incorporate for-profit actors. Some NGOs and 'socially responsible business' may be using humanitarian and cosmopolitan ideals instrumentally – as a marketing strategy to win grants or new customers. Indeed, much criticism of NGOs focuses on how they often have self-interested motivations of organizational survival. However, NGO advocates respond that at least their institutional survival interests are somewhat linked to humanitarian goals and the global public good, rather than profit alone.

The result of the above trends is a Human Security-Civil Society Complex that responds to the insecurity of the New Wars by trying to raise the level of protection for broad populations. This position claims to deconstruct the traditional 'Great Divide' between domestic and international politics, favoring instead a more global and universalist understanding of security – at least a rhetorical commitment to the security and welfare of the global population, especially the poor and oppressed. Rather than trying to shore up 'green zone' enclaves, post-statist idealism claims to make the 'red zones' of the world safer – a little greener, so to speak. By creating links with populations in the 'red zones', they aim to diffuse safety (through deploying aid workers, peacekeepers and diplomats) out from the core into the periphery.

Figure 4: Diffusion of Security to the Periphery in a Human Security-Civil Society Complex

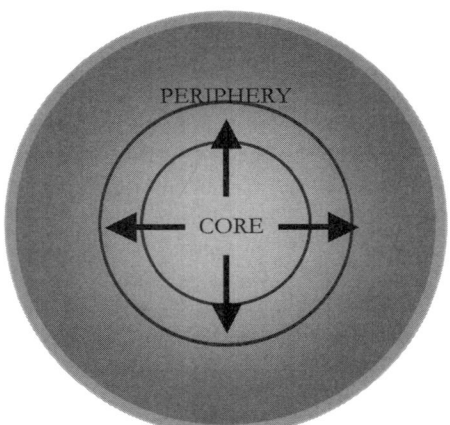

Unlike traditional idealism, the Human Security-Civil Society Complex sees this as a task not just for an intergovernmental club like the UN, but rather for partnerships between organizations of all types which share their mission. This is often expressed institutionally as a network of middle power states, UN humanitarian agencies and NGOs. To return to the metaphor of 'neo-medievalism,' the Human Security-Civil Society Complex hearkens back to the partnerships between medieval monarchs and the transnational, non-governmental and ideological institution of the Catholic Church.

Together, this partnership aims to assist those most in need through a commitment to aid for development, humanitarianism and peacebuilding. For example, the Scandinavian states have the highest levels of aid per GNI, much of which is allocated through NGOs and multilaterals. This is supplemented with the money NGOs raise privately. Moreover, this Human Security-Civil Complex dedicates significant effort to building peace in regions of conflict that may have low strategic importance. For instance, Norway, in partnership with NGOs, has spent considerable resources on peacebuilding efforts in Guatemala, Sri Lanka and Sudan.[121]

Akin to its traditional idealist cousin, the Human Security-Civil Society Complex has an aversion to the use of force, except when used to protect people from genocide and crimes against humanity or restoring stability to regions affected by conflict.[122] Unlike traditional peacekeeping operations, however, this Complex does not see peace processes as simply about an agreement between two states. Rather, it tries to build support for peace at the local and global levels and seeks protection for the victims of war, such as civilians and internally displaced persons.

At the global level, NGOs and middle powers work together to develop international law through advocacy and diplomacy. Working in partnership allows the Human Security-Civil Society Complex an opportunity to mobilize on several different levels. They can operate both through the state-centric traditional diplomatic channels, as well as through the 'Second Track' by using NGOs to build links to citizens and society.[123]

With its enthusiasm for the development of international law and its concern for the civilian victims of conflict, the Human Security-Civil Society Complex has campaigned for strong international restrictions against the use of landmines and cluster munitions, both through transnational advocacy networks and diplomacy. Demining and other mine action programs are conducted through humanitarian NGOs. Rather than prioritizing demining tasks according to strategic or commercial interests, this approach emphasizes those who need demining the most, either because of life circumstance or proximity to contaminated areas.

Figure 5: A Human Security-Civil Society Complex

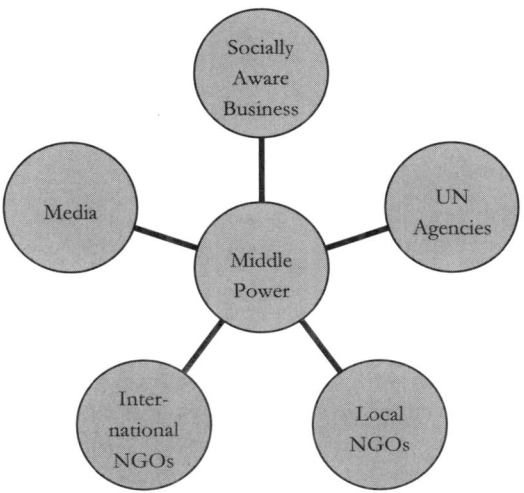

Like the Strategic-Commercial Complex, the Human Security-Civil Society Complex is a network of actors and thus requires some system of power relationships to coordinate action. However, it tends to rely on a different response to the principal-agent problem, called a 'principal-steward relationship,' in which the principal devolves implementation of a task to a trusted steward who believes that 'collectivistic behaviors have a higher utility than individualistic, self-serving behaviors.'[124] The principal may trust this steward for many reasons: a long track record of effective implementation, shared values or a shared mission. Most importantly, the steward's idealistic and professional commitment to the achievement of the principal's goals is trusted.[125] Therefore, the principal will control few of the operational details and largely trust that the job will be done effectively, perhaps with some basic monitoring and evaluation. This can actually reduce costs, as there is less need to rebid contracts or micromanage the steward. Moreover, the stability of the relationship acts as an incentive for the steward to increase the quality of programming by investing in increasing the capacity of their organization and staff, as they can expect long-term funding.[126] Because of the confidence of the principal in the steward, the steward also has more room than an agent to question the principal, especially if the steward is not wholly reliant on the principal's funds. Therefore, the steward may have a great deal of power in determining the actual nature of the tasks to be implemented, and how the work will be done. Indeed, principal-steward relationships create systems that 'facilitate and empower rather than…monitor and control.'[127] There is a more equal relationship between the principal and steward than between a principal and an agent.

Figure 6: The Principal-Steward Relationship

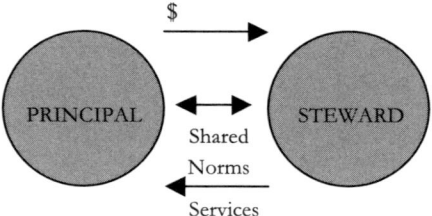

The potential problems with the principal-steward relationship are two-fold. Firstly, the principal effectively gives up some control over the precise nature and operation of the activities conducted by the steward. This means the steward may actually conduct activities the principal had not originally intended. Secondly, it is possible that the steward, by having more information about its work and the general context, may abuse its trusted position, hiding malpractice or corruption from the principal.

As will be shown in the rest of this book, many of the relationships in the Human Security-Civil Society Complex take the form of the principal-steward model. In the case of demining, a state donor will give a grant (rather than a tendered contract) to support the activities of an NGO, often over a period of time (such as a year), rather than specifying geographic areas to be cleared. Because there is often a shared mission and values between the principal and steward, they sometimes collaborate on advocacy issues such as calling for a ban on landmines or cluster munitions. Due to the mutual trust in this relationship, the steward may also have more room to challenge, shape or set the agenda of the principal, such as suggesting particular demining projects, influencing priorities and relaying community concerns upwards. There is thus more room for the steward to be receptive to its responsibility to additional actors, such as beneficiaries, other than the principal.

While many admire the motivations of post-statist idealism, it faces five main criticisms. Firstly, some critics argue that its good intentions cause it to underestimate the gravity of the situation and the very real security dilemmas involved. This means members of a Human Security-Civil Society Complex may be unaware of the ways in which their intervention interacts with the political economy of conflict or the real agendas of their local partners.[128] For instance, several commercial demining personnel in Afghanistan doubted the ability of demining NGOs to deal with violent threats from the Taliban. Moreover, those of a realist bent would argue that the impact such a complex could have is limited when confronted with armed groups willing to use lethal force. This is especially the case when domestic electorates may become dissatisfied with 'foreign adventures.' Critics say resolving the problem of failed and rogue states requires the political and military might of the great powers and a resolve to use violent force when necessary.[129] They believe post-statist idealism is not successful in diffusing security out to the peripheries, rather it allows the diffusion of insecurity (in terms of unwanted immigration, crime and terrorism) into the core.

Secondly, some have argued that the universalism of post-statist idealism fails to acknowledge differences in culture and national context. Proponents, for instance, of the so-called 'Asian Values' would argue that human rights is a 'Western' invention and should not be imposed upon other parts of the world.[130] Some have seen the involvement of the Human Security-Civil Society Complex in other countries as a violation of the principle of sovereignty. The Russian government has increasingly come to see NGOs, including local ones, as a sort of externally sponsored 'fifth column' that subverts the state's ability to safeguard national security (such as its ability to conduct military operations in Chechnya without international exposure).[131] For example, Russia has harassed the NGO HALO Trust's demining operations in Chechnya, accusing it of aiding the rebels.[132] Less state-centric criticisms argue that NGOs and international agencies have a tendency to act paternalistically – so convinced of the rightness of their cause that they adopt a patronizing or disparaging attitude to local culture, authority and society. This can be expressed in a reluctance to hand meaningful authority and capacity in aid operations over to local personnel and organizations.[133]

Thirdly, the enthusiasm of middle powers (and other industrialized countries) for international aid and peacebuilding has sometimes been criticized as a way to try to stem the flow of migrants from poor and war-torn countries. Many of the industrialized middle powers are undergoing significant internal political struggles over the issue of immigration. Aid and peacebuilding efforts are sometimes seen as ways to ameliorate conditions in the migrant-sending countries, in order reduce the number of people who want to leave. As such, some would argue that the Human Security-Civil Society Complex is at least somewhat motivated by 'containment' of international migration, rather than good intentions.[134]

Fourthly, there are technocratic criticisms that highlight the ineffectiveness and inefficiency of using NGOs as partners. NGOs have been criticized for confused mandates, poor management, amateurism and inefficient use of resources. Some neo-liberals argue that NGOs are simply a channel for rent-seeking and do not engage in 'productive' economic growth-enhancing activities like businesses. Milton Friedman, has argued that it is impossible to 'do good with other people's money.'[135] This has led some to wonder why donors often use NGOs to manage activities (such as construction of roads, schools and clinics) that would more suited to a commercial company. Competition for funding and poor communication also means that in comparison to state organizations, NGOs are continually dogged by failures to effectively coordinate their activities.[136] Many demining professionals interviewed by the author had similar criticisms of demining NGOs.

Finally, some argue that since NGOs may be funded by external donors and have no local electoral mandate, they have less potential for political accountability than local states. They often face the criticism that they are inherently undemocratic. NGOs operating in areas where state control is lacking can often become de facto state institutions, albeit unelected and not bound by a social contract to the citizenry[137] – 'Philanthrocracy,' as Maurice Amutabi called it.[138] With pressure to come up with success stories for donors and a lack of cultural and linguistic knowledge among expatriate staff, the Human Security-Civil Society Complex may have poor systems of needs assessment and program evaluation. Indeed, international NGOs are particularly prone to accusations of neo-colonial behavior as they are often 'too close for comfort' to external donors that might have interests in the country and have a tendency to bypass the concerns of local authority.[139] Moreover, international NGOs may sometimes be just as prone to enclavization as commercial companies. They create walled compounds to protect themselves and through the geographical fragmentation and local focus of their programming, can create 'islands of development' rather than an improvement of general well-being.[140] Therefore, some have criticized the Human Security-Civil Society Complex as just a whitewashed form of the neoliberal privatization of aid or another 'technology of social control' used by the north to dominate and exclude the south.[141] Both Mark Duffield[142] and William Easterly[143] have shown how there are shockingly similar parallels between the rhetoric of imperialism's 'White Man's Burden' and today's aid workers, 'peacebuilders' and Western diplomats. There is a danger then, that the 'petty sovereignty' held by the Human Security-Civil Society Complex over aid recipients is simply a form of neo-imperialism.

Conclusion

The era of sovereign and territorial *government* is giving way to an era of *governance*, diffused through global networks of authority below and beyond the state and incorporating polyarchical constellations of public and private actors.[144] Meanwhile, insecurity can no longer be adequately conceived of as the possibility of threat to the state by another state or a rebel pretender with statist ambitions. Insecurity has become de-territorialized, de-statized and no longer contained to bounded space and time. In the quest to govern these threats, many states, companies and individuals have abandoned faith in traditional realist and idealist understandings of security and have sought to mobilize protection through the new networked modes of social organization.

These new forms of governance are organized and constructed in different ways, depending on the motivations, understandings and institutional structures of the constituent members. Some of these networks are organized to protect the private space a privileged subpopulation, employing surveillance, force and securitized aid to separate safe enclaves from the zones of disorder around them. Such Strategic-Commercial Complexes engage the 'hard power' of coercion and economic incentives to overcome the 'principal-agent problems' that inevitably arise in such a diffuse and complicated organizational form. In essence they are Neo-Machiavellian, and draw on the belief that optimal outcomes can arise from the pursuit of national and/or commercial interest.

In contrast, there are coalitions of middle powers, multilateral agencies, NGOs, social movements, media and 'socially responsible' businesses that are drawn together by shared values of cosmopolitanism, humanitarianism and a broader understanding of interest as the global public good. Such Human Security-Civil Society Complexes try to raise the level of protection and services available to the whole population, or those perceived as most in need, through the diffusion of 'soft power' – persuasion, aid, rule of law and non-violent dispute resolution – out into the conflicted peripheries. This complex is generally not pacifist, but they are reluctant to deploy military force except to protect those in danger of genocide and crimes against humanity. In essence, they are Neo-Kantian or Neo-Durkheimian, and stem from assumptions that the pursuit of interest must be at least regulated and constrained by a commitment to humanistic values and ideals.

The real world is obviously more complicated than the simple dichotomy of ideal types presented in this chapter. Some actors will operate according to a Human Security-Civil Society model in one situation and adopt a Strategic-Commercial model in the next. Rather than ideal types, these kind of complexes may actually represent poles on a continuum, with real world examples falling somewhere along the line between the two extremes. For instance, an NGO receiving humanitarian funding from a middle power might still protect its compound with a private security force. Moreover, sometimes these models converge into complex blends or agents are embedded in multiple kinds of networks. It is rare to find any one actor wholly motivated by ideals or completely driven by narrow self-interest. Ideals can be expressed instrumentally, in an attempt to secure self-interest. The two complexes sometimes make an implicit division of labor; in Afghanistan, the US/NATO-led Strategic-Commercial Complex is engaging in counterinsurgency operations while the Human Security-Civil Society Complex concentrates on development, grassroots peacebuilding and humanitarian aid. The two types of complexes may also converge on similar goals – such as 'humanitarian intervention' in Northern Iraq, Bosnia and Kosovo – as the hard strategic interests of a superpower align with middle powers' and NGOs' calls for the protection of human rights. Nonetheless, for the purposes of this study, it is helpful, in the tradition of Weberian ideal types, to highlight – perhaps even caricature – the differences between two poles, to enable comparison and understanding of their strengths and weaknesses.

This book will not attempt to determine the overall advantages and disadvantages of these two complexes in every given situation. It will focus on describing how they have developed and the impact they had in one particular sector – demining. Specifically, it will try to determine the outcomes of each model as they attempt to protect populations from the threat of landmines and aid in the restoration of post-war political and economic systems. The next chapter looks at the role of two major mine action donor countries, the US and Norway, as nodes in these two types of networks.

Table 1: Responses to Insecurity

Model of Security	Threat	Security for Whom?	Motivation	Nature of Institutions	Internal Structure	Sources of Power
National Security/ Realism	Old War	Few/ The Nation-State	National Interest	Public	Unitary actor	Hard Power
Collective Security/ Idealism	Old War	All/ All states	Ideals and Global Public Interest	Public	Intl Law	Hard and Soft Power
Strategic Commercial Complex	New War	Few/ Privileged	Narrow Interest, Profit	Public-Private Partnerships	Principal-Agent Model	Hard Power
Human Security-Civil Society Complex	New War	All/ Vulnerable	Ideals and Global Public Interest	Public-Private Partnerships	Principal-Steward Model	Mostly Soft Power

3

DONOR POLICY MAKING IN THE US AND NORWAY

'The U.S. ... seeks to relieve human suffering caused by landmines and unexploded ordnance ... while promoting U.S. foreign policy interests.'
– US Department of State[1]

'It is in Norway's interest to ... improve the safety of people who are ... exposed to mines and other explosives... [and] defend an international system based on the rule of law and respect for human rights.'
– Jonas Gahr Støre, Norwegian Minister of Foreign Affairs.[2]

In his book *Impotent Superpower—Potent Small State,* Jan Egeland argued that while Norway and the US shared strong moralistic foreign policy motivations, a great power like the US is a 'more ineffective and inefficient human rights actor than the small state.'[3] This was because, as a superpower, US foreign policy was constricted by numerous strategic and commercial interests that hijacked its ability to shape policies according to humanistic ideals. By contrast, a small state like Norway had fewer transnational interests and more space to pursue a normative foreign policy. Indeed, such a policy might be the only way for a small state to achieve prominence on the global scene and build up international good will. This chapter will borrow Egeland's theoretical framework, applying it to a comparison of US and Norwegian mine action policy. As noted previously, these countries were selected for study because they are among the top bilateral mine action donors.

Table 2: Top Mine Action Donors in 2006.[4]

Absolute Terms		
Rank	Country	Millions
1	United States	$94.5
2	European Commission	$87.3
3	Norway	$34.9
4	Canada	$28.9
5	Netherlands	$26.9

Percentage of GNI		
Rank	Country	%
1	Slovakia	0.02635
2	United Arab Emirates	0.01922
3	Norway	0.01130
4	Denmark	0.00516
5	Netherlands	0.00385
...	...	...
18	United States	0.00070

This chapter will argue that Norway's mine action policy, driven largely by an idealist foreign policy tradition and rooted in close partnerships between the Ministry of Foreign Affairs and NGOs, approximates the ideal type of a Human Security-Civil Society complex. This is expressed through three key characteristics:

1. Championing tight regulation of mines, cluster munitions and other remnants of war,
2. Supporting demining programs prioritizing humanitarian concerns,
3. Channeling significant amounts of assistance through international NGOs.

In contrast, it will show that US mine action policy, shaped by significant strategic and commercial foreign policy interests that blunt its humanitarian elements, usually approximates a Strategic-Commercial Complex, with the following three characteristics:

1. Opposition to tight regulation of mines, cluster munitions and other remnants of war,
2. Supporting demining programs shaped by strategic and commercial interests,
3. Channeling significant amounts of assistance through commercial companies and military units.

However, it will show that when the US has fewer strategic interests at stake, it is able to act in a more Human Security-Civil Society mode.

Adapting Egeland for a Post-Statist Age

While this chapter is rooted in Egeland's comparison of the US and Norway, it also recognizes that, written in 1988, his theory needs to be updated to take into account trends of privatization and contracting in both US and Norwegian foreign policy. Therefore, this section will outline Egeland's theory in detail and update it for a 'post-statist' age.

Egeland believed the US was less able to form coherent and effective human rights policies partly because the size of its polity meant more complex decisionmaking processes, diverse internal interests and more inter-bureaucratic battles, leading to less coherent polices. More importantly, he argued that a superpower has considerable transnational military and commercial interests that tend to blunt the effectiveness of human rights promotion. Since the military and capital are generally self-interested rather than idealistic, Egeland believed they were more likely to lobby for '"constructive cooperation" with dictators and junta generals' than transformative human rights policies that risked disrupting trade or military collaboration.[5]

For instance, while the US State Department is supposedly the focal point of US foreign policy, one often finds the US military's regional commanders are more powerful than ambassadors.[6] Moreover, the US has become the world's leading arms supplier[7] and, as one former US Commerce Department official argued, its foreign policy 'has reflected an obsession with open markets for American firms.'[8] The strength of military and commercial interests has also shaped US foreign aid, with an estimated 70-80% of it 'tied to the purchase of US goods and services.'[9] Moreover, a review of US foreign aid argued it was often 'motivated more by … security concerns than by the objective needs' of recipients.[10] Indeed, the ongoing War on Terror has seen a further securitization of assistance, as 'fighting terrorism became the leading concern of American foreign aid.'[11]

In contrast, Egeland believed the small state's minimal economic and military power means it has fewer national interests to defend. As a result, a smaller nation has more space to conduct its foreign policy according to 'altruistic norms and principles.'[12] For instance, according to one historian, 'the belief in…a foreign policy infused by high ethical standards has been a trademark of Norwegian aid policies.'[13] As the steward of the Nobel Peace Prize, Norway has long seen itself as a 'bridge builder' or 'honest broker' in defusing international conflict and promoting peace.[14] It also had a long history of sending missionaries to the developing world[15] – a spirit which seems continue within the Norwegian aid community, much of which has close links to the Lutheran church.[16] Additionally, Norway has a strong internationalist labor movement and an internal consensus that the welfare of the poor and needy is the responsibility of society as a whole.[17] Therefore, notions of Third World solidarity and responsibility to share Norway's enormous oil wealth with the 'poor and oppressed' have been very popular. This has encouraged a 'race to the top' among Norwegian political parties, vying with each other to produce the largest aid budgets.[18] Therefore, Norway is one of only five countries that have surpassed the OECD aid target of 0.7% of GDP.[19]

Egeland did not argue that Norway had a more human rights-driven foreign policy because it was less self-interested than the US. He said that if key interests clash with norms, then a 'small state is normally not any more willing than the big state to sacrifice national interests for ideals.'[20] For instance, during the 1970s and 1980s, shipping interests delayed and hampered Norwegian sanctions against South Africa even while Norway funded the ANC.[21] Rather, Egeland said that in comparison with a superpower, these key 'red-line' interests were relatively few. Moreover, for a small state, becoming known as a champion of norms and ideals may be the only method by which it can gain prominence on the international scene.[22] Joseph Nye contended that because Norway lacks the resources for 'hard' economic and military power, it has built up sources of 'soft power,' using the symbolic power of its ideals and normative commitments to achieve its national interests.[23] Therefore, Norway has been an enthusiastic champion of multilateralism, disarmament, international law and collective security. Unable to protect itself in the event of an invasion by a great power, it is in Norway's strategic interest to tie great powers down to stable and predictable norms and to make sure its own voice is heard on the international scene. The Norwegian Ambassador to Sarajevo put it this way: 'small countries need to have some kind of international order…that's not altruism, it's a vital foreign policy interest.'[24]

As a result, Egeland argued that in terms of promoting human rights, Norway may actually be more powerful than a superpower because it is freer to act according to humanistic ideals rather than the constraints of expediency. Moreover, becoming known as an idealistic rather than self-interested actor may actually assist a small state's survival. Egeland's research heavily influenced his practice as a major political advisor in the Norwegian Foreign Ministry for much of the 1990s and has been behind part of Norway's drive to brand itself, somewhat hyperbolically, as a 'humanitarian superpower.'[25]

While already a trend when Egeland wrote his book, the past two decades have seen an increasing tendency of both the US and Norway to privatize the implementation of significant elements of their foreign policies. However, they have done so in different ways. The US has tended to look more to the for-profit sector. The US military now relies on an extensive network of private contractors providing everything from laundry and food services to guard duty, intelligence gathering and interrogation. Likewise, State Department and USAID foreign assistance initiatives are often carried out by consulting companies, foundations and international NGOs.[26] Therefore, US foreign and aid policy is now conducted by a complex network of both civilian and military actors, which are public, commercial and non-profit. However, with a strong emphasis on competitive tendering, the US generally has a 'principal-agent' relationship with its implementers. This means implementing agencies have little say in the overall policy direction and 'speaking out' may actually harm their chances of maintaining funding. Indeed, USAID has 'commercialized' its relationship with NGOs through competition and tendering in grant allocation. This 'marketization' of NGO aid functions to reduce NGOs' autonomy and independence. This process has been magnified by the growing 'securitization of aid' during the 'War on Terror.'[27] Many major US NGOs have felt uncomfortable being seen by the US government as 'a force multiplier' and 'an important part of our combat team….'[28] However, they are dependent on US government funds and took note when USAID administrator Andrew Natsios threatened to tear up their contracts if they failed to understand that they are 'an arm of the U.S. government' in Afghanistan and Iraq.[29] Therefore, the practice of contracting out has simultaneously multiplied the number of vested interests involved in shaping US foreign policy, while muffling the voices that might call for action contrary to narrow interests. One might argue that privatization has actually exaggerated the tendencies in US policymaking that Egeland highlighted in 1988.

In contrast, Norway has tended to form close partnerships with a small group of trusted NGOs – more akin to a 'principal-steward' relationship. Norway channels more aid through NGOs than any other OECD country.[30] Often called the 'Norwegian Model' of foreign policy[31] or 'Track 1 ½ diplomacy'[32] in the literature, these NGOs act as the eyes and ears of the Ministry of Foreign Affairs (MFA), gathering information about local conditions, and implementing programs designed with guidance from the MFA. In return, the MFA provides long-term grant funding (with little interference in day-to-day management), diplomatic cover and advocacy for their causes in international fora. There is also a great deal of staff rotation between the NGO community, the MFA and the UN, contributing to ideological continuity.[33]

Some might argue that the tightly integrated partnership between the MFA and NGOs is a hegemonic one, designed to dominate and control civil society. Moreover, some critics believe the 'Norwegian Model' allows the MFA to take credit for NGOs' success while denying their failures. However, the MFA is not always clearly the controlling partner – on multiple occasions the Norwegian NGOs have acted as the MFA's 'conscience' and set the humanitarian and development agenda. Because of the shared ideological commitments and personal connections spanning both the MFA and NGOs, NGOs seem more able in Norway than the US to lobby successfully for policy changes. This was illustrated by Norway's rapid changes in policy on both the landmine and cluster munitions bans. As a result, while the reliance on NGOs has added another layer to Norwegian foreign policy, and possibly a strong vested interest, the ideological cohesion between the MFA and the NGOs contributes to a continuing commitment to cosmopolitan and humanitarian norms.

Global Policy on Remnants of War

The general trends in US and Norwegian foreign policy outlined by Jan Egeland are reflected in their policies on international regimes controlling, managing and eliminating remnants of war. Focusing in particular on measures to control mines and cluster munitions, the following will examine how the US and Norwegian positions have evolved from domestic moratoria to three key international treaty processes – the Convention on Certain Conventional Weapons (CCW), the Antipersonnel Mine Ban Treaty and the Oslo Process on Cluster Munitions. It will show how in both countries, the military and 'realist' diplomats were resistant to greater control over mines and cluster munitions. However, in Norway, the power of the military relative to civil society was much less than the US. Its objections were thus overruled. In contrast, the US military and security interests have been much more able to resist the influence of the not insignificant pro-mine ban lobbies arrayed against them.

During WWII the US laid some 17 million mines,[34] but by the Vietnam War, it was dissatisfied with the slow process of manual mine-laying that left sappers exposed to enemy fire. Therefore, from the 1960s, the US developed systems to scatter mines remotely, by firing them as submunitions from artillery or air-dropped bombs. However, this had unintended consequences in the Vietnam War, when American soldiers often ended up 'retreating through areas that their own pilots had previously saturated with mines.'[35] Therefore, since the 1980s, the US has increasingly emphasized the development of self-destructing and self-neutralizing 'smart mines', intended to become inert after a given period of time.

US mines have been manufactured for both the Department of Defense and export by private companies, most notably Alliant Techsystems, a spin-off of Honeywell. While the US claims little responsibility for the landmine crisis,[36] investigations by human rights groups have shown that the US was a major exporter.[37] Moreover, by 1999, the US had stockpiles of around 12 million antipersonnel landmines.[38] The US has also been a major developer and user of cluster munitions. As described in the previous chapter, the US dropped some 256 million bomblets on Cambodia, Laos and Vietnam during the Indochinese Wars and has continued to use them in Kosovo, Afghanistan and Iraq.

That said, US landmine use has been relatively limited in recent years. The last time the US used antipersonnel mines was in the 1991 Gulf War. Until 1999, the US maintained minefields around the Guantanamo Bay Naval Base in Cuba, but these have now been cleared. While the minefields in the demilitarized zone in Korea are a major reason cited by the US to stymie landmine control efforts, these minefields are technically under the control of South Korea, though US war planning calls for sowing new minefields should North Korea invade.[39]

By contrast, Norway has never had a significant landmine production industry, though in the 1950s there was some 'minor military production...of some very primitive mines' and overseas subsidiaries of the Norwegian company Dyno-Nobel manufactured mine components. Nevertheless, landmines formed an integral part of their defense doctrine, with plans to lay minefields in the north in the case of a Soviet invasion. For this purpose, Norway stockpiled mines imported largely in the early post-WWII period. By the mid-1990s many of these stocks had been destroyed simply because they were out of date. That said, Norway had allowed the US, its NATO ally, to store antipersonnel landmines on its territory, the exact amount being classified. These were removed in 2002.[40] Until recently, Norway stockpiled artillery and air-delivered cluster munitions, but there were no reports of using them in combat.[41]

Unilateral Moratoria, 1992-1996

Most of the influential organizations that first called for a landmine ban were located in the US, notably Vietnam Veterans of America Foundation (VVAF), Human Rights Watch and Physicians for Human Rights. They found a sympathetic and powerful supporter in US Senator Patrick Leahy (Democrat, Vermont), who had been a champion for civilian victims of war after meeting a Nicaraguan boy in Honduras who had lost a leg to a landmine in the late 1980s.[42] Leahy drafted legislation in October 1992, that, despite lobbying from mine manufacturers, made the US the first country to unilaterally place a moratorium on the export of antipersonnel landmines. The law, later extended and formalized, also called on the President to negotiate an international prohibition of 'the sale, transfer, or export of anti-personnel landmines.'[43]

In September 1994, President Clinton, reading a speech Leahy helped draft, became the first world leader to call for the 'eventual elimination' of antipersonnel landmines. This led to a US-sponsored General Assembly resolution urging member states to impose export moratoriums 'at the earliest possible date' and encouraging 'further international efforts' to control antipersonnel landmines, 'with a view to their eventual elimination.'[44] By 1996, the US had also placed a one-year moratorium, beginning in 1999, on the use of antipersonnel landmines except along 'internationally recognized national borders' or in 'demilitarized zones' (i.e. Korea).[45] These legislative efforts and reports that President Clinton was personally dedicated to a landmine ban (keeping a defused antipersonnel mine on his desk to remind him of the issue's importance), led many to believe the US would prove a major champion of a ban treaty.

Norway's moratorium banning the use, production, trade and stockpiling of antipersonnel landmines in June 1995 followed the UN General Assembly resolution. This moratorium was much tighter than that passed by the US, but represented a major victory for Norwegian civil society over the resistance of the Norwegian government.[46] The government had initially considered a mine ban too drastic and unrealistic, arguing instead for strengthening the CCW Mines Protocol. A former Norwegian military officer recalled that 'there was certainly resistance' within the officer corps to the idea of a ban.[47] As a result of lobbying by the Norwegian Campaign to Ban Landmines, Parliament requested the government to seek a total international ban.[48]

The 1996 CCW Amended Mines Protocol

In the 1970s, Norway was among the seven smaller powers, led by Sweden, which had called for tight CCW restrictions on cluster munitions and landmines. As a result, it was among the first countries that called for a strong international regime to govern conventional weapons. However, despite the Norwegian Parliament's endorsement of a total ban in 1995, a year later the Foreign Minister continued to argue for a relatively weak compromise within the CCW.[49]

The US has had an ambivalent attitude toward the CCW. On one hand, it has generally sought to use the veto the consensus process gives them to seek the weakest and least onerous restrictions. For instance, in the 1970s, the US, among other NATO countries, treated the Swedish proposal with disdain and made sure the 1980 CCW restrictions were as weak as possible. One US negotiator said the US 'entered the CCW negotiations as a holding action.'[50] Even then, it did not ratify the CCW until 1996, as a result of Leahy's urging.[51] On the other hand, the US has consistently argued that any new international norms regarding conventional weapons must be negotiated through the CCW or UN Conference on Disarmament processes, where they retain a veto. Thus, US investment in the CCW is arguably an attempt to ensure that negotiations produce minimal restrictions, reflecting a broader US tendency to privilege strategic concerns in disarmament and humanitarian law treaties. Throughout the post-WWII era, the US has been consistently skeptical of extending the laws of war, failing to accede to the 1977 Geneva Protocols and actively working to undermine the International Criminal Court. This is driven, in part, by a fear that international treaties erode American sovereignty and thus the freedom of the electorate to determine policy.[52]

Therefore, in the negotiations following the CCW review conference in 1994, the US, though rhetorically supportive of a ban, tried to make the resultant Amended Mines Protocol as weak as possible and exempt the self-destructing and self-neutralizing mines that made up the bulk of its stockpiles. While the ICBL had initially been impressed with Clinton's rhetoric on mines, they were disappointed with what they saw as an 'obstructionist' US attitude to the negotiations.[53]

When the 1996 CCW conference concluded the Amended Mines Protocol, which placed moderately tighter controls on the use of mines, those in favor of a complete ban were disappointed. Therefore, in October that year, Canada hosted a conference in Ottawa of states who were still interested in seeking a ban. On the last day of the conference, Canada's Foreign Minister Lloyd Axworthy shocked the participants by calling for a maximalist ban on anti-personnel mines to be negotiated outside the CCW framework by the end of 1997. This came to be known as the 'Ottawa Process.'[54]

The 1997 Antipersonnel Mine Ban Treaty

Despite its initial reluctance to accept a ban, intense civil society pressure made the Norwegian government an early and strong supporter of the 'Ottawa Process.' It came to believe that the CCW's 'consensus practice' and 'lack of external exposure resulted in non-committal and weak solutions with little relevance to the situation of people living in mine-affected areas.'[55] Norway became part of a 'core group' of pro-Ottawa states supporting activities aimed at encouraging the involvement of African countries. In September 1997, it hosted the negotiation of the treaty in Oslo and its 'generosity and rapidity' in responding to the 'enormous diplomatic and organizational challenge' of running the conference were credited as a major contribution to the Ottawa Process.[56]

During the negotiations, a Norwegian foreign minister later observed, the 'strategic partnership' between middle powers such as Norway and international NGOs, 'injected an internal dynamism' into the Ottawa Process that turned the incentive structure 'upside down.' Instead of a 'race to the bottom', the state participants found themselves in a process where they were continually challenged by civil society actors to take a maximalist position.[57] With decisions made on the basis of a supermajority instead of consensus, the negotiations were also not held hostage to the most conservative members. The result of the Oslo conference was a treaty text that was a straightforward and relatively loophole-free 'prohibition of the use, stockpiling, production and transfer of anti-personnel mines.' The treaty also required states to clear all minefields on their territory in ten years and called on donor countries to contribute to demining, survivor assistance and awareness programs.[58] Since the treaty came into effect in 1999, Norway has seen the Mine Ban Treaty as the sole framework for its policy on antipersonnel landmines; the ICBL has praised the country as 'a key promoter of the universalization and effective implementation' of the Ottawa Convention.[59]

In contrast, US officials were angry at what they saw as a unilateral move by Canada. It appears US diplomats misjudged the level of support for Ottawa initiative, dismissing it as a 'pep rally' and announcing in early 1997 that they would, instead, seek to negotiate a ban in the UN Conference on Disarmament.[60] Meanwhile, opinion in the US government was divided. Senator Leahy was an active champion of a ban and found significant support within the State Department for participating in Ottawa Process. However, the Department of Defense was not in favor of doing so, and Clinton, burned by Somalia and clashes over gays in the military, wanted to avoid alienating the Pentagon.[61] As a compromise, Clinton sent diplomats to the negotiations in Oslo but with a 'package' of non-negotiable demands aimed at 'modifying the treaty to accommodate existing US policy.'[62] The US insisted on 'a geographic exception for Korea, a change in definition to exempt certain US 'smart' mines, and a nine-year delay in the effective due date of the treaty.'[63] To this end they 'cajoled and arm-twisted with great vigour,' alienating many sympathetic countries.⁶⁴ Unsurprisingly, in the end they decided not to sign the treaty.

Disappointed by his misjudgment of the ability of the Ottawa 'core-group' and the ICBL to win a complete ban, Clinton developed an alternative unilateral landmine policy. In the 1998 Presidential Decision Directive 64, he committed the US to stop using antipersonnel mines outside Korea by 2003. If suitable alternatives to mines were found, by 2006, the US would give up antipersonnel mines altogether and sign the mine ban treaty.⁶⁵ The Bush Administration took the US even further from the Ottawa Convention, dismissing it as an 'absolutist formulation' whose 'simplicity' is its 'greatest weakness.'⁶⁶ It argued that joining the treaty would require the US 'to give up a needed military capability' ⁶⁷ required to 'protect our forces, saving the lives of our men and women in uniform and of those civilians they defend.' ⁶⁸ They felt the treaty was inadequate as it does not address anti-vehicular mines (though it is unclear why this should be a reason against joining the Ottawa treaty and seeking other agreements on anti-vehicular mines). They argued the treaty 'commits its adherents to the costly and unnecessary act of clearing every last mine' – leading to diminishing returns on donor's money.⁶⁹

Therefore, in February 2004, the Bush Administration outlined the current US landmine policy, claiming to make both 'humanitarian and military goals... fully compatible.'⁷⁰ The new policy was to increase mine action spending by 50%⁷¹ and commit 'to leave no mine behind of any type on any battlefield anywhere in the world.'⁷² However, this rhetoric was misleading, as the following details of the new policy showed a continuing reluctance to give up mines:

1. 'Persistent' antipersonnel landmines will continue to be stockpiled for use within Korea until 2010.
2. Until 2010, 'persistent' anti-vehicle mines can still be used if specifically authorized by the President.
3. After 2010 the US will continue to stockpile and maintain the right to use 'non-persistent' or 'smart' mines. ⁷³
4. The US will invest more in developing landmine systems that incorporate 'enhanced self-destructing, self-deactivating technologies and control mechanisms....'⁷⁴

This policy was criticized by mine ban campaigners for four major reasons. Firstly, the self-destruct or self-neutralization mechanisms in so-called 'smart' mines do not always function properly, meaning that humanitarian deminers have to treat 'smart' mines as if they were still dangerous. Secondly, such mines remain active for weeks or months before self-destructing, still posing a threat to civilians. Thirdly, the major commitments in this policy were only to take place in 2010, after Bush left office. Finally, it was seen as an excuse for the US to develop two new landmine systems, the Spider Networked Munitions System and Intelligent

Munitions System (IMS).⁷⁵ If these systems go into production, they will join the existing US stockpiles of some 10.4 million antipersonnel mines and 7.5 anti-vehicular mines, third in size only to China and Russia.⁷⁶

Post-Ottawa Negotiations in the CCW

Following the Ottawa Convention, negotiations have continued in the CCW as a means of tightening up regulation of other remnants of war. In these discussions, the US and Norway have demonstrated attitudes and positions similar to those they held concerning preceding treaties.

Despite resistance from Senator Leahy and the US Campaign to Ban Landmines, the US has continued to position itself rhetorically as a champion of tighter regulation, while simultaneously trying to limit actual restrictions on its freedom to deploy weapons according to 'military necessity.' For instance, it signed the new Protocol V on Explosive Remnants of War, but this, as outlined in the previous chapter, has many loopholes.⁷⁷ Even with these 'escape clauses' the US has not yet ratified the Protocol. Since 2001, the US has championed the idea of a new CCW protocol restricting anti-vehicle mines that do not have 'effective self-destruction or self-neutralization mechanisms.'⁷⁸ However, by the time such a Protocol would come into effect, according to the US landmine policy, it would have given up such 'dumb mines' anyway – banning a weapon they already consider antiquated and dispensable.

For its part, Norway has generally stood for a maximalist CCW, having ratified the new protocol on Explosive Remnants of War and called for further restrictions on anti-vehicular mines and cluster munitions.⁷⁹ However, the Norwegian MFA has become increasingly skeptical of the CCW, feeling that the consensus process effectively gives great powers a veto on the maximalist position. This has brought it into conflict with the US over the issue of cluster munitions. The US had argued that cluster munitions were already adequately regulated by Protocol V and by 'strict implementation of international humanitarian law.'⁸⁰ In contrast, by 2006, the Norwegian MFA, scandalized by Israel's use of cluster munitions in Lebanon, decided there needed to be a 'ban on cluster munitions that have unacceptable humanitarian consequences' and introduced its own unilateral moratorium as a step 'towards developing an international norm....'⁸¹ At the November 2006 CCW review conference, the Norwegian MFA felt that such a ban was impossible within the CCW and invited other pro-ban states to start a course of action outside the CCW similar to the Ottawa Process.

Oslo Process on Cluster Munitions, 2007-Present

In February 2007, Norway held a conference in Oslo where 46 countries signed a declaration calling for a ban on cluster munitions causing 'unacceptable harm to civilians' outside the CCW.⁸² Intense treaty negotiations followed, with some 100 countries signing a final treaty, the Oslo Convention, banning the production, stockpiling, transfer and use of cluster munitions in December 2008.[83]

The US, however, saw this as a unilateral move by Norway. Going outside the CCW framework, a State Department representative argued, was 'not healthy for the development of widely adhered to rules of international humanitarian law.'[84] While it has continued to play down the humanitarian impact of cluster munitions as 'episodic and limited in scope, scale and duration', [85] the US has changed its earlier position and now says it is open to discussing further restrictions on cluster munitions within the CCW – because it 'takes into account both military and humanitarian considerations.'[86]

Interestingly, elements within the Norwegian government, particularly the military, were also unenthusiastic about a ban on cluster munitions. They considered them useful weapons and were concerned about issues of NATO interoperability with US forces. However, it is indicative of the nature of the two countries' polities that anti-ban resistance was shouted down by civil society in Norway and pro-ban resistance was largely ignored by the Bush Administration.

Emergence of Support for Humanitarian Mine Action

Comprehending the historical roots of policies is essential to a full understanding of them. Therefore, before looking at the current structure of US and Norwegian mine action funding, the following explores how the two countries initially became involved in supporting demining programs. The circumstances from which these programs arose shaped the way in which the two counties later institutionalized support for mine action.

The US originally supported 'humanitarian' demining as an extension of support to military or paramilitary operations in support of strategic objectives. Operations were conducted either by military units or a commercial company; humanitarian motivations tempered by strategic considerations. Chapter one already outlined how early US 'humanitarian' demining efforts in Indochina were rather small and only conducted when a clear strategic advantage was present. One saw a similar pattern when the US deployed military personnel to work with the Egyptian army clearing mines and UXO from the Suez Canal zone in the aftermath of the 1973 Arab-Israeli War.[87] The enormous strategic and commercial importance of the Suez Canal almost certainly played a role in initiating this program. Other than these brief and limited efforts in Indochina and the Middle East, however, the US largely avoided involvement in demining efforts until the close of the Cold War.

The late 1980s saw a massive increase in the proliferation of mines, caused by the 'Reagan Doctrine' proxy wars, in which the US supported anti-government insurgencies in socialist countries of the Third World (such as Afghanistan, Angola, Cambodia and Nicaragua). It was the recognition that mines hampered the military efforts and political legitimacy of its proxy armies that contributed to the US's nascent interest in sponsoring demining programs for civilian benefit. For instance, in 1986, US Army Special Forces trained Honduran Army sappers to clear agricultural land mined by Nicaraguan factions. While the US claimed the focus was 'humanitarian, rather than military, mine clearance,' this operation was part of a series of joint US-Honduran military exercises and counterinsurgency training.[88] Similarly, the US Army helped the Royal Thai Army develop a mine detection dog unit to demine areas blocking US proxies' access across the Cambodian border.[89]

This model became most developed in Afghanistan where the US was engaged in 'the biggest paramilitary affair in CIA history',[90] pumping some $2.75 billion worth of arms and aid into the mujahideen struggle against the Soviet occupation and Afghan communist regime.[91] Coordinating with the CIA, USAID used 'humanitarian' assistance to develop the mujahideen's logistical capacity, rehabilitating roads, building bridges and importing pack mules.[92] While these programs were officially for humanitarian purposes, they also facilitated the movement of arms.[93] Posing a major threat to this logistical system, the USSR littered the border area between Afghanistan and Pakistan, where the majority of mujahideen bases were located, with thousands of mines.[94] By 1988, USAID began to realize that keeping the mujahideen supply lines open required a demining capacity.[95] Therefore, USAID had its logistical contractor, RONCO Consulting Corporation, develop a demining program in collaboration with the mujahideen parties. The US also seconded military officers to the nascent UN demining program that started around the same time, partly motivated by a desire to learn about Soviet mine warfare.[96]

These military and commercial-led models of demining continued to dominate US mine action assistance in the early 1990s. In Somalia, US soliders engaged in limited demining to support Operation Restore Hope. In Cambodia, military PSYOPs specialists developed mine awareness materials and Special Forces trainers trained Cambodian military deminers. The US Army also trained Latin American military deminers at the School of the Americas. In Mozambique, USAID, State Department and the Department of Defense contracted RONCO to clear priority roads in Mozambique.[97] Consolidating these efforts, in 1993, the US formally established its Demining Assistance Program, a joint endeavor involving the Department of State, Department of Defense and USAID, which would give priority to mine affected countries based primarily on balancing considerations of 'US interests' and the 'severity of the problem.'[98]

In contrast, Norway's early involvement in mine action stemmed more from a commitment to UN peacebuilding missions and was implemented largely through an international NGO. Norway has long been an enthusiastic supporter of the UN. Despite its small population, Norway is the UN's seventh largest financial contributor[99] and one of the few rich countries that consistently provides troops to peacekeeping missions.[100] The post-Cold War era also saw Norway take on an increasingly internationalist role, participating in NATO operations and playing the role of the third-party negotiator in a variety of peace processes. Therefore, in contrast to the strategic underpinnings of early US funding for demining, the genesis of Norwegian support was rooted in the post-Cold War rise of 'New Interventionism' by the UN, Western powers and humanitarian organizations in conflict zones. The early 1990s saw the UN move beyond traditional peacekeeping to 'rebuilding the institutions and infrastructures of nations torn by civil war and strife.'[101] The large UN program of assistance to Afghanistan from 1989 on was an early harbinger of such programs. Norway seconded military officers to the early UN Afghan demining program, one of whom became its head for a time.[102] Later, Norway contributed financial resources to the UN-led, local NGO-implemented Afghan demining program.

When the UN launched its Transitional Authority in Cambodia (UNTAC), (which ran Cambodia as a sort of 'neo-trusteeship' from 1992-1993) it became clear landmines posed a major threat both to post-war reconstruction and UN peacekeepers. As part of its UNTAC contribution, Norway asked the NGO Norwegian People's Aid (NPA) to recruit former Norwegian military ordnance experts to serve as deminers in Cambodia under UN auspices.[103] Since then, NPA has continued to provide technical assistance to the Cambodian Mine Action Center (CMAC), preferring to build local capacity than run its own operations.[104]

An analogous pattern occurred in Mozambique, where the 1992 peace agreement mandated a similarly interventionist UN mission (UNOMOZ). Funded by Norway and UNHCR, NPA started the country's first demining program in 1993, concentrating efforts on priority refugee return areas.[105] Unlike many of the other mine clearance agencies in Mozambique, NPA tried to integrate demining into broader community development programs including health, education, water supply and microcredit.[106] Similarly, following the 1994 Angola peace agreement, the UN asked NPA to clear a road to facilitate World Food Program operations.[107]

The Current Structure of US Mine Action Funding

The US Humanitarian Mine Action Program, a conglomeration of programs within the Department of Defense, Department of State, USAID and a few other government agencies, has become by far the largest global contributor to mine action in absolute terms, spending over $1.1 billion on demining, mine survivor assistance and mine risk education since 1993.[108] However, critics point out that US funding for mine action is actually much smaller than many other countries, including Norway, if measured as a proportion of GNI.[109]

The primary reason cited by the US for its major commitment of resources to mine action is 'to relieve human suffering caused by landmines and unexploded ordnance (UXO).'[110] Indeed there is significant public commitment in the US for mitigating the humanitarian impact of mines, as evidenced by the large private contributions to mine action NGOs. However, the former 'Humanitarian Demining Strategic Plan' explicitly acknowledged that one of its major goals was to 'Promote U.S. foreign policy, security, and economic interests.'[111] Indicating the importance of security matters, US mine action programs answer to a committee chaired by a representative of the National Security Council and that includes the CIA.[112] Looking at Table 3, one can see the influence of strategic considerations in the allocation of funds. Together, Iraq and Afghanistan received more funding than the rest of the top ten recipients put together. While both these countries are among the most heavily mined in the world, they only received such large funding allocations after the US became strategically interested in them, as Figure 7 displays:

Table 3: Top Recipients of US Mine Action Funding FY2004-2006. [113]

Rank	Country	Million
1	Iraq	$60.6
2	Afghanistan	$40.5
3	Angola	$17.1
4	Lebanon	$16.9
5	Cambodia	$12.6
6	Azerbaijan	$10.2
7	Bosnia and Herzegovina	$9.7
8	Vietnam	$8.9
9	Sudan	$7.8
10	Laos	$7.2

Table 4: US Mine Action Appropriations, 1999-2006 (Million US$). [114]

Budget Line	1999	2000	2001	2002	2003	2004	2005	2006
Department of State (NADR)	35.0	39.5	39.9	40.0	49.0	48.7	59.0	55.4
Department of State (ITF)	12.1	14.0	12.7	14.0	10.0	9.9	9.9	9.9
Department of Defense (OHDACA)	16.0	28.9	16.6	16.8	6.2	2.1	4.0	6.5
Department of Defense (R&D)	18.2	18.2	12.6	13.2	12.6	12.8	13.2	13.8
Supplemental Appropriations (Afghanistan, Iraq and Lebanon)					3.0	15.4	35.8	9.0	22.6

Figure 7: US Demining Funding to Afghanistan and Iraq, 1996-2005 (Million US$)[115]

The US also uses its demining funding as a way 'of deflecting attention' from its refusal to join the Ottawa Convention.[116] Unlike most bilateral mine action donors, the US does not organize its funding within the framework of the Mine Ban Treaty, dismissing attempts to do so as 'absolutist' and 'not grounded in practical realities.'[117] Citing the 'law of diminishing returns,'[118] State Department officials argue that the treaty's commitment to 'destroy...all antipersonnel mines in mined areas' in ten years[119] is a 'losing proposition' resulting in 'expanding expenditure of resources' for 'decreasing humanitarian impact.' The US, they say, 'can not afford the opportunity costs of trying to find the last 'million dollar' landmine.'[120] Therefore, contrary to the Ottawa Treaty's requirement that countries become 'mine free', the US argues that funding programs should aim to make the country 'mine impact free' – that is, free from mines 'that pose a high or medium threat to the ability of people to safely live normal lives....'[121] Indeed, the State Department's demining chief described the Ottawa Treaty's article on mine clearance as 'inspiring politically, but operationally...literally insane.'[122] Therefore, fitting into overall US foreign aid policy, US mine action programs are driven by tightly interlocking humanitarian and strategic interests and are implemented by a complex network of public and private actors.

It is telling that the vast majority of mine action funding has been placed not under USAID, but under the Department of State Bureau of Political-Military Affairs, which 'administer[s] military and other security-related programs.'[123] Through its Office of Weapons Removal and Abatement (WRA), this Bureau manages two primary accounts contributing to mine action, with some additional funding coming from emergency supplemental (in addition to the regular budget) appropriations (for Iraq, Afghanistan and Lebanon). WRA acts as the lead agency in determining the policy framework for US mine action.

The majority of WRA mine action funding comes from the Nonproliferation, Antiterrorism, Demining and Related Projects (NADR) appropriation.[124] WRA oversees this directly, allocating about half through prime commercial contractors and half through grants to NGOs, universities or multilateral bodies. The decision of whether to allocate money through contracts versus grants is determined according to 'the comparative advantages of both the private sector and not-for-profit sector'[125] and 'the best method in a given country and given situation to achieve U.S. objectives.'[126] Former WRA director Richard Kidd explained, 'resource allocation decisions are based on business models: …perceived return on investment, contractor value, contractor behavior, principal-agent relationship, etc.'[127]

Commercial contracting, using 'price competition considerations,' is used when WRA is 'procuring a supply or service' in exchange for money. They are used 'when the U.S. is conducting direct bi-lateral engagement with a sovereign state, or to oversee the actions and develop the capacity of local NGO's.'[128] Thus 'contractors are expected to act as agents of the United States…a principal-agent relationship.'[129] In other words, contractors are used when the State Department wants greater control over the final product produced. From FY2003 to FY2004, there was only one primary contractor, the US-based demining company RONCO Consulting. Since then, WRA has selected an additional two prime contractors, the private security and military contracting firms ArmorGroup and DynCorp. When WRA has a new task to contract out, each of these three companies can bid on the tender.[130]

Grants are used when WRA receives 'only indirect, intangible benefits' in return for its funding, but wishes 'to facilitate continuation of the grantee's mission.' Reflecting a principal-steward relationship, 'A grantee is already doing good things. … [We] give them a little more money; and they will do more of those good things.'[131] The primary six grantees have largely remained consistent.[132] They are:

1. Three mine clearance NGOs: HALO Trust, Mines Advisory Group and Norwegian People's Aid.
2. The Organization of American States, which funds military demining programs in Latin America.
3. Cranfield University, which provides training to mine action program managers.
4. James Madison University, which runs a Mine Action Information Center and edits the trade publication, the *Journal of Mine Action*.[133]

Being the primary funder of the leading mine action periodical gives WRA significant power over mine action discourse as it has final editorial say over the publication, and makes sure any 'political' or 'anti-American' language is edited out.[134] That the State Department has been able to effectively wield the censor's pen over the publication of an academic institution, shows how even grantees with a strong code of independence can be politically reined-in and controlled when operating the context of a Strategic-Commercial Complex.

In addition to these main recipients, WRA has made 'a dedicated effort to expand the number of organizations receiving U.S. support' through a 'Request for Application' process. Therefore, the number of WRA grantees has risen from 19 in FY2003 to 42 in FY2007. During that period, there was also an increase of funding allocated through grants rather than commercial contracts. However, WRA asserts that this should not be interpreted as a policy change or 'necessarily be taken as predictive of future U.S. allocations.' [135] Rather, the reduction in overall funds to commercial companies is due more to such contingent factors as a decrease in the previously large appropriations to Iraq and the recent completion of several contracts for programs in Lebanon and Sri Lanka.[136]

Table 5: WRA Modes of Fund Dispersal.[137]

		Mode of Funding				
		Contracts	Grants	ITF[1]	Other[2]	TOTAL
2003	Million	$29.7	$29.0	$10.0	$1.9	$70.6
	%	42	41	14	3	
2004	Million	$61.2	$43.3	$9.9	$4.5	$118.9
	%	51	36	8	4	
2005	Million	$32.4	$34.8	$9.9	$0.8	$77.9
	%	42	45	13	1	
2006	Million	$20.1	$40.5	$9.9	$2.9	$73.5
	%	27	55	13	4	
2007	Million	$21.5	$33.5	$8.6	$1.2	$64.8
	%	33	52	13	2	

In addition to the NADR funding, WRA also oversees a direct congressional appropriation to the Slovenian International Trust Fund for Demining and Mine Victim Assistance (ITF), the primary funder of mine action projects in the Balkans. The US provides a matching grant to the ITF, to act 'as a draw for other international donors…'[138] As will be described in more detail in later chapters, the ITF allocates its funding to both NGOs and commercial companies largely through a tendering model. Not included in the NADR and ITF appropriations is

---

[1] The International Trust Fund contribution is classified neither as a grant or a contract, though ITF's competitive tenders include contracting and granting to implementing agencies.
[2] All other WRA obligations, including 'voluntary contributions, administrative expenses (including salaries and travel), fund cite transfers to other operating units for obligation and expenditure, purchase orders, and other such obligations as the Department may enter into.'

the US assessed contribution to the UN peacekeeping budget, which pays for just over 20% of the budget for the UN Mine Action Service.

In addition to the State Department, Department of Defense (DoD) runs three activities under the rubric of 'humanitarian mine action': humanitarian demining, mine awareness and technical research. In addition to humanitarian demining, the various branches of the US military contract demining and explosive ordnance disposal tasks required for their mission (such as clearing military bases) to commercial companies.[139]

The humanitarian demining and mine awareness activities are funded by the DoD Overseas Humanitarian, Disaster, and Civic Aid account and overseen by the Defense Security Cooperation Agency (DSCA). The DSCA is 'the primary DOD body responsible for foreign military financing and training programs'[140] and also organizes bilateral arms sales.[141] The programs are implemented by US Special Operations Forces – a form of 'tied aid' as a significant amount of the budget goes into training Special Forces at the Humanitarian Demining Training Center at Fort Leonard Wood, Missouri.[142] As US troops are forbidden by law from actually engaging in humanitarian mine clearance themselves, these soldiers train local troops in mine clearance methods.[143] The mine awareness program is carried out by Psychological Operations units. Sometimes these troops are aided in their work by private contractors such as RONCO. Beneficiary countries are selected by the regional military commands, in consultation with the State Department.[144]

As can be seen in Table 4, in FY2006 the DoD's humanitarian mine action activities were worth $6.5 million. However, in past years, this figure was much higher, for example, $28.9 million in FY2000. The reduction is due to Special Forces being tied down with combat in Iraq and Afghanistan, leaving fewer surplus troops available for humanitarian demining.[145] While this training can undoubtedly play an important role in a local mine action program, the DoD sees its mine action efforts as a way to 'counter ideological support for terrorism', 'provide access to regions where traditional military-to-military engagement is virtually impossible', provide 'training opportunities in remote and austere environments' and enable troops to 'learn about the host nations' economy, culture, and hone their foreign language skills.'[146] Special Forces mine action is seen as a 'first foot forward' into a country that may be suspicious of the US military, opening the way for further cooperation through joint exercises, partnership and eventual contribution to NATO or other alliances. A program administrator pointed out that mine action training had opened up dialogue with Laos and Vietnam and was an important tool in leading to counterterrorism cooperation with Yemen.[147]

In his review of mine action, Stuart Maslen heavily criticized the US tendency to 'put its military at the heart of its mine action strategy,' arguing that the training efforts have had 'uneven' results and that civilian control is more effective. He is particularly critical of conceptualizing mine risk education as a form of psychological operations, saying that concentrating 'excessively on the high-tech production of printed media...[is] an unsustainable and typically inappropriate approach in a developing country.'[148]

The Pentagon also provides funds – about $13.8 million in FY2006 – to the Humanitarian Demining Research and Development Program within the US Army's Night Vision and Electronic Sensors Directorate (NVESD) for 'research and development of promising mine detection and removal technologies.'[149] The NVESD is also mandated with research into combat countermining. Again, this is a form of tied aid, because the money generally goes to research institutions and contractors based in the US that are often engaged in defense research. For all the millions of dollars spent on developing high-tech solutions to mine contamination, the broad consensus in the mine action community is that the most reliable demining methods are still low tech – human beings or dogs. Some have suggested, therefore, that DoD funding for demining is more concerned with developing new defense technology, especially technology that could be used to cross minefields during combat, than making a significant contribution to humanitarian mine action.[150]

Though it is the primary conduit of non-military US foreign assistance, and is on the steering committee for US mine action, USAID plays a relatively minor role in mine action funding. That said, when sponsoring major infrastructural programs in mine-affected countries, such as Afghanistan and Bosnia, it has funded its prime contractors' subcontracts with demining companies. In both those countries, the infrastructure programs were closely tied to US strategic objectives and often incorporated significant involvement of the military in influencing priorities.

USAID also oversees the Patrick J. Leahy War Victims Fund ($14.4 million in FY2005).[151] This fund 'contributes to improving the mobility, health, and social integration of the war disabled, including landmine survivors.'[152] This program seems to be less influenced by strategic priorities than others. It was started by Senator Leahy in the late 1980s after he met a boy in Honduras during the Nicaraguan war who was a landmine survivor. Leahy asked him which faction had laid the mine but the boy did not know as both sides had used mines. Leahy was moved by this experience and set up the fund for civilian victims of conflict.[153] This is an indicator that the US is not a 'carbon-copy' of the Strategic-Commercial Complex ideal type – there are still considerable humanitarian influences. However, the Leahy Fund represented only some 15% of US mine action funding in FY2005.

Several other US government agencies have made a minor contribution to mine action, including the Centers for Disease Control and Prevention,[154] the Department of Education and the US Department of Agriculture. However, the role of these agencies is rather limited.

Primary Implementing Partner: RONCO Consulting Corporation

Among the many organizations implementing US mine action projects, RONCO Consulting Corporation has had a particularly prominent position. Founded in 1974 by Ronald Boyd and Stephen Edelmann, who 'wanted to do something profitable and humanitarian',[155] RONCO's early projects managed agribusiness development programs and logistical pipelines for USAID. After pulling out of Afghanistan in 1994, where it had set up the first major US demining program, RONCO seems to have specialized in running programs treading the fine line between US politico-security functions and developmental objectives. In Serbia, RONCO played an important role in surreptitiously providing USAID supplies to the democratic opposition to Milosevic.[156] During Operation Enduring Freedom in 2001, RONCO returned to Afghanistan, demining US and Afghan military bases and clearing up unexploded munitions.[157] It has recently found a burgeoning market for its services in Iraq.[158] In addition to demining, RONCO developed a plan to restructure the Iraqi Army in 2003.[159] RONCO's work in Afghanistan and Iraq appeared to have further militarized their approach, with their employees in Iraq carrying arms.[160]

RONCO has continued to be a favored US demining contractor – indeed, for a long time RONCO was the only contractor doing US operational demining. In addition to State Department-funded contracts with the Slovenian International Trust Fund, RONCO won a five-year worldwide demining contract worth up to $250 million from WRA in October 1999.[161] Though RONCO continues to hold a prime position in US demining contracting, its position as market leader has been challenged in recent years by several other companies. UXB International has won contracts with USAID and the US military. Likewise, RONCO is no longer the sole holder of the primary WRA contract. Since 2005, it has shared this with two large military contracting firms, DynCorp and ArmorGroup.[162] These two companies, which have major private security operations, are significantly more militarized than RONCO, perhaps indicating the trend toward increasingly blurred lines between commercial demining and private security operations.

RONCO itself has been drawn into closer integration with the private security market, following its acquisition (along with ArmorGroup) in April 2008 by Group 4 Securicor (G4S), one of the world's largest security companies.[163] RONCO is now a subsidiary of Wackenhut Services Incorporated (WSI), itself a subsidiary of G4S. WSI provides 'high-end armed and unarmed security personnel, paramilitary protective forces, law enforcement officers' and other security related services.[164] Both G4S and WSI have controversial reputations and have been accused of human rights abuses in their work.[165]

The Current Structure of Norwegian Mine Action Funding

In contrast to the array of government agencies involved in US mine action funding, Norway's support for mine action is led almost exclusively by the Ministry of Foreign Affairs (MFA), with no more than five key staff involved. While for a time NORAD was involved in providing some funding, and Norwegian troops have participated in some demining operations in Kosovo, Afghanistan and Macedonia, the vast majority of funding is now dispersed by MFA. Mine action funding structures and policy are therefore much more coherent and simple than in the US.

Since 1997, when Norway pledged $120 million over five years to mine action, its support has been rooted in the Ottawa Convention's articles on stockpile destruction, demining and survivor assistance and concentrated in highly mine-affected countries party to the convention.[166] In 2002, Norway's five-year pledge ended, but it has indicated that it will continue to provide similar levels of funding. In 2006 Norway provided almost $36 million for mine action programs, making it the third biggest donor both in absolute terms and relative to GNI.[167] Twenty percent of this budget generally goes to survivor assistance.[168]

Norway's commitment to the Ottawa Convention means that it supports making countries 'mine free' rather than just 'mine impact free.'[169] While many see the 'impact free' formulation as an economically sensible approach, Norway sees it as a rhetorical device to undermine the Ottawa treaty. Firstly, it is not clear that budgeting money for mine action necessarily takes it away from other foreign aid budgets. Therefore, Norway believes the US argument that the diminishing returns from demining takes money from other humanitarian efforts is inaccurate. Moreover, they would argue that the 'impact free' formulation gives donors like the US an excuse to declare success while leaving the job unfinished – with countries that are supposedly 'impact free' still having a significant mine contamination.

In addition to technocratic criteria of efficiency and coherence, Norway's mine action funding is allocated on the basis of three key principles 'humanitarianism,' 'partnerships,' and 'national ownership.'[170] Each of these is considered in detail below.

Reflecting the normative concerns of Norway's broader foreign policy, humanitarianism – the allocation of resources according to need, rather than political, economic or strategic priorities – is one of the most important guiding principles of Norwegian mine action funding. Therefore, Norway supports demining in areas that are 'worst affected by mines,' specifically where they 'constitute unacceptable threats to the lives and livelihoods of civilians' or 'seriously hamper the return and resettlement' of displaced persons.[171] Hence in the table below, one sees that the countries receiving the most mine action funding from Norway are not necessarily strategically important.

Table 6: Top Recipients of Norwegian Mine Action Funding, 2003. [172]

Rank	Country	Million
1	Iraq	$3.0
2	Bosnia and Herzegovina	$2.5
3	Mozambique	$2.2
4	Angola	$2.1
5	Croatia	$1.9
6	Ethiopia	$1.5
7	Eritrea	$1.4
8	Lebanon	$1.2
9	Afghanistan	$1.0
10	Sri Lanka	$0.9

This principle also guides the types of partner organizations selected to implement projects (see subsection below). Norway generally picks implementing agencies that at least claim to act on behalf of the poor and needy, rather than profit or national interest. Norwegian diplomats explained that implementing partners should 'have some values which coincide'[173] with the MFA and that 'We think humanitarian work should be done according to humanitarian principles....'[174] They have also made a concerted effort to push implementing agencies to take gender issues seriously, because they feel this angle has been overlooked and ignored by other donors and mine action agencies.[175]

The one political conditionality that does impact their allocation of funding is a preference for countries that have ratified the Ottawa Convention. However, this has not been applied rigidly (for example Lebanon and Sri Lanka are not signatories, nor was Iraq at the time of the above statistics), and Norway would argue that this conditionality is actually in place for the larger humanitarian ideal of eradicating landmines altogether.

Its emphasis on humanitarianism means Norway is critical of efforts to 'securitize' or militarize mine action, saying the use of aid for 'hearts and minds' builds 'mistrust' in local population, reducing 'humanitarian space.'[176] This means they are also displeased with the power the Department of Peacekeeping Operations has over UN mine action efforts. Norway would prefer OCHA or UNDP, rooted in the humanitarian and development communities, to have more control.[177]

Reflecting Norway's commitment to multilateralism and close cooperation between the MFA and civil society organizations, 'Norway structures its support for mine clearance as partnerships' with host states, NGOs and UN agencies.[178] Most of the assistance for demining programs is channeled through Norwegian People's Aid (see subsection below), though the MFA also funds other NGOs such as MAG and the HALO Trust. Much of the assistance to mine accident survivors is channeled through the Tromsø Trauma Care Foundation, another

Norwegian NGO, though it also supports groups like Landmine Survivors Network. None of its funds are dispersed through commercial-style tendering.[179] Therefore, in comparison with US-issued tenders and Requests for Proposals there is much less competition for funding. As one Norwegian diplomat explained:

> We don't like to provide assistance money to commercial companies…they are there to make money and for them making money means that they will try to provide the least amount of service in order to maximize their profit. … We think that NGOs have a tendency to work somewhat differently, that there's a stronger sense of idealism, that they're more committed to the task at hand, rather than being committed to the bottom line and the statement of accounts and shareholder responsibilities.[180]

As a result, Norway has a principal-steward relationship with its implementers. It trusts them with long-term grants (at least a year) and implicit guarantees of funding over long periods of time. In comparison with US funding, there is much more flexibility and program design and reporting requirements are much less onerous. As a one diplomat explained, 'Even though we have a dialogue with these organizations we don't ask them to go to particular areas…or clear the mines along this road or that road. … We're not really interested in exactly where they are, as long as they do a decent job.'[181]

Table 7: Top Implementers of Norwegian Mine Action Funding, 2003.[182]

Rank	Organization	Million
1	Norwegian People's Aid	$13.0
2	ICRC and Norwegian Red Cross	$4.2
3	UN Development Programme	$3.3
4	HALO Trust	$1.2
5	Trauma Care Foundation/ Tromsø Mine Victim Resource Center	$0.9
6	Landmine Survivors Network	$0.7
7	Mines Advisory Group	$0.7
8	International Campaign to Ban Landmines	$0.7
9	Organization of American States	$0.6
10	Geneva International Centre for Humanitarian Demining	$0.5

The MFA also grants money to progressive and technical thinktanks, such as the Peace Research Institute, Oslo (PRIO), the Institute of Applied Social Science (FAFO), the Geneva International Centre for Humanitarian Demining (GICHD) and advocacy organizations, as a way of moving the agenda and discourse on mine action in directions it feels to be particularly important. For instance, it has funded research on incorporating peacebuilding, national ownership and community participation concerns into mine action.[183] It has also contributed to the International Campaign to Ban Landmines and the Cluster Munition Coalition.

As the Ottawa Convention gives primary responsibility for mine action to the countries affected by mines, Norway gives priority to governments 'that demonstrate and mobilise the political will necessary to tackle their own problem.'[184] This does not necessarily mean the country has to divert significant resources to mine action, but rather that the government is able to develop a national plan and establish local priorities. This concern for local ownership of mine action planning is a major reason why Norway has taken its funding for Afghanistan out of the UN Trust Fund, arguing that the UN has not done enough to build national governmental structures to regulate demining.[185]

This principle also means that Norway gives priority to 'reinforc[ing] rather than undercut[ting] existing capacities.'[186] Implementing partners are expected to build local capacity and the excessive use of expatriate staff is discouraged. Nevertheless, some argue that Norway's tendency to fund international NGOs actually contradicts this principle and that Norway should funnel more of its resources through local government or local NGOs. Indeed, one Norwegian diplomat admitted that 'It's easier to explain to the Norwegian parliament that we're providing all this money for mine clearing, if it's a Norwegian entity involved.'[187]

Primary Implementing Partner: Norwegian People's Aid

Like the US, Norway has a primary implementing partner – Norwegian People's Aid (*Norsk Folkehjelp*) – that it consistently relies upon to carry out most of its mine action assistance. Indeed, NPA received more funding in 2003 than the rest of the top ten recipients put together. Unlike RONCO, however, NPA is driven by its strong ideological and leftist roots, rather than commercial profit.

Created from the amalgamation of an organization that had supported the Republicans in the Spanish civil war and a worker's health program, NPA is the humanitarian wing of the Norwegian labor movement. It has a long-standing relationship with the Norwegian Labor Party, which has guaranteed access to significant public funds – in 2006 about 60% of NPA's income came from Norwegian government agencies.[188] Its first program sent aid to Soviet-occupied Finland in 1940. The German occupation of Norway prevented NPA from engaging in major operations during the war, but after 1945 it channeled aid for the reconstruction of Norway and began advocacy for greater social welfare spending. It also provided aid to war victims in Austria and contributed to UNICEF's work throughout Europe.[189]

Reflecting its roots in the labor movement, NPA says its humanitarian and development programs are motivated by 'solidarity' with the poor and oppressed.[190] Therefore, its programs are largely targeted at groups considered particularly needy and are often linked to broader advocacy and awareness raising campaigns. However, it has often taken its notion of 'solidarity' to much more militant conclusions than most international NGOs. Throughout its history, NPA has believed that it 'is not possible to be neutral' when working in conflict zones – 'Not taking a stand against oppression is also to take sides.'[191] As a result, it was the primary conduit of Norwegian support to the African 'liberation movements'[192] and historically maintained close links to the Sudan People's Liberation Army, PLO, Eritrean government, Tamil Tigers and Sandinista regime. An independent evaluation of NPA found that their explicitly political stance:

> facilitates access to guerilla zones...[and] has the advantage of allowing the population to own and evaluate aid programmes. But...NPA always risks to legitimise violence and political forces which are not necessarily representative and can divert humanitarian logistics to military ends.[193]

As outlined above, NPA's involvement in mine action evolved out of requests that it support UN peacebuilding missions in Cambodia, Mozambique and Angola in the early 1990s. Since then it has continued to expand its operations to a range of mine-contaminated countries, especially to places where it has felt demining could be an act of 'solidarity' with a particularly oppressed group. For instance, it began a program in Iraqi Kurdistan in 1995, as a way of showing support to the Kurds in their struggle against the Iraqi government.[194] Similar sentiments have motivated NPA programs in Western Sahara (a secessionist region of Morocco), Kosovo, Lebanon and Sri Lanka. Indeed, a 1999 NPA strategy paper asserted that its mine action programs should be 'implemented according to our ideological foundation and principles' including 'human dignity, equality, solidarity, unity, peace and freedom.'[195]

Likewise, NPA has linked its demining operations to advocacy for bans on landmines and cluster munitions: 'To NPA, field operations and advocacy are mutually reinforcing activities, both aimed at changing realities on the ground.'[196] Therefore it is 'engaged politically... to prevent the use of landmines and increasing political awareness among governments and the general public....'[197] NPA played a critical role in lobbying the Norwegian government to take negotiations for a cluster munitions ban outside the CCW process and start the Oslo process.[198]

Nevertheless, the more conservative outlook of the primarily former military personnel staffing NPA's demining programs has not always meshed well with the organization's leftist and activist roots. Employees explained that there was a 'culture clash' between NPA's traditional 'anti-militarist, anti-war, pacifist people' with backgrounds in sociology and anthropology and the ex-army people who came 'crashing in' with the mine action program.[199] For instance, NPA's mine action personnel coined the epithet 'batik skirts' to refer to the traditional leftist members of the organization. As a result, there has been significant intra-organizational conflict over the role and nature of NPA's mine action programs and how they fit into the broader framework of the organization's work. For example, as part of a general effort to diversify its funding base, NPA's mine action unit took contracts from Norwegian oil companies to provide mine clearance support in both Iran and Angola.[200] This has not sat well with some of NPA's other staff. During the early years, the mine action programs were hived off from the rest of the organization and operated almost as if they were an 'NGO within an NGO.' However, in recent years, there has been both a pragmatic shift within the organization as a whole, and a concerted effort to integrate mine action into NPA's broader strategic goals and principles.

Conclusion

Max Weber argued that people are driven by material interests, but similar to a train running along tracks, the specific direction they go in is determined by the ideas that channel and shape their action. Interests are the engine of the train, ideals guide it.[201] This seems an apt description of the way in which interests and ideals combine to shape US mine action policy. If the US government had no concern for the humanitarian effects of mines and UXO, it, like Russia and China, would not bother committing major funding to mine clearance or negotiating arms control measures. However, the strength of US strategic interests mean that these ideals merely shape the manner in which mine action policy is packaged and implemented – they are not the driving force. As a result, US policy on international law regulating mines and cluster munitions has shown a 'significant gap between rhetorical leadership and policy realities.'[202] While the US has shown greater openness to humanitarian concerns than other great powers (like China and Russia), mirroring Egeland's thesis, the US has been constrained by its strategic national security concerns. Therefore, some argue US mine and cluster munitions control policies are focused more on defending an institution of the state – the military – than the human security of civilians. When pushed by public and international pressure, the US has been open to negotiating restrictions on mines and cluster munitions, but only in fora where it has control over the final outcome and as long as the treaty does not threaten the freedom of the Pentagon to deploy weapons according to 'military necessity.'[203]

Moreover, US mine action assistance is shaped and constrained by the country's strategic interests as a military power and growing trends of privatization. Therefore, a significant portion of US mine action takes the form of military-to-military training or is implemented through private security companies. Therefore, it often roughly represents a 'Strategic-Commercial Complex' of mine action. Nevertheless, it would wrong to completely pigeonhole US mine action. When strategic concerns are less important and the US feels it is less necessary to control the outcome, it is able to take a more idealistic approach. For instance, the Leahy War Victims Fund derives from compassion for civilians affected by conflict and WRA provides considerable funds to mine action NGOs. However, it is difficult to gather the same level of political mobilization within US governmental structures without the 'fuel' of a major strategic or commercial interest.

By contrast, Norway's foreign policy seems to flip Weber's metaphor. For Norway, the missionary impulse of their ideals and concern for the global public good do genuinely seem to be the engine of their interaction with the rest of the world, though it is constrained by the few genuine national interests they have. Concerning mines and UXO, the Norwegian government has been open to civil society pressure, overruling objections from its military in favor of becoming a leader in efforts to ban mines and cluster munitions. Similarly, following the 'Norwegian Model' of foreign policy, Norway's aid for mine action is implemented through a network of NGOs and humanitarian agencies. Its primary grantee, NPA, is rooted in a leftist ideological project of 'solidarity with the poor and oppressed.' Since it is less likely to be motivated by concerns of material gain than companies like RONCO, NPA has the independence necessary to push the government to adopt more progressive stances. Therefore, Norway's mine action programs rough represent a 'Human Security-Civil Society Complex.'

4

IMPLEMENTATION IN AFGHANISTAN, BOSNIA AND SUDAN

This chapter provides historical background to mine action in three mine and UXO-affected countries, showing how different political structures of demining developed internally and how donors, especially the US and Norway, selected which systems to support. The analysis of the case histories focuses on the development of two broad typologies of mine action. First, the author found in each of the countries a civil society-driven approach, implemented generally by international NGOs that prioritized the humanitarian and socio-economic needs of communities and espoused a cosmopolitan politic. This approach was supported by Norway and fits into its broader political vision of collective security outlined in previous chapters. A second approach was driven by commercial tendering and was more integrated into the strategic and 'national security' politics of self-interested actors, both locally and internationally. This approach was often supported by the US, echoing the broader *modus operandi* of its foreign policy. In the case of Sudan, this approach was adopted more by the UN, which had significant politico-military interest in the country.

The chapter also examines three exceptions in Norwegian and US behavior: a) Norwegian support for Bosnian military demining b) US support for UN-led, local NGO-implemented demining in Taliban-era Afghanistan, and c) US support for international NGO-led demining in Sudan after the Comprehensive Peace Agreement. These exceptions are accounted for by arguing that when the US is strategically disengaged from a region it is more likely to frame its involvement as 'humanitarian' rather than political and diplomatic. When this happens, the significant, though weak, idealistic influences in US policy became ascendant, though underfunded.

Before continuing, there are a couple important points to note. This is, in effect, a political history of demining. Thus it focuses on the differing socio-political modes of US and Norwegian-supported demining, rather than any difference in technical methods (such as machines, dogs, prodding, metal detectors, etc.). Moreover, while the subsections analyzing the US support to mine action are generally longer than those about Norway, this is not because of bias or lack of information. It is because the US programs were generally more complicated, involving more donors (such as the State Department, USAID and Department of Defense) and implementing agencies.

Afghanistan[1]

Afghanistan is considered the birthplace of humanitarian mine action. In addition to being the oldest operation, the Mine Action Program for Afghanistan (MAPA) is also the biggest, employing almost 10,000 people at its peak. This section tells the story of three distinct Afghan demining programs that began at the end of the 1980s, how the UN-led program became ascendant and absorbed the others, and how the aftermath of September 11 led to a major shake-up in Afghan demining. In particular, it focuses on structures supported by the US and Norway.

The roots of Afghan demining lie in the significant international humanitarian and strategic interest in Afghanistan at the end of 1980s. Following the Soviet invasion in 1979, battles between Afghan communist government forces and the US, Saudi and Pakistan-backed rebels were devastating to life, infrastructure and livelihood and created staggering levels of mine contamination.[2] While demining had always been considered a military issue, the high casualty rates and socio-economic blockages caused by mines became too big for humanitarian agencies to ignore.

Early US support to demining in Afghanistan developed in the context of the USAID Cross-Border Humanitarian Assistance Program (CBHA), which smuggled aid from Pakistan into mujahideen-controlled regions of Afghanistan.[3] While cloaked in language of humanitarianism and providing only 'nonlethal assistance', the CBHA was closely integrated with the CIA effort supporting the mujahideen. The former CBHA director described USAID, the CIA and the Pakistani Inter-Service Intelligence (ISI) as 'one big happy family.'[4] During the course of supplying American pack mules to the mujahideen to transport humanitarian cargo and ammunition over the mountains, USAID became concerned about the impact mines had on their supply lines. Therefore, in 1989, with the assistance of the CIA station in Bangkok, the CBHA persuaded the Thai military to help train the mujahideen in the use of demining dogs.[5] RONCO Consulting Corporation, the contractor responsible for CBHA procurement and logistics, was running an

Animal Holding Facility in Peshawar, Pakistan to train and care for the pack mules. At USAID's request, RONCO expanded this to include a Mine Dog Center (MDC), training mine dog handlers and sending them into Afghanistan to clear roads and airstrips.[6]

The US continued to support MDC directly through RONCO until 1994, when it was reorganized as a local NGO and incorporated into the fold of the UN program (described below). The MDC is now the biggest mine detection dog organization in the world.[7] Because it was set up just as US interest in Afghanistan began to decline, MDC became a relatively depoliticized part of the larger UN program. However, its roots lay US covert action in Pakistan and Afghanistan. While ostensibly a humanitarian program, it had also acted in support of a paramilitary campaign and was run by a commercial contractor.

The many international NGOs involved in providing aid to the Afghans in the 1980s were also fully aware of the threat mines and UXO posed both to their own workers and the people they served, and began seeking ways to mitigate it. The International Committee of the Red Cross, which was running hospitals for war-wounded, provided medical care and developed prosthetics for mine accident survivors.[8] Handicap International also worked with mine survivors both in Afghanistan and in the refugee camps in Pakistan.[9] Several international NGOs also began incorporating mine awareness education into other programs like health and education. In early 1988, despite accusations of imprudence from other NGOs, Rae McGrath, then country director of World Vision International and former British serviceman, organized one of the first NGO-implemented mine clearance programs in the context of a rural rehabilitation project.[10] He then set up Mines Advisory Group (MAG), later a founding member of the International Campaign to Ban Landmines, with the purpose of assisting the UN in developing a mine clearance program in Afghanistan.

However, only one international NGO, the Hazardous Area Life-support Organisation (HALO) Trust, developed a sustained, long-term mine clearance program in Afghanistan. Now one of the largest demining organizations in the world, the HALO Trust was set up by ex-British Army officers Colin Mitchell and Guy Willoughby, who had both worked in Afghanistan and were shocked at the humanitarian impact of mines. Rather than leaving it to the military, they felt demining should be 'an act of charity.'[11] Following an assessment in 1988, they recruited Kabuli doctor Farid Homayoun and set up an office in communist-controlled Kabul.[12] With ex-British Army volunteers, many of whom learned mine-clearing in the Falklands/Malvinas War, HALO began its nascent demining program in 1989.[13]

Since HALO was the only demining group and one of very few western NGOs that coordinated its efforts with the communist government, some NGOs saw their work as 'aiding the enemy.' However, Willoughby, who was influenced by the 'impartiality and neutrality' doctrine of the Red Cross, felt that since the area around Kabul was the most heavily mined, HALO should focus where it could have the most humanitarian impact. This contrasted with the US-supported MDC and the UN-created NGOs that only worked in mujahideen-controlled regions. In order to bolster its credibility as a neutral third party, in 1990 the UN decided to also support HALO's demining in communist-controlled areas and HALO has continued to be a UN implementing partner ever since. It has since grown into the biggest demining agency in Afghanistan, with some 2,800 staff.

In 1988, the US and USSR signed the Geneva Accords, aimed at ending hostilities in Afghanistan and the Soviets withdrew in 1989. Anticipating large-scale refugee return, the UN Office for Coordination of Humanitarian Affairs (OCHA) launched a massive program of assistance to Afghanistan, dubbed Operation Salam, which included a Mine Clearance Programme.[14] Initially, this consisted of military personnel from several Western countries, including the US and Norway, training Afghan refugees to clear mines voluntarily when they returned home.[15] However, this effort failed miserably, marked by 'a good deal of confusion', 'lack of communication' and a 'lack of planning at all levels.'[16] After a 'radical rethink,'[17] OCHA developed a formal and specialized institutional structure to coordinate and implement demining in Afghanistan. OCHA would fund, coordinate and supervise the program and provide expatriate technical advisors to several local NGO implementing partners and the HALO Trust, through a Pakistan-based office called the UN Mine Action Center for Afghanistan (UNMACA). These NGOs and UNMACA, their supervisor and funder, were known collectively as the UN Mine Action Program for Afghanistan (MAPA).

The early expatriate military officers seconded as technical advisors to MAPA were from non-Communist states and were interested in learning about Soviet mine warfare.[18] Moreover, some of the Afghan NGOs, for a time, were associated with political factions.[19] However, as mujahideen parties began fighting each other, following the collapse of the communist government, OCHA was determined to maintain neutrality from factions.[20] Using NGOs for implementation was seen as a better alternative to 'carv[ing] it out for international commercial companies to make a lot of money out of'[21] and a way to keep the program out of the sphere of political contestation until an internationally recognized government sat in Kabul.

After the communist government fell in 1992 and the Soviet Union collapsed, the US lost interest in Afghanistan as a strategic priority. Therefore, the CBHA pulled out in 1994. MDC was reorganized as a local NGO and incorporated into the MAPA. The US continued to fund Afghan demining at a level of about $1 million to $3 million a year from 1994 to 2001, but channeled and coordinated all its assistance through the MAPA.[22] Unlike the partisan spirit of the CBHA, the MDC, like the other Afghan NGOs, 'established a working relationship with all sides'[23] in the ongoing conflict between the mujahideen parties. The HALO Trust, since it was largely funded through the UN system, also became subsumed under the MAPA umbrella, though it was 'highly resistant to coordination by UNOCHA.'[24] An uneasy truce developed, analogous to a 'federal' system, where HALO was under the UN umbrella, but had more autonomy that the local mine action NGOs. Thus, from 1994 to 2001, the system remained largely unchanged and unchallenged.

The rise of the Taliban in 1997 posed some difficulties; they objected to the employment of women as mine awareness instructors and harshly persecuted some non-Pashtun deminers. However, most mine action agencies had tolerable relations with them. Security for deminers improved considerably as the Taliban cracked down on warlordism and violent crime. The Taliban, like many Afghans, viewed demining as a continuation of the jihad against the Soviets and were largely supportive of demining, even donating land to some of the NGOs. The Taliban also used few mines and in 1998 publicly backed a ban on them. The Northern Alliance, fighting a defensive war against the Taliban, continued to use mines extensively, but they too were largely supportive of the demining program in areas of low strategic importance.[25] The MAPA's perceived neutrality was critical in maintaining its ability to work extensively all over the country.

However, due to the international isolation of the Taliban regime, the UN found it difficult to expand the program. Afghanistan had fallen off the strategic maps of the great powers. Donors were reluctant to engage too deeply with an Islamist regime linked to international terrorism. The political climate following the bombings of US embassies in East Africa in 1998 and subsequent US missile strikes on Afghanistan made it especially difficult to secure funding. Nevertheless, both the US and Norway maintained small commitments of funds to the UN-led MAPA throughout this period.

After seven years of isolation, September 11, 2001 put Afghanistan back on the US geo-political map. Unlike in the 1980s, when the US mainly operated through Pakistani interlocutors from the safety of Islamabad, US agencies now involved themselves directly in Afghan politics. While they were relatively hands-off in comparison to their participation in Bosnia, Kosovo or Iraq, US involvement was unprecedented as far as Afghanistan was concerned. Other developed countries, including Norway, followed America's lead, increasing their foreign aid contributions to Afghanistan substantially – though Norway's demining aid allocations changed less drastically (see figures 8 and 9).

As one might expect, the sudden new international interest in Afghanistan led to the explosive growth of the MAPA. Figure 10 shows how quickly the program grew post-9/11. At the same time, funding modalities for demining multiplied. The staggering increase and diversification of funds led to demands for reform and greater scrutiny.

Figure 8: US Aid to Afghanistan, General and Mine Action, 1991-2005.[26]

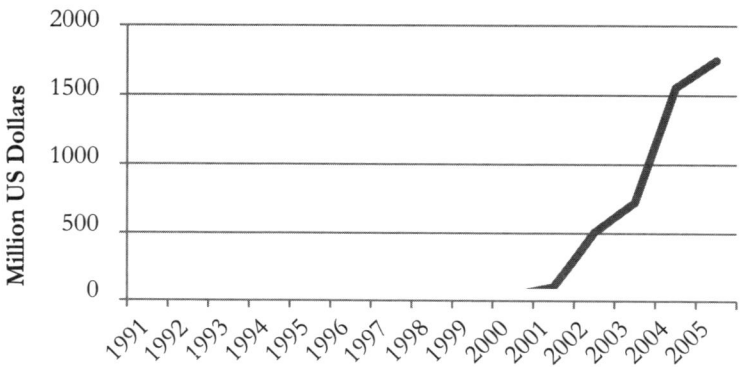

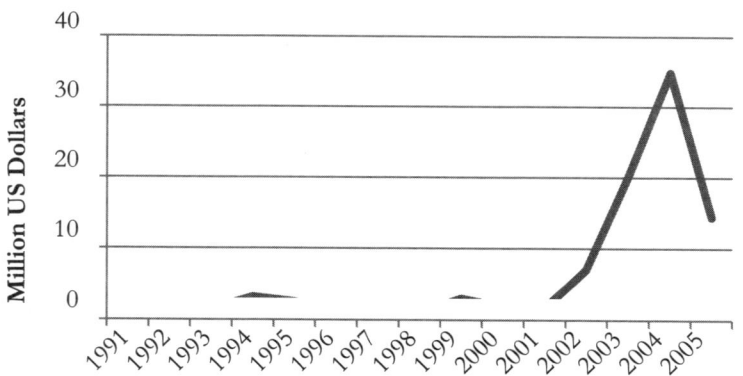

Figure 9: Norwegian Aid to Afghanistan, General and Mine Action, 1991-2005.[27]

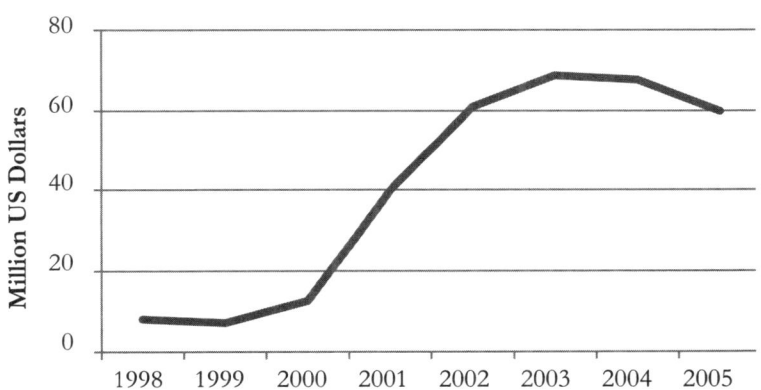

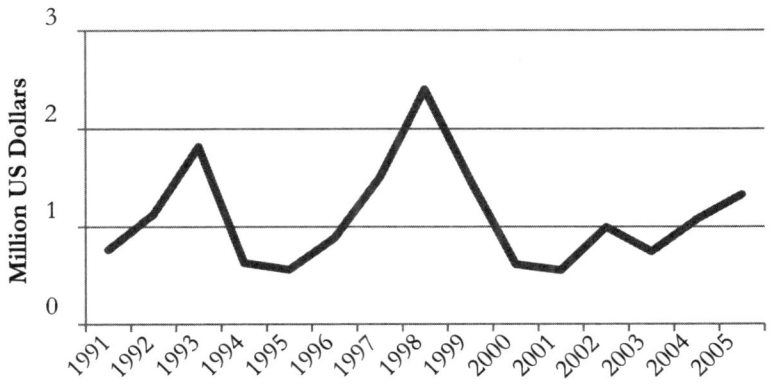

Figure 10: Contributions to UN Mine Action, in Millions of US$, 1991-2005.[28]

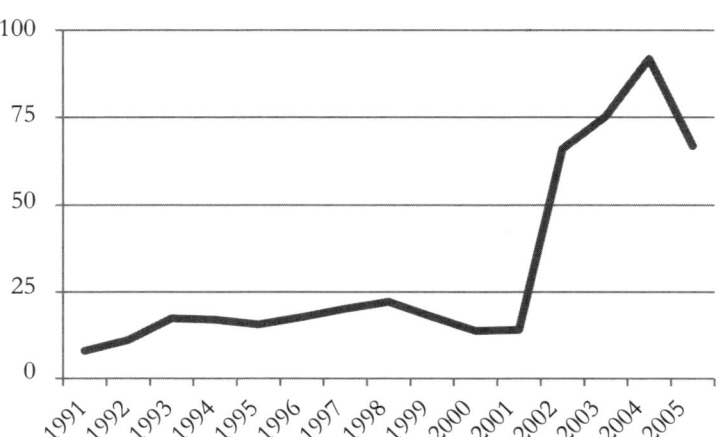

One of the major policy shifts, in 2004, was an UNMACA decision to allow the accreditation of commercial companies interested in the new demand for military and commercial reconstruction tasks. The rationale given by UNMACA and donors was that the new Afghan NGO Law forbade NGOs from bidding directly on commercial reconstruction tasks.[29] However, some have argued that donors could have circumvented this problem by closely coordinating grants for demining with reconstruction work. It is more likely that the decision was made as a result of pressure from the US and that UN Mine Action Service (UNMAS), located in the Department of Peacekeeping Operations, took over management of UNMACA from OCHA in June 2002. OCHA, rooted in the humanitarian aid world, had a bias toward using aid agencies. UNMAS, on the other hand, has tended to contract out its operations through the UN Office of Project Services (UNOPS) and was more open to commercial demining.

Despite its efforts, however, neither the US nor the Norwegian government were fully satisfied with UNMACA's reform. USAID's demining coordinator, Dean Hutson, said he saw UNMACA as a 'broken' and 'failed organization.'[30] Both the US and Norway chose to direct their demining funding outside the UN system, but for different reasons and in different directions. The divergence in the US and Norwegian post-9/11 approaches derives from their differing interests in and perspectives of the situation in Afghanistan.

US Post-9/11 Support for Afghan Mine Action

The US foreign policy community sees Afghanistan as a location of immense importance in the Global War on Terror. Thus US aid to the country has bolstered broader objectives of promoting political stability, countering the blooming narcotics trade and building support for counterinsurgency efforts. The first paragraph of USAID's current Strategic Plan for Afghanistan asserts that it is 'a critical partner' in the 'fight against terrorism and tyranny.'[31] The US post-9/11 strategic interest in Afghanistan has also shaped its demining funding. Initially, USAID channeled its new funding for reconstruction demining through the UN system and RONCO was contracted by the State Department to build the capacity of the Afghan NGOs.[32] However, a parallel structure also began to emerge, with a growing commercialization, politicization and militarization of US demining funding.

Firstly, the US military worried that mines could hinder its mission. When occupying the massive former Soviet military bases at Bagram and Kandahar, US forces found high levels of mine and UXO contamination but lacked demining capacity.[33] In April 2002, four US soldiers were killed in explosive ordnance disposal operations in Kandahar.[34] Thus, the US State Department asked RONCO, which returned to Afghanistan in 2002 after an eight-year absence, to assist military engineers in demining Bagram and Kandahar air bases. Moreover, RONCO helped the US Army develop its first mine dog teams deployed since the wars in Indochina.[35] Though RONCO lost its position as the sole commercial demining company in Afghanistan in 2004, it has continued to do military demining at bases and airfields, winning a $16.4 million three year contract with the US Army in 2007.[36] Several other companies also do military-related mine/UXO tasks. For instance, UXB International, on sub-contract to DynCorp, provided demining and explosive ordnance disposal services to US-funded forced poppy eradication efforts.[37]

Secondly, the US was concerned about the enormous caches of unsecured munitions, including mines, scattered around the country – a potential bonanza for the insurgency, terrorist networks, drug traffickers and warlords. As part of its State Department contract, RONCO developed explosive ordnance disposal (EOD) teams to clear strike zones and arms caches.[38] RONCO cites the following anecdote as one of its success stories:

> When the coalition airfield at Kandahar came under numerous rocket attacks in early 2004, regional authorities requested RONCO deploy its teams to clear up munitions sites in the area surrounding the base. Clearance operations lasted five months, and since their completion, there have been no significant rocket attacks.[39]

After August 2005, this contract went to DynCorp (and subcontractor UXB) and included support for destroying small arms, anti-aircraft rockets and other ordnance.[40]

Thirdly, increasing US demining support reflected a general growth in US funding for reconstruction, aimed at building US political capital in the country.

USAID's 'largest project category' is road building,[41] seen by the US government as a way to increase NATO and US access to remote regions. As one US commander put it, 'where the roads end, the Taliban begin.'[42] USAID decided in July 2006 that it would no longer fund UN-led, NGO-implemented demining, instead prime reconstruction contractors would subcontract needed tasks to commercial demining companies. In September 2006, USAID awarded a five year $1.4 billion energy, water and transportation infrastructure contract to Louis Berger Group, Inc. and Black & Veatch Special Projects Corp[43] and in August 2007, issued a Request for Applications for a further $400 million to construct 'Strategic Provincial Roads' in South and East Afghanistan – those areas most affected by insurgencies.[44] It is understood that a sizable portion of both these contracts will be sub-contracted for demining and battle area clearance. This arguably erodes the needs-based prioritization of demining assets in favor of strategically important tasks.

Finally, the commercialization of demining reflects a broader trend in US security and development policy. In Iraq, Bosnia, Kosovo and Afghanistan, the US military has outsourced many of its core functions to private contractors. Moreover, USAID's overall economic strategy for Afghanistan relies on encouraging private sector-led growth through liberalization and privatization.[45] This is applied both to Afghan institutions and to USAID's own services, which have shifted from grants to contracts.[46] Privatizing the functions of public service institutions like the MAPA was seen as a way of correcting dysfunctions, by introducing standards of competition and transparency. USAID demining coordinator Dean Hutson said:

> We don't want to babysit this country for the next twenty years, we want this country to stand up on its own, with its own business, its own private enterprise. ... We don't want this to be a non-stop, continuing welfare operation where we just hand out money to do it.[47]

Therefore, while the US State Department continued to fund and provide technical assistance to the Afghan NGOs through DynCorp in 2005, part of this program's emphasis shifted to assisting the local Afghan NGOs spin off commercial operations.[48] Moreover, the US State Department has agreed to allow equipment donated to local NGOs to be used by local commercial companies, if the NGOs have a surplus of equipment.[49] Meanwhile, several private corporations, such as the mobile telecommunications firm Roshan, have begun to contract commercial clearance. As a result of this liberalization and deregulation, at the time of the author's fieldwork there were seven international commercial companies involved in Afghan mine action and three local start-ups. Many of the international companies had other divisions that were providing security services and several companies' staff also carried arms.

Norwegian Post-9/11 Support for Afghan Mine Action

Norwegian interest in Afghanistan also increased precipitously after 9/11, prompted largely by its commitment to NATO. Afghanistan, according to the Norwegian Minister of Defense, is Norway's 'most important international operation.'[50] For the first time in its history, Norway deployed troops outside Europe on a non-UN mission, facing real combat. Since 2001, Norway has contributed both to the US-led Operation Enduring Freedom[51] and the NATO International Security Assistance Force (ISAF). From February 2006, all its troops came under ISAF command and were deployed largely in northern Afghanistan, centered around Mazar-i-Sharif. It retained a small Special Forces counterterrorism capacity around Kabul.[52]

Like the US, Norway supported the UN-led demining program until 2002, but soon became frustrated. Its objections to UNMACA, however, were very different. It felt that the massive expansion of UNMACA's international staff after 2002 undermined the possibilities of transferring the responsibility for demining to a local institution – a policy Norway felt would be more democratic, more sustainable and less expensive.[53] At the same time however, Norway felt uneasy about the transparency of the local Afghan NGOs. Therefore, in 2003, Norway began giving its Afghan demining funds to the HALO Trust, which is largely run by Afghans, but conforms to international standards of accounting and reporting.

The Norwegian ISAF contribution has included a mine clearance unit which did some limited demining efforts for military purposes[54] and HALO's main area of operation – Kabul and northern Afghanistan – corresponds to the deployment of Norwegian troops and where most Norwegian aid is directed. However, Norwegian diplomats insist that they maintain a clear separation between the humanitarian nature of demining and the politico-military objectives of the NATO mission. Multiple officials denied HALO was supported because they were working in similar areas as Norwegian troops, and denounced the tendency to blur the lines between humanitarian and security objectives. Other Norwegian aid projects are also kept separate from military efforts, with implementation left to the World Bank, UN and NGOs.[55]

Thus, it is likely that Norway supports the HALO Trust for reasons other than potential strategic gains. The Norwegian government prefers giving demining funds to international NGOs since they tend to have greater standards of financial accountability. Other than Danish Demining Group, which is new to Afghanistan, HALO Trust is the only international in the country. Moreover, HALO, as the biggest operational demining organization in Afghanistan, has a good reputation and is able to leverage other funding and gains from economies of scale. Though HALO is an international NGO and rotates a few expatriate technical advisors through the program, the rest of its 2800 staff, including the country director, are Afghans. It thus embodies the kind of local ownership the Norwegians wish to see UNMACA adopt.

Finally, HALO's organizational culture is in line with the kind of humanitarianism Norway wishes to support in Afghanistan. In contrast to the private demining companies, HALO works hard to cultivate 'humanitarian space' for mine clearance: 'The important thing is to keep neutrality, so you are not seen as pro this or against that group,' said Dr Farid. 'That's the key thing – having a neutral, impartial humanitarian organization.'[56]

In sum, the origins of Afghan humanitarian mine action offered three alternative models of organizing demining:

1. Demining in support of broader US security objectives, contracted through a private company (the RONCO/MDC program),
2. International NGO mine action claiming political neutrality (the HALO Trust),
3. UN coordinated mine action, implemented by local NGOs and also claiming political neutrality (the UN Mine Action Program for Afghanistan).

From 1994 to 2001, the UN-led model was ascendant and supported by both Norway and the US. However, US re-engagement in Afghan politics after 9/11 saw it return to a 'securitized' and commercialized model. In contrast, objecting to the massive, top-heavy growth of UNMACA, Norway switched to funding the HALO Trust in order to support an Afghan-led civil society actor espousing traditional humanitarian values.

Bosnia and Herzegovina[57]

Following the 1995 endgame of Bosnia's war, the frontlines dividing Bosnian Government territory from its separatist statelets – the 'Republika Srpska', 'Herzeg-Bosna' and the 'Bihac pocket' – were littered with extremely high levels of mine and UXO contamination.[58] The Dayton peace agreement that ended the war required factions to clear all their minefields in 30 days.[59] This was a ludicrous deadline, given the heavy contamination and lack of accurate minefield maps. It quickly became clear that demining required a long, intensive effort and significant international funding. In early 1996, a Bosnian Mine Protection and Removal Agency (MPRA) was set up to oversee mine action. This was soon embroiled in political contestation and was replaced by a new system. The highest Bosnian mine action authority became the Demining Commission, with three Commissioners, each representing a 'constituent ethnic group' (Bosniak, Serb and Croat). This was advised and assisted by an internationally staffed UN Mine Action Center (UNMAC), set up in May 1996, which also gathered data and conducted quality assurance inspections. UNMAC was localized in 1998 as the Bosnia and Herzegovina Mine Action Center (BHMAC), which became the coordinating body, accreditation and quality assurance authority and data depository for Bosnian mine action, with counterpart Entity Mine Action Centers (EMACs) in the Federation and Republika Srpska. In 2002, following the passage of a Law on Demining, BHMAC was reorganized to incorporate both the EMACs.[60]

The vast majority of mine action funding was provided by international donors and implemented by a diverse set of local and international actors. As per their obligations under the peace agreement, the militaries of both of Bosnia's decentralized 'entities' – the Federation (populated mostly by Bosniaks and Croats) and Republika Srpska (populated mostly by Serbs) – have performed a significant proportion of demining in Bosnia. Local civil protection units, responsible for disaster preparedness and civil defense, have taken responsibility for small tasks and explosive ordnance disposal. Private actors, international and local commercial companies and NGOs, have also played a large role. This section on Bosnia will not cover the history of all these actors and their programs; it will focus solely on those funded by the US and Norway.

US Support for Bosnian Demining

Though through most of early 1990s the US carefully avoided getting involved in the wars of Yugoslav succession, growing pressure from human rights groups and concerns about the credibility of the US and NATO's 'security blanket' in Europe provoked the US into leading robust NATO intervention in 1995.[61] The Bush administration has slowly disengaged from the region, prompted by military overstretch elsewhere and replaced by a growing EU presence. However, during the late 1990s, the US staked a great deal of its superpower credibility on the success of the peace agreement it had brokered, deploying 16,000 troops to Bosnia and establishing one of the largest CIA stations in the region.[62] As in Afghanistan, the US strategic stake in Bosnian stabilization shaped its support for demining programs.

From a military perspective, landmines posed a major threat to US troops stationed in Bosnia – the first US casualty of the NATO peacekeeping mission fell to a landmine.[63] Therefore, the US military saw mines as 'the greatest threat to force protection and the success of the mission.'[64] US troops gathered information about mines in their area of operation and were mandated by Dayton to monitor the entity militaries' efforts to meet clearance obligations.

Cooperation with the entity armies in mine action was also a potential way transform the Bosnian militaries from a security threat into potential future NATO contributors. The US and NATO saw demining and explosive ordnance disposal as a possible contribution the Bosnian military could offer NATO, if it became a member.[65] While some countries were nervous about directly training the entity armed forces in anything, the US decided to train and equip the Bosnian military demining units.[66] In 1997, 55 US Special Forces troops, supported by RONCO, trained 450 personnel from the entity armed forces in basic demining.[67] Both Johnson[68] and Priest[69] have described how US Special Forces training opens doors for the US to gain influence with the host country armed forces.[70] The next year, the US Defense Department expanded its efforts, training 71 military demining instructors and building three military demining academies (one for each 'constituent people')[71] – 'because they didn't want to be trained together.'[72] With encouragement from the US, Bosnia deployed a multiethnic military demining unit to Iraq as a part of the 'Coalition of the Willing' in 2005, showing that US investment in training the Bosnian military had reaped a small strategic return.[73]

The US poured millions of dollars into Bosnian reconstruction efforts, attempting to create noticeable benefits from a 'peace dividend.' Much of this focused on creating conditions for the return of displaced persons. The return process was a keystone of Dayton's legitimacy because without it, the agreement would appear to be a crude ethnic partition. In terms of economic reconstruction, the US encouraged the privatization and liberalization of Bosnia's centrally-planned economy, as it did throughout Eastern Europe and the Former Soviet Union. As a result, most US support for Bosnian demining focused on return areas and infrastructure projects and operated in a variety of commercial tendering frameworks.

For example, USAID funded demining through its infrastructure programs, which 'focused on restoring electric power, water, sanitation, shelter, schools and other public services'[74] in minority return areas. Many of these projects were 'developed in cooperation with the U.S. Army and implemented in the U.S. Army SFOR Area of Operation.'[75] Parsons, an engineering firm acting as prime contractor for these projects, subcontracted management of any necessary demining work to UXB International, an American demining company. UXB subcontracted much of the work to local companies. UXB also did some demining for corporate clients, including the energy company Energoinvest.

The local firms dominant at the beginning of Bosnian mine action were created in 1996 and 1997 when the US State Department contracted RONCO to train three 'demining units' – one Serb, one Croat and one Bosniak. These units were transformed by local entrepreneurs into the companies UNIPAK (Serb), SI/OKTOL (Bosniak) and DECOP (Croat). In 1999, the State Department also assisted in the creation of three local NGOs, again along ethnic lines: STOP Mines (Serb), BH Demining (Bosniak) and PROVITA (Croat). In addition to small grants to help them start up, they were given sole-sourced (non-competitive) contracts from the State Department through the ITF for the first few years of their operation, though these were later turned into competitive tenders.[76] Many other local commercial companies and NGOs soon developed.

From late 1996, the World Bank, funded in part by the State Department, contracted 14 commercial companies to implement its demining program in Bosnia. Over 60% of the project funds went to two international contractors – Mine Tech International and RONCO Consulting Corporation.[77] Mine Tech teamed up with UNIPAK and worked in the Republika Srpska; RONCO worked with SI/OKTOL and DECOP in the Federation. Mine Tech left Bosnia in 1998, and RONCO took over as the main intermediary between international donors and the three main local companies until 2000. RONCO claims credit for shaping SI/OKTOL, UNIPAK and DECOP into viable commercial operations and ran its demining operations through them.[78] Actual contracting and tasking was handled by Project Implementation Units (PIUs) in each of the entities, staffed by the entity government, which administered World Bank money.

This system became extremely controversial, with reports of local 'cowboy companies' using government connections to gain contracts.[79] Indeed owners of the local companies, several of whom had links with war criminals and organized crime, had deep familial and political ties to key persons in the government.[80] The World Bank, concerned with the extent of corruption, terminated its demining funding in 1999; an internal evaluation of deemed it 'Highly Unsatisfactory.'[81] In summer 2000, Bosnian Financial Police raided the offices of several demining companies, including RONCO. As a result of these investigations, the three Demining Commissioners were removed from office in October 2000 by the Bosnia's international High Representative, citing 'misuse of office,' 'breach of public trust' and 'widespread conflict of interest.'[82] This was followed by a major expose of the demining sector in the local newspaper *Slobodna Bosna*.[83] RONCO left Bosnia later that year, with bitter complaints of a 'smear campaign.'[84]

In 1999, the US began channeling its funds through the International Trust Fund for Demining and Mine Victim Assistance (ITF), set up by the Slovenian government. The State Department promised to match funds raised by ITF up to a certain level. This moved contracting out of the country, away from the PIUs, thus making it less susceptible to nepotism and corruption. However, until 2003, the ITF continued to award US-funded tenders to the same companies that had been at the center of the World Bank scandals.

The US finally moved against demining corruption in mid-2003. Renewed international efforts to tackle organized crime and networks supporting indicted war criminals prompted the State Department to blacklist demining organizations allegedly linked to the network hiding war crimes fugitive Radovan Karadzic. The blacklist, which barred organizations from receiving US demining funds but was not publicly disseminated. The blacklist ran into criticism for several reasons. Firstly, it was nontransparent. There was no way to publicly determine the reasoning behind placing a company on the list, which did not officially exist. Secondly it was reactive, rather than proactive. Organizations were placed on the list when links to such clandestine networks were discovered. There was little effort to investigate rigorously the backgrounds of other organizations.[85] Thirdly, it seemed to focus solely on those companies linked to Serb organized crime networks, neglecting those operated by other ethnicities. Eventually, in early 2007, the State Department conducted thorough investigations and organizational audits of the local Bosnian demining organizations. The result was a continued commitment to ITF commercial tendering, but with one key modification: in April

2007, the State Department released a shortlist of organizations eligible to be US-funded contractors.[86]

Why did the US continue for so long to support organizations uncovered as corrupt and linked to organized crime? Though there were warnings as early as 1997 that the whole demining system was in danger of being captured by private interests,[87] State Department officials involved at the time claim that they did not know of the backgrounds and suspect practices of the main local companies until 2003. It is possible that this was simply due to naiveté. One official said the early passive State Department policy was a 'function of information'; when they finally found out about links to war criminals, they took action. This explanation is implied by the State Department's embarrassed muddling over the local company MEDECOM, which was blacklisted in August 2005, but received a US-funded contract for a clearance project in Serbia the very same month. Investigations by the author and Hugh Griffiths suggested this was due to bureaucratic inertia preventing the news of MEDECOM's blacklisting from reaching the US Embassy in Belgrade.[88]

However, a 1996 document written by the above State Department official who denied knowledge of Kojic's alleged nefarious connections until 2003, reported to his superiors that Kojic had 'close ties' to Karadzic.[89] Indeed, as Keen[90] and Clay and Schaffer[91] warn, observers should be careful not to prematurely judge a policy as a failure. It can be helpful to ask what functions and unstated objectives might be disguised by a policy's rhetoric. The key is to bracket questions of motivations and intentions (notoriously difficult to discern) and focus on how actions function. It makes sense then, to ask whether contracting demining to nationalistic and criminalized elite in post-conflict Bosnia also served the political and security objectives of the US.

Getting the buy-in of powerful nationalists was crucial for ensuring the stability of the country. In the early days, the nationalists still had secret arms caches and the power of mass mobilization, while the international community lacked the political will to tackle them head-on. Providing lucrative contracts may have encouraged potentially dangerous men to engage with the reconstruction process rather than shoot at US troops. This may be a necessary evil, though could be corrosive to the political system. It is conceivable then that the $1 billion of aid to Bosnia reported to have disappeared by 1999,[92] and the estimated $6.7 million of demining funding unaccounted for,[93] might have functioned (whether intended or not) as a pay-off for elites who stood to lose from the emergence of peace. It was perhaps a political 'transition cost' of moving from one social system (war) to another (peace).[94] Indeed, in an interview with the author, former High Representative Wolfgang Petrisch, said his order firing the Demining Commissioners, was the only occasion during his tenure when the US ambassador attempted to stop a High Representative's order. Petrisch believes the US Embassy was acting to protect RONCO and the State Department from a public scandal.[95]

Even if individual State Department bureaucrats were unaware of the level of corruption and shady pasts of its primary Bosnian contractors, a lack of knowledge about war crimes often reflects political choices over intelligence priorities, as Samantha Power has argued.[96] Information is often available in public sources – indeed the involvement of organized crime in demining had been covered by several newspapers and was well-known by demining personnel. However, information is only processed into actionable data when enough political pressure builds to analyze it in a systematic manner – as happened when the US finally moved against demining corruption. Therefore, even if the US funneling of money to the 'nationalist-criminal nexus' was a complete accident, such a policy decision operated in a Darwinian framework (the fittest policy survived). As the World Bank says, 'An institution [or policy] exists in part because some constituencies gain from its existence and so have the incentives and influence to support it.'[97] Accidental policies only survive if they are well-adapted to the broader political context. In the early post-war period, the US policies that survived were those of accommodation and collaboration with the nationalist status quo in the interests of stability.[98] Therefore, 'Nationalist, and in many cases criminal, politicians [became]... the key interlocutors for the international community and their principal entry points into local communities.'[99]

In sum, US support to Bosnian demining took two forms: a) support to the entity militaries, as part of a larger effort to transform them from a threat to European security into future NATO contributors, and b) a commercial tendering system that for several years was captured by a criminalized ethno-nationalist elite. This may have functioned as part of a broader US 'passive policy' toward such elites, aimed at getting their buy-in to the Dayton process.

Norwegian Support for Bosnian Demining

During the Bosnian war, Norway contributed to the UN Protection Force. Following the peace agreement, Norway provided troops to the NATO and EU peacekeeping missions. Since 1992, at least 9,000 Norwegian troops have rotated through Bosnia, five of whom lost their lives.[100] This was a major military commitment for a small state. In addition to its usual support for peacekeeping missions, Norway felt it was obligated by NATO membership to provide maximum support for NATO's Bosnian missions. Therefore, like the US, Norway provided assistance to the entity military demining teams, including insurance coverage to deminers,[101] training in the use of a Bozena mechanical mine clearance machine[102] and a technical advisor to the entity armies' Humanitarian Mine Action Coordination Cell, seconded from Norwegian People's Aid (NPA).[103] However, support to the military deminers remained limited.

Norway also provided significant humanitarian assistance to Bosnia during the war, much channeled through Norwegian People's Aid (NPA). NPA worked largely in Bosnian government territory, following its custom of working 'in solidarity' with the people it views as most oppressed in a social conflict. NPA focused on the area around the northern city of Tuzla, whose municipal government tried hard to maintain a cosmopolitan and multiethnic polity. Following the peace agreement in 1996, NPA supported several youth and media programs aimed at encouraging inter-ethnic integration and grassroots peace and democratization processes. NPA also had a major shelter and reconstruction program facilitating the return of refugees and displaced persons, by building more than 5,600 houses.[104]

NPA began preparing for a demining program in October 1995, by giving mine action training to Bosnian refugees in Norway.[105] In 1996, supported by the Norwegian Ministry of Foreign Affairs (MFA), it deployed to the Tuzla area and began demining tasks in support of the shelter and reconstruction programs.[106] In 1997, it also seconded staff to UNMAC, as a way to develop the central coordination capacity in Bosnia.[107] In addition to its humanitarian and development value, the MFA saw mine clearance as a way to support several political peacebuilding objectives, such as restoring freedom of movement and facilitating refugee return – so that 'ethnic cleansing was not to be seen to succeed.' Through this, hoped to support broader global aims of strengthening international legal norms such as the Refugee Convention and the Ottawa Treaty.[108]

Since the advent of the ITF, the MFA has channelled its assistance through the Slovenian body, earmarked for NPA. This has been to attract US matching funds – 'working together with allies…is always a good thing' – and because it 'provides some basic kind of oversight over the money spent.'[109] Nonetheless, the MFA gave NPA significant freedom in determining the scope and operation of the program. A former manager said the MFA never interfered nor 'tried to dictate, in any way where we work.'[110]

NPA relocated its operations to Sarajevo in 1997 to support UNMAC's Mine Action Plan for Sarajevo, since the city's population density and wartime legacy left it one of the most heavily mine-impacted areas. As a result of the wartime siege, the city was ringed by a belt of minefields that, in a sense, perpetuated the siege into peacetime. UNMAC recommended that NPA be the primary implementing partner for the plan because it felt the priority tasks (such as houses and high-rise buildings) in Sarajevo were extremely complex and difficult to parcel into commercial tenders. UNMAC also felt an NGO would be 'better placed to provide assistance with staff training and capacity building.'[111] NPA's work was thus 'particularly time consuming and laborious'[112] and very treacherous, but was vital for restoring public safety in Sarajevo.[113] During the winters in Sarajevo (and previously in Tuzla), when the weather became untenable for demining, NPA moved its operations to the Mostar area, where the climate is warmer.

After 2003, as downtown Sarajevo became relatively mine-free, NPA shifted the locus of its operations to the Brcko District. Brcko, a strategically important and divided municipality, was designated a 'special' autonomous district under international supervision in 2000. A site of brutal 'ethnic cleansing' and heavy fighting during the war, it was one of Bosnia's most heavily mine-contaminated regions.[114] However, with heavy-handed international intervention, it was the first municipality to reintegrate its police force and school system and had the highest rate of return of displaced persons in the country. The Brcko District was seen by international actors as a 'showcase' of how Bosnia could return to a multiethnic polity, and demining was seen as a 'crucial condition for sustainability of returns, improving both the economic and social environment.'[115] NPA's demining was closely integrated with efforts to reconstruct returnee villages in the District and was seen as a way to support 'one of the key areas for return and further normalization and development of the country.'[116]

Initially, NPA's program was very expatriate-heavy. However, in a similar fashion to the Sudan program (see below), they slowly trained and promoted talented Bosnians, who replaced their expatriate managers. NPA's in-country operations, employing 147 people, are now completely run by Bosnians of several ethnicities. Their country director, Darvin Lisica, former deputy director of BHMAC, is a widely respected authority on Bosnian mine action, and a main intellectual author behind BHMAC's attempts to encourage community participation in mine action priority-setting.[117]

In sum, while Norway gave some token assistance to military deminers, the vast majority of its assistance was channeled through an international NGO that employed a multiethnic staff and supported efforts to build a cosmopolitan polity in Bosnia.

Sudan[118]

At war since 1983, Sudan has an as yet undefined landmine and UXO contamination problem, caused largely by fighting between the northern government and Southern rebels, which has resulted at least 3,700 mine and UXO casualties.[119] Most of this problem is concentrated in the South, around former government garrison towns and along major roads. The level of contamination appears initially to have been dubiously overestimated, fueling a massive expansion of demining efforts in recent years. However, the ongoing Landmine Impact Survey (LIS) suggests that the level of contamination is probably much lower than previously expected.[120]

Mine action began in Sudan in 1996, when the southern rebel group, the Sudanese People's Liberation Army (SPLA), declared a unilateral moratorium on the use of mines. This prompted the Government of Sudan to sign, though not ratify, the Antipersonnel Mine Ban treaty in 1997. The SPLA then created two national 'NGOs' (more like parastatals) – Operation Save Innocent Lives (OSIL) to coordinate clearance and Sudan Integrated Mine Action Service (SIMAS) to coordinate mine risk education and victim assistance. In the north, the Sudanese Red Crescent Society and the Sudan Campaign Ban Landmines (SCBL) successfully lobbied the government on landmine issues and by 1999, both the

SPLA and the government re-affirmed a commitment not to use mines. That same year, OSIL, which received some limited assistance from NGOs and UN agencies, reported that it had cleared a total of 236 miles of roads, 2,179 mines and 20,740 UXO from the south. The Government of Sudan Army also reported doing limited clearance. However, these efforts were not subjected to any independent quality assurance nor conducted to the UN-recognized International Mine Action Standards.[121]

Large-scale international intervention was prevented by the ongoing conflict. Few mine action donors were willing to put money forward while there was the possibility that factions could begin re-mining demined areas or threaten the security of demining agencies. Moreover, while the SPLA signed the Geneva Call Deed of Commitment (an equivalent of the Antipersonnel Mine Ban Treaty for non-state actors) in 2001, many bilateral and multilateral donors were reluctant to fund mine action in a country whose government had not ratified the Treaty.[122]

In 2002, the combined diplomatic efforts of the US, Norway and Switzerland achieved a ceasefire between the Government of Sudan and the SPLA in the Nuba Mountains region. This was seen by the international community as a major opportunity to build confidence between the two parties and demonstrate a potential 'peace dividend,' in socio-economic assistance. The Joint Military Commission (JMC), created to monitor compliance with the ceasefire, saw mine survey and clearance of roads as critical to its mission and a possible means to build cooperation across lines. Therefore, Landmine Action, which had been involved in organizing 'crosslines' meetings between mine action groups from the North and South, began training teams of deminers and surveyors from both sides to work on a project called Sudan Landmine Information and Response Initiative (SLIRI). Meanwhile, DanChurchAid (DCA), supported by Norway, set up a similar program working in cooperation with JASMAR, a mine action NGO with ties to the northern government, and OSIL. The US also funded RONCO to provide demining support to JMC, which itself had significant US involvement.

Progress on the ground in the Nuba Mountains contributed to political progress toward mine action cooperation. Three years before the signing of the Comprehensive Peace Agreement (CPA), the northern government, SPLA and United Nations Mine Action Service (UNMAS) signed a memorandum of understanding laying out the framework of a mine action program for Sudan. In 2003, the Sudanese government ratified the Mine Ban Treaty, making it attractive to donors. The international involvement in Nuba Mountains mine action presaged the trend toward greater internationalization as the general peace process between the Government of Sudan and the SPLA improved. In 2004, WFP and UNMAS began survey and clearance operations in the south. The next year, the Comprehensive Peace Agreement between the SPLA and the northern government, brokered in large part by the US and Norway, allocated responsibility for mine action technical assistance and coordination to a UN Mine Action Office (UNMAO) within the UN Mission in Sudan (UNMIS).

The vast majority of demining in Sudan has been done on contracts tendered by WFP or UNOPS (on behalf of UNMAS). Implementers bid on specifically defined tasks and contracts generally went to the lowest bidder, with some consideration of experience and organizational capacity. The UNMAS/UNOPS contracts were funded from the UN's assessed peacekeeping budget and focused on clearing major roads to open access, mostly in South Sudan, for UN peacekeepers, as well as commerce, returning refugees and humanitarian aid. Since DPKO's priority has been to gain rapid access for peacekeepers to all parts of the country, the UNMAS/UNOPS contracts have only cleared roads to a width of eight meters.

In addition to the UNOPS contractors, UNMAO also coordinated several UNMIS military peacekeeping contingents doing demining. This is another example of how UNMIS has integrated mine action into the politico-military objectives of the UN peacekeeping mission. At the local level, UNDP has assisted in the creation of Joint Integrated Demining Units (JIDUs), incorporating Government of Sudan and SPLA soldiers and under the joint management of the northern and southern Sudanese demining authorities. However, as neither the US nor Norway support the international peacekeeping military demining units nor the JIDUs, this chapter will not explore Sudan's military 'humanitarian' demining in more detail.

The WFP contracts are part of a larger road rehabilitation project started in 2004 as a way to reduce the cost of transporting humanitarian aid to the south. Since WFP's focus has been on long-term infrastructural rehabilitation – including the verges and drainage systems of roads – WFP contracts cleared roads to a width of 25 meters. The contracts were funded through member states' voluntary contributions to WFP; both the US (USAID and State Department) and Norway have contributed.[123] WFP's original projects in Sudan were contracted to an NGO, Swiss Foundation for Mine Action (FSD), on a 'cost-plus' basis – FSD was paid a premium on top of costs incurred in conducting operations. However, WFP felt the cost-plus contract, though useful for starting up the program rapidly, incentivized overspending. WFP thus switched to a tendering model similar to the UNMAS/UNOPS system, contracting both international NGOs and commercial companies. A WFP manager explained that they switched to tendering because they wanted 'to gain as much control over the operation as possible.'[124]

UN officials believe the increased competition between agencies in the tendering process encouraged efficiency. Moreover, they felt tendering gave them more control, enabling them to stipulate precisely which areas should be cleared, in what order they should be tackled and to punish – through penalty clauses in the contracts – poor quality or slow work. It is also possible that, as seen in Afghanistan, the earlier shift from OCHA and UNDP toward UNMAS/UNOPS managing the bulk of UN demining has led to a greater UN openness to commercial demining. For instance, UNMAS officials in Sudan have talked about wanting to 'do a Kosovo,' where the UN had tight control over coordinating mine action and tendered most contracts to commercial operators.[125]

Early US demining efforts in Sudan followed a similar model of commercial demining in support of a military peacekeeping force. Immediately after the Nuba Mountains ceasefire, the State Department contracted RONCO to use its Mozambique-based Quick Reaction Demining Force to begin clearance work 'as a confidence enhancing and peacebuilding effort.' In 2004, it contracted RONCO to expand its operations to work throughout the Nuba Mountains 'in direct support of the Joint Military Commission' which monitored the agreement.[126] A RONCO supervisor explained that they cleared 'roads to make the job easy for the JMC monitors, so they could reach the isolated villages.'[127] In the same year, the State Department also contributed to the WFP commercial tenders for 'high priority roads in South Sudan.' As will be explained below, however, the State Department abandoned the commercial model of funding Sudanese mine action in 2005.[128]

Nonetheless, USAID appears to be moving back in that direction. Despite having been a major donor to the WFP road rehabilitation program since 2004,[129] in mid-2006, (at about the same time it pulled its money out of UNMACA's trust fund for Afghan demining) USAID decided to contract reconstruction directly. It thus announced the award of a five-year South Sudan infrastructure contract worth up to $700 million to Berger Group.[130] One of the first projects on this contract will probably be the rehabilitation of a road from Juba to Nimule, and Berger will subcontract RONCO for at least $1 million, to provide demining support.[131]

Beginning with Landmine Action's operations in the Nuba Mountains, international NGOs have been major players in Sudanese mine action, largely supported by the European Commission and bilateral donors, though some, especially FSD, have also competed for UN tenders. Unlike other UN agencies, UNHCR supported DDG's unexploded ordnance disposal in returnee areas through annual grants. Most of these international NGOs worked in partnership with local NGOs.

Following the CPA, the US State Department switched from commercial mine action to supporting international NGOs and building the capacity of local mine action authorities. Contradicting its commonly stated position that commercial mine action is faster and more efficient (also espoused by UNMAO), the State Department explained its shift in policy by saying 'the use of grants to NGOs enabled a more rapid response to the emergency landmine/UXO threat.' It also said the grants aimed to 'provide the necessary infrastructure and management/technical training for the Sudanese to 'own' the program themselves and hopefully develop the indigenous capacity.'[132] The State Department supported the mine risk education, survey, clearance, EOD and small arms destruction efforts of Mines Advisory Group (MAG), Landmine Action and the HALO Trust, largely in South Sudan. MAG, which works in partnership with OSIL and JASMAR, is one of the largest demining NGOs in Sudan and works both in the North and South of the country. While MAG has done considerable work on the roads, it is beginning to focus its efforts on local communities and integrating mine action operations into broader development efforts. In keeping with these priorities, MAG has led the gathering of data for the Sudan Landmine Impact Survey, which will clarify the extent of mine contamination and how it impacts public safety and socio-economic concerns.[133]

Landmine Action, and its programmatic successor HALO, however, have not had as much success. Following a review of their programs in 2006, Landmine Action determined that low productivity levels and fraught organizational politics were beyond their capacity as a small mine action organization. They pulled out and handed over some of their South Sudanese projects to the HALO Trust. However, the HALO Trust, despite being one of the world's largest demining organizations, with experience in some of the most complex political situations in the world, was not able to report any concrete successes from their $3 million, year-long program. Following an ordeal of endless strike action, staff members threatened and subjected to physical violence and crude attempts to seize their funds and equipment, HALO was finally expelled from the country by the South Sudanese government in late 2006.

The State Department has also supported DCA and NPA, the same organizations that Norway has used to fund Sudanese demining. Like MAG, DCA works with JASMAR and OSIL, but in the Nuba Mountains. The aim of the program has been to build 'crosslines' collaboration between deminers from the two sides. Intended to 'contribute to the peace and confidence-building effort... Every aspect of the programme is carried out together and joint capacity building has led to increased trust and reliance on each other.'[134]

NPA, in contrast, is one of only two international NGOs that does not work in partnership with local NGOs. With a 20-year legacy of humanitarian operations in South Sudan, NPA has a close relationship with the SPLA leadership. This may be why it has escaped pressure from the local authorities to work with a local partner. From its base in the far southern city of Yei, NPA's mine action team has focused its efforts on roads, especially the Juba to Yei road. However, using the Task Impact Assessment methodology it also used in Bosnia, NPA has begun retargeting its efforts toward local communities.

In sum, US support to demining between the Nuba Mountains ceasefire and the CPA followed the commercialized and securitized patterns seen in Afghanistan and Bosnia. The more recent USAID contract to RONCO through Berger Group continued in this vein. However, following the CPA, the State Department concentrated on funding international NGOs and local government capacity building. Indeed, its funding of the Scandinavian NGOs Norwegian People's Aid and DanChurchAid overlaps with Norway's choice of implementing partners. This contrasts with the UNMIS model of commercializing and integrating mine action into the politico-military objectives of the peacekeeping mission, which, actually resembles the US strategic-commercial approach in Afghanistan and Bosnia. The next section of this chapter will analyze the reasons why US and Norwegian demining funding modalities have occasionally converged, as they did in Sudan.

Accounting for Convergences

The case study analysis shows that US and Norway's funding of mine action often took divergent paths. Norway consistently supported NGOs, generally international ones, while the US often opted for a more militarized and commercialized approach. However, as the analysis has indicated, there are three important exceptions where the two countries' policies converged. First, Norway, like the US, supported entity armed forces demining in Bosnia. Second, from 1994 to 2001, in Afghanistan the US supported the same UN-led, NGO-implemented program as Norway. Third, following the Comprehensive Peace Agreement in Sudan, the State Department has supported demining by NGOs, including the two groups supported by Norway. The following will explore possible reasons for these convergences.

Norway and Bosnian Military Deminers

Norway's provision of equipment and insurance for Bosnian military demining was quite unusual. It is possible that this divergence from Norway's ordinary mine action policy stems from two factors. First, the entity armed forces were required by the Dayton Peace Accords to engage in landmine clearance. Norway, with its commitment to international peacemaking may have felt it was important to encourage the entity armed forces to meet their obligations. Second, the NATO intervention in Bosnia was the first non-UN peacekeeping mission undertaken by Norway (other than the multinational force in the Sinai mandated by the Camp David Accords), its first in Europe and its first overseas NATO deployment. While their military operations in Bosnia were consistent with its history of peacekeeping, it also represented a break with the past, toward a more strategic deployment in support of its NATO allies. As the Norwegian ambassador to Sarajevo said,

> ...[W]e have interests in the success of NATO and therefore we support NATO actions. So that's not altruism, it's a vital foreign policy interest. We want a strong NATO because a strong NATO is the best guarantor for Norwegian and collective security.[135]

This may have influenced its willingness to more readily engage with the security structures of a non-NATO foreign country. However, this seems to be an exception in Norwegian practice. Though it has supported military demining in Nicaragua, where there are no other organizations operating,[136] and has occasionally deployed its own military demining units in the context of peacekeeping units (such as in Lebanon[137]), these seem to be isolated occurrences.

US Relations with Afghanistan, 1994-2001

Following the end of the Cold War, the US saw little need for major commitments in Afghanistan. By 1994, US aid to the country, both overt and covert, had plummeted. Rather than controlling its humanitarian aid contributions directly, or through commercial contractors, it began giving them to NGOs and the UN. Though Afghanistan's civil war continued with great ferocity, it had shifted in US policy from being constructed as a politico-military problem within a Cold War strategic framework, to a 'humanitarian' problem. According to Barnett Rubin, for the US 'Afghanistan became an object of charity and neglect, not necessarily in that order.'[138]

US interest, accompanied by limited covert funding, began to shift following the Taliban ascendency, a growing interest in oil pipeline politics,[139] Al Qaeda's bombing of the US embassies in East Africa and the US missile strikes in 1998. However, it was not until 9/11 that pressure for a full strategic re-engagement with Afghanistan returned to US foreign policy. It is likely then, that in the absence of clear political objectives in Afghanistan, the US felt it unnecessary to control demining efforts directly, preferring to delegate to the UN. That the US continued to provide Afghanistan with demining aid at all points to the existence of humanitarian elements in US mine action funding. However, it took renewed post 9/11 strategic interest, rather than a sudden rise in charitable concern, for the US to increase its demining funding by a factor of 10.

US Foreign Policy in Sudan

Post-Cold War US foreign policy in Sudan is driven in multiple directions by the competing interests and ideals of different branches of the US government and society.[140] The CIA, though concerned about the northern government's previous links to terrorism also sees this as an opportunity to gain important sources of information on Al Qaeda (Osama Bin Laden lived in Sudan between 1991 and 1996). Thus, it has built a cooperative information-sharing relationship with government security organs.[141] In contrast, Congress is aggressively pro-south and pro-Darfur, influenced by a unique coalition of four powerful pressure groups – human rights groups, African-Americans (originally concerned about slavery in Sudan), Jewish groups (concerned about genocide) and churches, including the evangelical religious right (which sees the SPLA as a Christian institution fighting Islamic persecution).[142] American petrochemical businesses, on the other hand, are interested in Sudan's abundant natural resources and the US Treasury has been 'favourably impressed' by its rapid economic growth.[143] The State Department, while sympathetic to the south and Darfur, wants to avoid the renegotiation of borders in Africa or give encouragement to separatist movements elsewhere.[144] Thus the State Department's guiding policy is 'making unity attractive to the south' to prevent it from declaring independence.[145] Adjudicating between all these competing groups, the Bush White House, found itself torn between a commitment to the War on Terror (thus needing Sudanese intelligence information) and its core political support from the evangelical churches.

During the 1990s, the complex tangle of US strategic and humanitarian interest in Sudan shaped a policy in which the US tilted toward the SPLA, but without a making major commitment: 'supporting the rebels, but not openly, and not enough to enable them to win the war.'[146] The US criticized Sudan's human rights record, imposed sanctions and even launched a cruise missile strike in 1998 following the East African embassy bombings. However, the US was not willing to dedicate significant overt or covert assistance directly to the SPLA. Instead, it became the largest donor of humanitarian aid to the South, tacitly understanding that this would indirectly support the SPLA, through fungibility and diversion. Thus humanitarian aid 'with its reputation of neutrality and its moral appeal concealing a fundamental vulnerability to all sorts of manipulation' became the 'main channel of the US's Sudan policy.'[147] The US also directed significant amounts of its assistance to NGOs operating outside the UN framework of Operation Lifeline Sudan (OLS), which was required to coordinate all its activities with the northern government.[148] The non-OLS NGOs were far more likely to be close to the SPLA and take a pro-southern rather than neutral stand on the conflict.[149] By 2001, USAID was also funding 'civil administration training' for the SPLA to 'transform itself into a robust and internally democratic political party.'[150]

NPA was a non-OLS NGO and has been one of the favored US recipients of assistance for Sudan. NPA's role in Sudan during the North-South war was very controversial and reflected its history as the conduit of Norwegian funding to African 'liberation movements.'[151] From the beginning of their program, they took a partisan position 'in solidarity' with the SPLA. Egil Hagen, their first country director and former Norwegian ski commando and counterespionage officer, was nicknamed 'The Rambo of Relief' for his militant support for the south. 'Relief in war situations *is* politics,' he said. 'I am one hundred percent with the SPLA. ... I do the maximum to see that they get the material aid they need apart from weapons.'[152] A later country director, Dan Eiffe (nicknamed 'Commander Dan'[153]), was a major apologist for the SPLA and became very close to John Garang, the SPLA founder and leader for over two decades.[154] A Norwegian government evaluation in 1997 found that NPA workers turned a blind eye to the ways their work aided and abetted SPLA operations.[155] NPA was even accused of ignoring weapons smuggling on airplanes they used to transport humanitarian relief.[156] While NPA has tempered its rhetoric in recent years, and has tried to influence the SPLA in a democratic direction, a recent NPA public statement in the *Sudan Mirror* called Garang a 'giant hero' and reaffirmed 'NPA's continued commitment and support to the realization of his visions....'[157] In light of its historical relationship with NPA Sudan, it is not surprising that the US would direct some of its demining assistance to them. It is possible, also, that the US use of NGOs to conduct demining in South Sudan is driven by path dependency, following a long history of using NGOs as intermediaries in the Sudanese conflicts.

While it may be a coincidence, it appears that when US strategic interest in Sudan has been higher, it has reverted to the commercialized model of demining. In 2002, not long after the 9/11 attacks, masterminded by a former resident of Sudan, the US put a great deal of pressure on the Sudanese government and SPLA to sign the Nuba Mountains Ceasefire.[158] It then played a major role in the Joint Military Commission, and unsurprisingly, supported RONCO's commercial mine clearance in support of the JMC objectives. As the North-South war subsided with the CPA, attention shifted to Darfur and the war in Iraq took strategic precedence, the US continued to support mine clearance, perhaps kept alive by the humanitarian strains in US foreign policy, but switched to funding NGOs. Interestingly, the decision by USAID to return to supporting commercial demining coincided with renewed US attention in Sudan and the Horn of Africa. A USAID 2006 report declared Sudan the 'highest…priority country in Africa' due to 'counterterrorism and regional stability' concerns.[159] Moreover, in October 2007, the US military set up a new Africa Command (AFRICOM) in recognition of Africa's 'growing strategic importance.'[160]

The Impact of Strategic Interest on Demining Funding

Observing the above three convergences in US and Norwegian demining policy, it appears that when a donor has a higher strategic interest in a region, it is more likely to commercialize and militarize its demining funding. That the UN DPKO adopted such an approach in Sudan, where the UN does not want to see another failed peacekeeping mission in Africa, shows that the impulse to control and securitize the process when there are more interests at stake may not be limited to the US.

When strategic interest is lower, donors seem more willing to grant their demining funding to NGOs. This could be due to the tendency among Western countries to frame conflicts of low strategic importance as zones for 'humanitarian' rather than politico-military intervention.[161] NGOs and the UN are seen as tools for managing, rather than tackling head on, the political problems that exist in such countries.

Conclusion

In conclusion, just as the differing Norwegian and US visions of security impact their global policy on mines and cluster munitions, their different foreign policy motivations have impacted the manner in which they fund mine action on the ground in the mine-affected countries Afghanistan, Bosnia and Sudan. The case studies show that Norwegian foreign policy is not entirely idealistic or humanitarian, especially when called to uphold its NATO commitments. Norway's national security is so intrinsically linked to NATO that in Afghanistan and Bosnia, one can detect some impact of national security concerns upon its demining funding. However, with a few small exceptions, Norway was remarkably consistent across all three case studies in its funding of civil society-driven and non-militarized demining.

US funding converged with this style of demining when it had lower strategic interests in a region. In these situations, US-funded demining was kept alive by the humanitarian strains in US foreign policy, but at a low level. However, when the US engaged politically and strategically in the case countries, it opted for a more militarized and commercialized approach. The fact that the UN also adopted this approach in Sudan, suggests that other donors may also seek greater control over the demining process and integration into politico-military objectives when they have more strategic interests at stake.

Therefore, in the remaining chapters, the Norwegian-funded programs, and to a certain extent the US funding of NGOs in Sudan, will be used as rough proxies to measure the impact of Human Security-Civil Society Complexes. US funding of commercial companies and the UN tendering system in Sudan will be used as proxies for Strategic-Commercial Complexes. Table 8 summarizes how differing levels of strategic interest may shape demining programs.

Table 8: Demining Complexes

	High Strategic Interest	Low Strategic Interest
Level of Funding	• High	• Low
Implementing Agencies	• Commercial companies • Private security companies • Military demining units	• International NGOs • Local NGOs • UN Humanitarian Agencies
Tasks	• Military Bases • Securing Weapons Caches • Access Roads Important to Military • Large-Scale Infrastructure Projects	• Community-Based Priorities • Areas of High Human Impact • Focus on Marginalized Populations

5

COMPARING THE PERFORMANCE OF TENDERS AND GRANTS

'[The] insistence on the magical power of market forces seems at best misplaced and at worst irresponsible.'
– Timothy Donais.[1]

The two types of demining complexes described in this book produce different outcomes. This chapter examines the performance of both complexes in terms of the demining process, showing how the two modes of operation produced differing results in terms of price, speed, task complexity, quality and safety. Ultimately, the differing outcomes suggest the divergent conceptions of risk and resource allocation that underlie the two complexes, as well as the impact of competition on public service provision.

As proxies for the two complexes, this chapter compares organizations that received the majority of their funding from tendered contracts with those receiving longer-term grants. First, it compares performance in terms of the price and speed of demining. It will show that, as Eddie Banks earlier argued,[2] commercially tendered mine action was often cheaper and faster than that done on grants to international NGOs or by public bodies. This may be because commercial contracting encourages frugality and innovation. However, the chapter also argues that this should not be taken at face value; sometimes commercial demining may be cheaper simply because it does easier, thus cheaper, tasks. Secondly, using quality inspection and demining accident data, the chapter will show that NGO demining funded by grants was conducted at a higher level of quality and safety than commercially tendered or government-implemented demining. It will argue specifically that cut-throat competition erodes a public service ethic, increases tolerance of risk to deminers' lives and fragments accountability, leading to corner-cutting and a 'race to the bottom.' Donors had higher ethical expectations of NGOs, leading them to take fewer risks with the safety of their deminers. With less pressure to be profitable, NGOs could concentrate on quality and invest in training and capacity building. Finally, the chapter will close with an attempt at generalizable model-building, showing the impact of competition and profit-seeking on demining and the relevance of the principal-agent and principal-steward models. While it is obviously unwise to believe in a watertight generalizable model derived from only three cases, there is broad agreement between the three countries, which exhibited vastly different cultures, geographies, climactic conditions and socio-political contexts.

Price

In conventional wisdom, contracting out public services increases competition, leading to improvements in efficiency and price. In many cases, across many different sectors, this is indeed true.[3] Anecdotal evidence in Afghanistan, Bosnia and Sudan all suggested that demining done on a tendered contract was cheaper than that done by agencies who received most of their funds from grants. However, this is difficult to test systematically because in all three cases price data was scarce, due to the reluctance of demining agencies to release the size of their contracts. Moreover, complicating factors include the difficulty of establishing comparability of the type of areas cleared and the absence of data separating the cost of minefield clearance (which is much more expensive and time-consuming) from battle area clearance.

In Afghanistan, the author was able to obtain only vaguely comparable and reliable data for clearance at Kabul International Airport. The NGOs ATC and OMAR (supported by UN grants) were clearing civilian areas of the airport while several commercial companies were clearing adjacent military areas (on contract to the US Air Force). Thus the terrain was similar and on both sides most of the contamination was from UXO rather than mines. In this very limited case, NGO clearance, funded by grants was indeed significantly more expensive. ATC and OMAR's costs ran to over $9 a square meter, compared to UXB and S3Ag (both companies), both clearing for under $2 a square meter.[4] However, to make a generalization from this one particular place to a very large country, especially since the NGO clearance data was for the year prior to the commercial data, would be unwise.

In Bosnia, two earlier studies concluded that commercially tendered demining was cheaper than grants to NGOs or public agencies. As cited earlier, Banks' analysis argued that commercially contracted demining was much cheaper than that done by international NGOs and government agencies.[5] A later study by Lisica &

Rowe calculated that an average commercial team in Bosnia demined for 72% of the cost of one employed by an NGO (see table below). They believed that commercial companies were more exposed to competition, forcing better productivity and lower prices.[6]

Table 9: Price of Demining in Bosnia, by Sector.[7]

Type of Organization		Average Demining Price ($/m²)	
		2002	2003
Private	Commercial	1.70	1.61
	Nongovernmental	2.47	2.25
Public	Governmental	2.00	2.82
Country Average		**2.09**	**2.22**

This general picture in Bosnia appears to hold when specific comparison is made between the price of demining by Norwegian People's Aid and the average price in contracts let by the Slovenian International Trust Fund (see Figure 11 below). NPA's cost per square meter was almost three times that of organizations (both commercial companies and local NGOs) working on ITF tendered contracts. Though not displayed above, a similar picture is seen when comparing cost per mine cleared. The Geneva International Centre for Humanitarian Demining estimated that the cost of clearance by the Bosnian militaries was somewhere between $4.80 and $14.39 a square meter in 2002. One can thus say with a high degree of confidence that the rigors of competition (lowest for the military, highest for ITF contracts, NPA and other international NGOs in between) did contribute to lower prices of clearance in Bosnia.[8]

The author found it difficult to obtain precise cost data to allow for a proper statistical comparison of costs between the tendering and granting models of funding in Sudan because many organizations were wary of releasing what they saw as proprietary information. There was, however, some anecdotal information. WFP changed from a long-term cost-plus contract with the NGO FSD to a competitive tendering system because they felt FSD had had incentives to overspend.[9] That said, one evaluation found that Landmine Action, an NGO funded largely by grants from the European Commission 'was probably one of the most efficient programmes in northern Sudan' in terms of clearance costs.[10] Similarly, a MAG employee said that, unlike commercial companies, they often incorporated areas around UN-tendered tasks if asked by local communities. Moreover, when they finished tasks early or under budget, they used this time and money to take on other priority tasks pro bono. Therefore, he said that the international NGOs' lack of profit motive can give donors better value for money.[11]

Figure 11: Price of Demining in Bosnia: Comparing NPA and ITF-Funded Clearance[12]

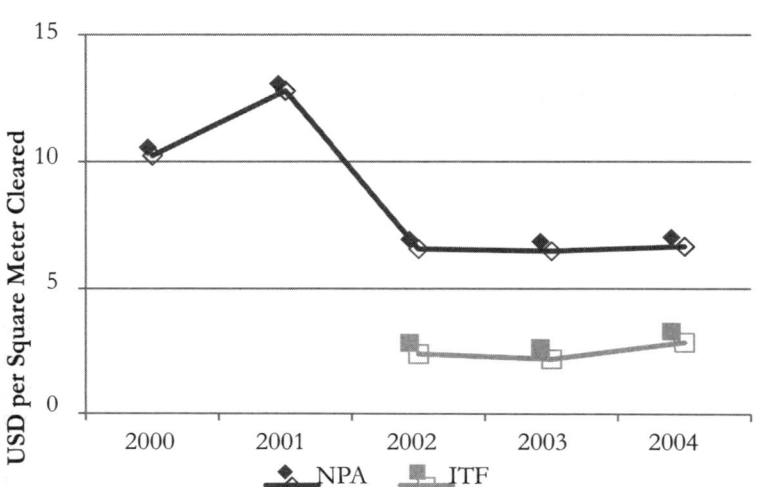

Table 10: Average Price of Demining in Mozambique, 1993-1998.[13]

Organization	Organization Type	US$ per Square Meter
Handicap	International NGO	41.83
CIDEV	International Company	4.81
ADP	Local NGO	3.05
NPA	International NGO	1.61
HALO	International NGO	0.63
RONCO	International Company	0.57
SCS	International Company	0.44
Mechem	International Company	0.23
Mine Tech	International Company	0.02

While not one of the cases studied, price data from Mozambique for 1993-1998 confirms the overall picture from the other cases. The four lowest priced organizations were commercial companies while four of the five highest priced were NGOs (see Table 10).

In sum, it appears that commercially tendered demining is generally cheaper than that done by NGOs on grants. However, the paucity of data means that one can make only the most tentative of generalizations. Thus it may be helpful to look for other measures of efficient performance. The following section will look at speed of demining.

Speed

In all three cases, commercial companies claimed a higher speed of clearance than NGOs and governmental agencies. Companies argued that profit motive forced them to use labor more efficiently and use new technologies such as mechanical clearance. As one RONCO manager said, 'The quicker you do that job, the better your profit is; the longer it takes you ... the more it eats into your profit.' He estimated that RONCO Afghanistan deminers worked three more hours a day than local NGO deminers, due to shorter breaks and beginning the work day when deminers arrive on site, rather than upon leaving their base camp.[14]

Again, systematic comparison was difficult in Afghanistan, as the author was not able to gain sufficient data to make comparisons. The only usable data was from Kabul International Airport, and gave a muddled picture. UXB (a company) was clearing over 180,000 square meters per month, compared with local NGO OMAR's 39,000. However, S3Ag (another company) was clearing 44,000 and ATC (another NGO) was clearing 77,000. The data from Afghanistan is thus far from conclusive.[15]

In Bosnia, the author found comprehensive speed data only for 2002. Figure 12 shows that competition may have had some positive impact on speed, though the margins of error are quite wide and one must be careful of reading too much into the statistical significance. Moving from left to right along the x-axis roughly approximates an increase in competition for funding and the prominence of profit-seeking as a motivation. The local government and military units, on the left, were funded through public budgets and bilateral contributions. International NGOs largely relied on grants, especially NPA (which is shown below disaggregated). Local NGOs received funding from a mix of ITF tenders and bilateral grants and had 'almost the same behavior as the commercial companies.'[16] Finally, the local and international commercial companies, on the right of the x-axis, received all their funding through competitive tenders.

Figure 12: Deminer Productivity Rates: Square Meters Cleared per Deminer in Bosnia and Herzegovina in 2002[17]

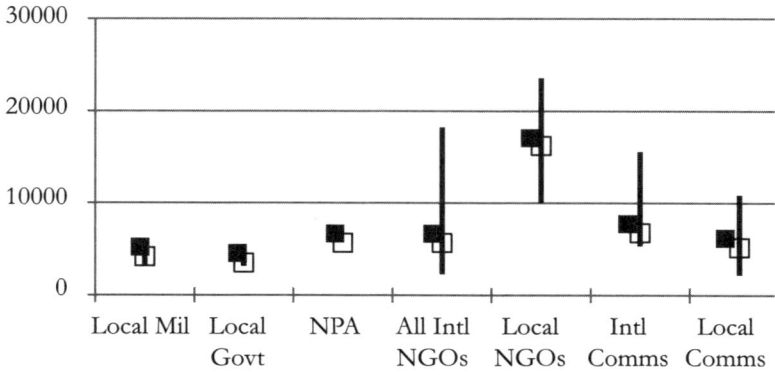

Figure 13: Speed of Clearance (Square Meters Cleared per Work Hour) by Organization's Dominant Funding Mode in Sudan, 2005-July 2007[18]

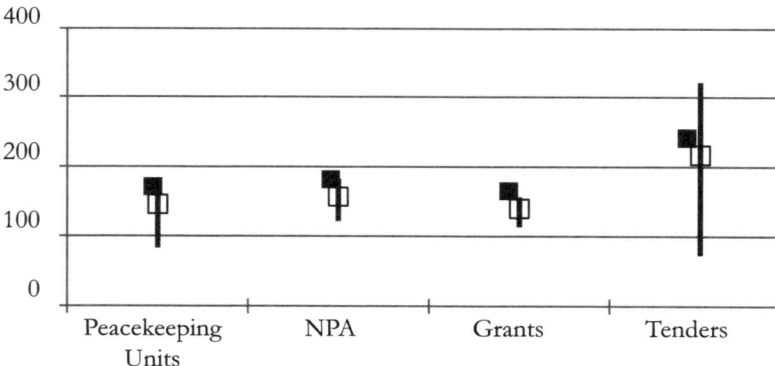

The best data on comparative demining speed was from Sudan. Agreeing with the tentative results above, UNMAO statistics showed that organizations receiving the majority of funding from tenders (including one NGO), cleared more square meters per work-hour than peacekeeping units or NGOs funded by grants. In Figure 13, the x-axis moves roughly from low to high competition and profit-seeking. Again, one must, of course, be cautious as to statistical significance, given the large margins of error.

In sum, though the data is again rough, commercially tendered demining is sometimes loosely correlated with a slightly higher speed of clearance. However, this should not lead immediately to hard and fast conclusions. At this level of aggregation, it is difficult to determine whether commercial companies and NGOs are clearing comparable tasks (which the following subsection will analyze in more detail).

Cost-Cutting, Innovation and Comparability of Tasks

Why did the commercial contracting model appear to yield lower prices and a higher speed of demining? And what impact did motivations of humanitarianism versus profit have? The most obvious reason why commercially tendered demining might have lower costs is the profit motive forces cutting of excess costs. In all three case studies, commercial companies were far more likely than international NGOs to cut back on retirement and health insurance costs and reduce the size of administrative staff. One company in Afghanistan even saved on secretarial salaries by having deminers do their own paperwork. By contrast, the international NGOs' public service ethic and security of funding often led them to spend more. While they sometimes were not able to pay as high salaries as the commercial companies, they were more likely to pay fringe benefits and offer longer-term contracts. For instance, several NGOs in Sudan paid for expensive HIV/AIDS medication for deminers. International NGOs also tended to spend more money and effort in training and capacity building for local staff. Year-around employment meant that NGOs could ensure that staff maintained their training in the off-season.[19]

Moreover, in all three cases international NGOs were more likely to integrate their demining into a wider 'package' of services, such as community liaison and mine risk education. For instance, in Bosnia, before demining begins, NPA community liaison teams conduct task impact assessments (which include basic socio-economic studies) for every demining task to identify the level of impact of mine contamination, the likelihood the land will be used productively in the future and how it will impact the local political economy – to 'make sure we're not demining the back garden of the mayor of Sarajevo.'[20] During demining operations, NPA community liaison teams meet with local authorities and visit residents bordering the minefield to listen to their concerns and educate them about the nature of the process and residual threat. Likewise, in Sudan, Landmine

Action and DanChurchAid's operations in the Nuba Mountains were probably made more difficult by their attempts to incorporate demining into a peacebuilding framework, integrating deminers from both sides of the conflict. Unlike RONCO (a commercial company), which swept into the Nuba Mountains with Mozambican deminers and was able to move fairly quickly, Landmine Action and DCA spent significant time negotiating local political issues and disputes between deminers. In contrast, one RONCO executive argued that 'Trying to integrate all aspects under one organization just can't happen, and, if it does, it is under extraordinarily high cost. ... Demining is demining, as far as we are concerned.'[21]

In addition to the predictable cutting of 'excess costs', the profit motive may spur innovation. Commercial companies in both Afghanistan and Sudan were more likely than some of the NGOs to adopt new detection and clearance technologies (though NPA and HALO were also adept at adopting technology). Moreover, by broadening the market through deregulation and commercialization, there is greater potential for different actors to specialize according to their respective comparative advantages. If one company is particularly good at manual demining and another in canine demining, each can concentrate on excellence in their specific niche. This gives added value to the client, who is able to pick and choose specific service providers according to the needs of the task.

However, there is a danger of overstating the impact of competition on the price and speed of demining because it is very difficult to establish the comparability of tasks. It is possible that NGO clearance is slower and more expensive because commercial companies are doing simpler tasks. This has been noted in the general literature on public service privatization.[22] Across the case studies, it appeared that NGOs were deliberately selecting or being assigned more difficult tasks, often areas with a high humanitarian impact. In contrast, the more strategically-driven donors often wanted large areas cleared quickly in order to gain a rapid strategic dividend, either preparing the way for military deployment, or winning 'hearts and minds' through kickstarting infrastructure projects. This means the tasks given to commercial contractors may not be the most complex ones.

For example, because of liability concerns, some US-funded reconstruction companies in Afghanistan were contracting commercial demining even when there was little chance of contamination. The NGOs were less likely to see these tasks as a useful allocation of resources. Paul Molam of the British/Zimbabwean firm Mine Tech said,

> A lot of the work the commercials are doing, ... there's nothing there, but before the [US Army] Corps of Engineers will allow a [construction] contractor onto an area, you've got to go and put your assets over the ground although you know there's nothing there.[23]

The graph in Figure 14 shows that, in the last couple years, NGOs have been clearing a larger amount of ordnance from their tasks than commercial companies in Afghanistan. This is sometimes a rough, though not foolproof, proxy for the difficulty of a task. Though mines per square meter is a more accurate measure of difficulty than UXO per square meter, both are displayed below. According to Guy Willoughby of HALO Trust, this is deliberate. HALO chose to work in some of the most heavily mined and high impact areas in Afghanistan because they saw this as a humanitarian imperative.[24]

In Bosnia, Per Breivik, former program manager of NPA, claims NGO average costs were higher because they were driven by humanitarian concerns to demine difficult tasks like apartment blocks, rather than flat open fields.[25] In contrast, a former senior official at the UN Mine Action Center in Bosnia (UNMAC) said RONCO, a commercial company, chose to clear 'nice easy areas that had no one living there.' Figures 15 and 16, illustrate how government agencies and international NGOs (primarily funded by public budgeting or grants) took on more complex tasks than companies and local NGOs, (both funded by competitive contracting in Bosnia). Note that here, especially with regards to areas cleared without ordnance, the statistical significance is probably higher than with the speed data.

The same pattern was found in Sudan (Figure 17). Organizations funded by grants cleared areas with a much denser concentration of mines. This is probably because most tenders have been for roads, while NGOs (generally funded by grants) have concentrated on high-impact areas. Furthermore, when NGOs have participated in UN-tendered demining, they have often expanded the tasks to include contaminated areas outside the initial contracted area when requested to do so by local communities.[26] Commercial companies, driven by their bottom lines, have stuck strictly to the parameters of their contracts.

That the commercial companies may be doing simpler tasks should not necessarily be taken as a criticism of them. In fact, it may be a wise division of labor to focus commercial demining on tasks that can be done quickly and allow humanitarian NGOs to focus on slower, more difficult ones. Having NGOs do tasks where there is minimal risk of contamination would be a waste of humanitarian resources. However, to then use the price and speed data to make firm assertions of commercial preeminence, as Banks does, is misleading. Future mine action researchers may wish to consider looking into the issue of targeting and task difficulty, perhaps using data from the Landmine Impact Surveys and correlating it with clearance data.

Figure 14: Comparing the 'Density' of NGO and Commercial Clearance Tasks in Afghanistan, January 2005-November 2006.[27]

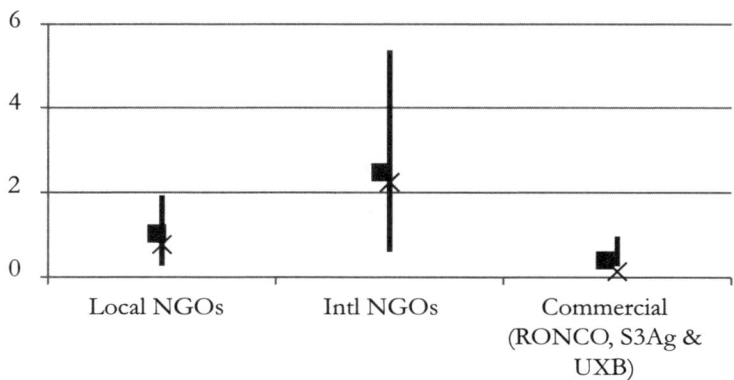

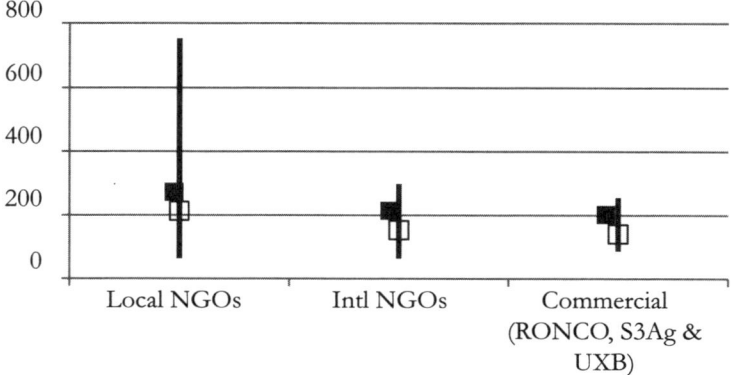

Figure 15: Task Complexity in Bosnia and Herzegovina.[28]

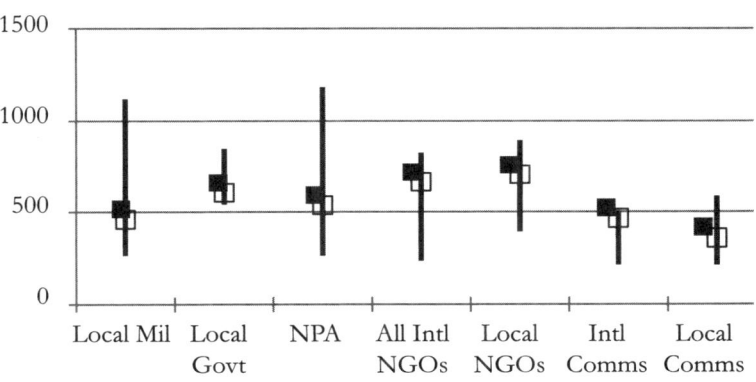

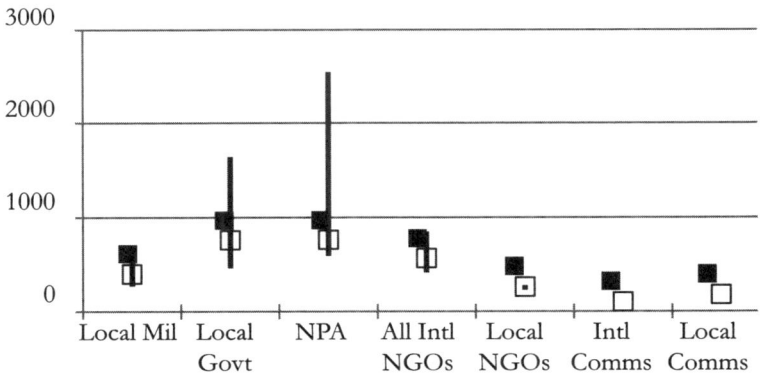

Figure 16: Task Complexity in Bosnia, Ctd.

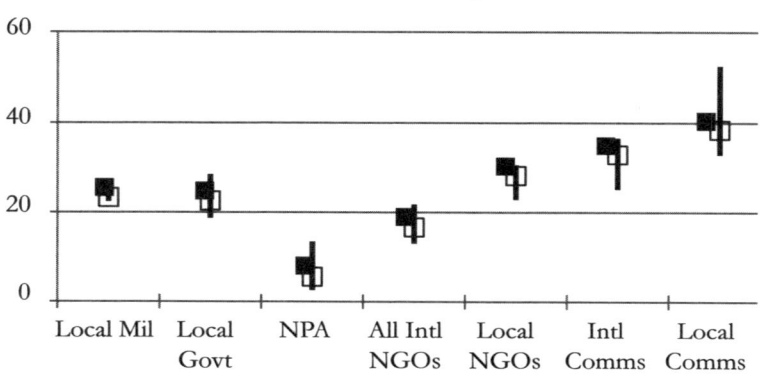

Figure 17: Density of Minefields Cleared in Sudan (Mines per Million Square Meters Cleared) by Organization's Dominant Funding Mode, 2005-July 2007.[29]

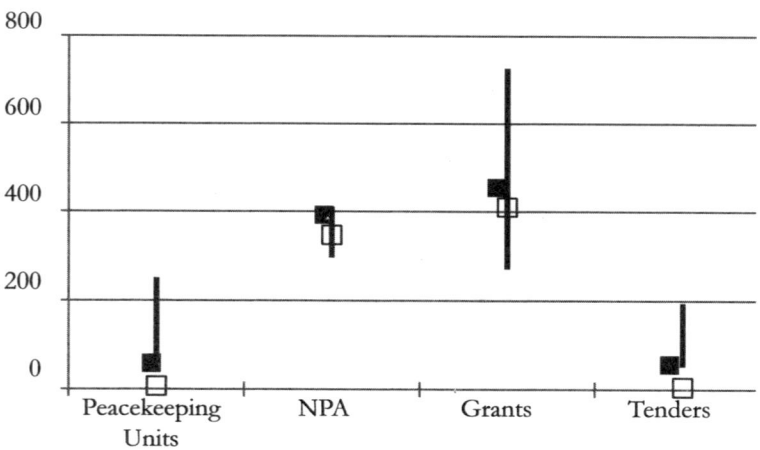

Quality and Safety

While there may be efficiency benefits to commercialization, simple price comparisons do not account for potential human costs, such as lowered safety. Indeed, the literature on public service privatization has generally found that cost savings in contracting out often come at the risk of reduced quality and/or safety. As one demining manager said simply, 'A high level of quality means more costs.'[30] While slightly different systems were used, in each case country public quality assurance teams visited demining sites and issued reports indicating whether the demining agency was following safety and quality regulations. Data from these inspections thus measures the quality of the demining process, which obviously influences the quality of the product. In all three cases, agencies receiving grant funding had better quality assurance records (measured by comparing the percentage of good versus bad quality assurance reports) than those working on tenders.

In Afghanistan, if an organization's site complied with mine action standards, the quality assurance team issued a Conformity Report. If there were small infractions, they issued a Minor Non-Conformity Report, indicating errors observed and suggested corrections. For infractions that threatened lives – of either deminers or future users of the land – the quality management team issued a Major Non-Conformity Report. Unfortunately, at the time of the author's fieldwork the Quality Assurance reports were not all centralized at UNMACA headquarters, due to the decentralized nature of data collection and some technical problems setting up a central database. Therefore, a country-wide comparison of quality assurance reports was not possible. However, the author obtained the paper records for 2005 and 2006 (till September) for the Central AMAC, covering Kabul, Kapisa, Parwan, Bamyan, Wardak and Logar provinces. The Central AMAC region is a good case study for comparison as some 36% of mine affected communities are located there[31] and around 42% of all Afghan clearance (in terms of square meters) in 2005 and 2006 occurred there.[32] Every demining actor, NGO and commercial, carried out demining and battle area clearance in the region in 2005 and 2006. The Central AMAC is thus relatively representative of Afghan demining as a whole. Moreover, this dataset is probably the most reliable of the quantitative data from Afghanistan used in this chapter.

Figure 18 shows the percentage distribution of Central AMAC Quality Assurance reports, by organizational type. It shows that commercial companies had a significantly higher rate of Major Non-Conformity. That is, according to the Central AMAC quality management specialists, the commercial companies were more likely to engage in activities posing a direct threat to the lives of deminers or future users of cleared land.[33] As in previous sections, the x-axis moves roughly from low to high competition (on the left the UN-funded local NGOs effectively have guaranteed funding, in the middle, international NGOs have to compete for grants given by bilateral donors, on the right, commercial companies compete for tenders).

Figure 18: Distribution of Quality Assurance Reports by the Central Area Mine Action Center, Afghanistan, 2005-September 2006.[34]

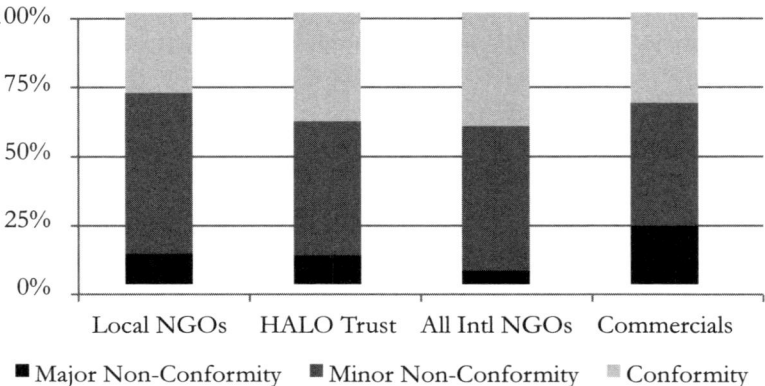

The author was able to get quality assurance data covering the whole of Bosnia, from 2000 onwards. While the methodology judging quality changed in June 2003, both the old and new methods display a similar picture. Organizations funded through tendered contracts (commercial companies and local NGOs) were much more likely to deviate from safety procedures than those receiving grants or funded through public budgets. NPA's record is particularly good. Looking at Figure 20, for June 2003 onwards, one can see that the NPA is between six and eight times less likely to commit a 'Critical Error' – one that could risk someone's life – than an organization receiving money from the US-funded ITF.

While the difference is less stark than in Afghanistan and Bosnia, the tendering model appears also correlated with a slight reduction in quality of demining in Sudan (see Figure 21 above). Organizations funded by tenders were three times more likely to receive an order to stop work due to work practice that posed a risk to life (a 'Fail-Stop' rating), and received less 'High-Good' ratings than those funded by grants. The consistency of commercial underperformance in quality assurance inspections across the three case studies highlights the inadequacy of the simple price and efficiency approach to examining demining privatization taken, for instance, by Banks.[35] While low quality work is a serious problem in any industry, in demining it can literally be fatal. Therefore, in cost-benefit analysis of contracting-out, one must account for 'human costs' in additional to financial ones.

Figure 19: Percentage Non-Compliance with Standing Operating Procedures in Bosnia, 2000-2003.[36]

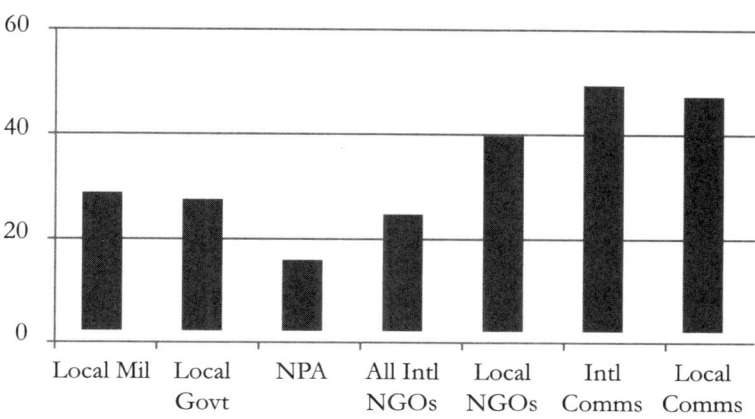

Figure 20: Percentage Non-Compliance with Standing Operating Procedures in Bosnia, 2003-2006.[37]

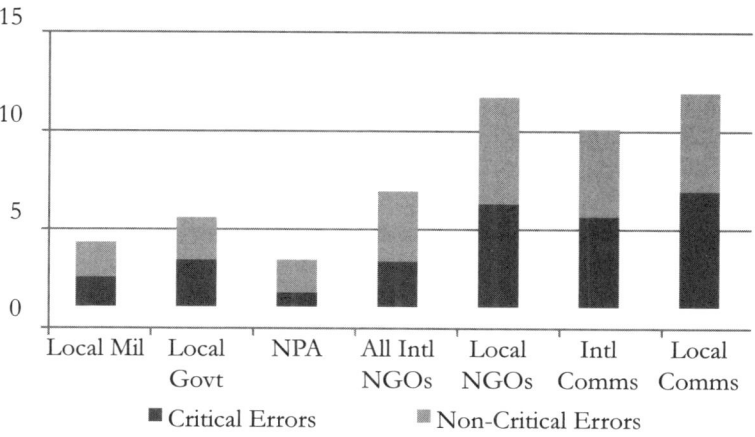

Figure 21: Quality Assurance Record, by Organization's Dominant Funding Mode, Sudan 2004-July 2007[38]

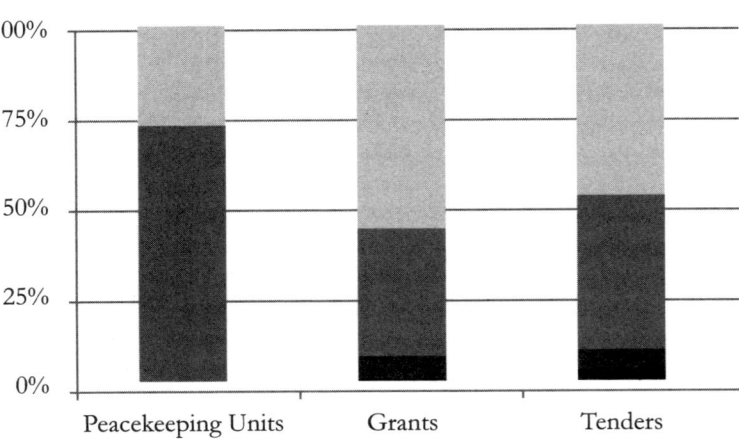

Unsurprisingly, given their poor quality assurance performance, organizations working on tendered contracts seem to have more demining accidents than great-funded NGOs and some public agencies. Since more than 80% of demining accidents are avoidable, caused by lack of proper management, supervision or training,[39] the poor safety record of commercial deminers may be a further indication of the poor quality of their demining. For instance, a comparison of accidents per area or ordnance cleared shows that commercial companies had a significantly poorer safety record in Afghanistan than international and local NGOs (Figure 22).

Some have suggested that the commercial companies' bad run of accidents in Afghanistan is due to their recent arrival in the country – that it takes time to get used to the conditions. However, one sees a similar pattern in Bosnia (Figure 23), where local military and NGO demining units consistently outperformed commercial companies and local NGOs that received significant portions of their budgets from tendered contracts. NPA's safety record is particularly impressive. While the Sudan accident data was too scarce to make meaningful comparisons, the Afghan and Bosnian experiences show that while a competitive tendering process may reduce financial costs of demining, there are high human costs – accidents – which appear correlated with the contract model of funding.

Some commercial demining companies claim that their increased speed of demining makes up for their higher accident rate. They say that speeding up the process reduces hazardous areas thus increasing the number of civilian lives saved, making up for the possibility of demining accidents. This hypothesis, of course, can be empirically tested with data from the field. Unfortunately, the only adequate set of data gathered from the three case studies was from Bosnia in 2002. During that year, the average square kilometer of contaminated area injured or killed 0.35 people. This figure is used below (see Table 11), along with accident data, to determine how many people an agency saved versus how many deminers' were hurt or killed during clearance. Though they cleared slower, and may have thus saved less people from injury or death than agencies funded by tenders, the international NGOs had no accidents in 2002. By contrast the agencies funded mostly by tenders harmed 2.10 people per square kilometer cleared. On balance, then, international NGOs saved more lives.

These calculations would be improved with an analysis of how the tasks done by each type of organization differed in level of hazard to the community. As described earlier, international NGOs seemed to have been clearing more contaminated tasks, thus suggesting they may have been saving even more lives per square kilometer. Nonetheless, this is only data from one country for one year and should not be considered any more than a possible indication and a suggested method for a future researcher to try in other contexts.

Table 11: Lives Saved versus Harmed, by Organization's Dominant Funding Mode, in Bosnia, 2002.[40]

		Agencies Receiving Majority of Funds from Grants (Intl NGOs)	Agencies Receiving Majority of Funds from Tenders (Comms & Local NGOs)
A	Area cleared (km²) per 100 deminers in 2002	0.952	1.297
B	Estimated people saved from injury or death as a result of A.	0.033	0.045
C	Deminers injured or killed per average km² cleared in 2002	0	2.096
D	Deminers injured or killed as a result of clearing A (D=C*A)	0	2.719
E	**Balance (people saved from injury or death minus deminer casualties, E=B-D)**	**0.033**	**-2.674**

Figure 22: Accident Record by Organization Type in Afghanistan, January 2005 to November 2006.[41]

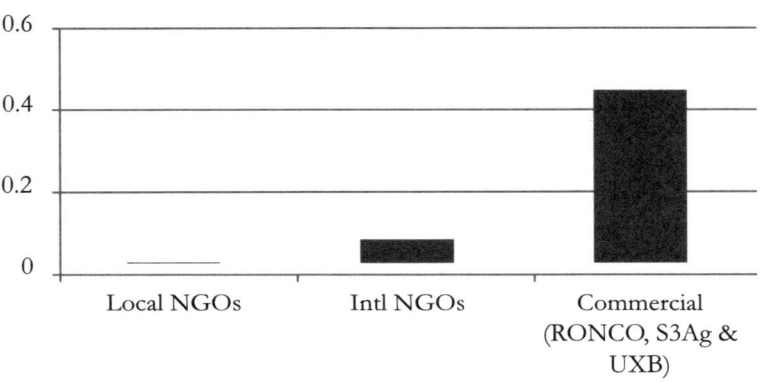

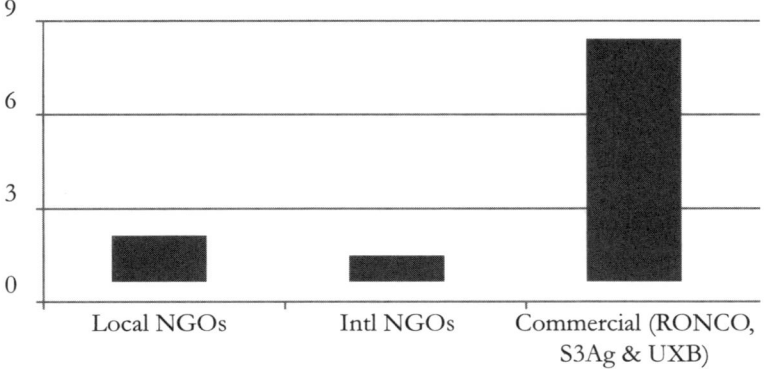

Figure 23: Safety Record by Organization Type in Bosnia, 1999-2006.[42]

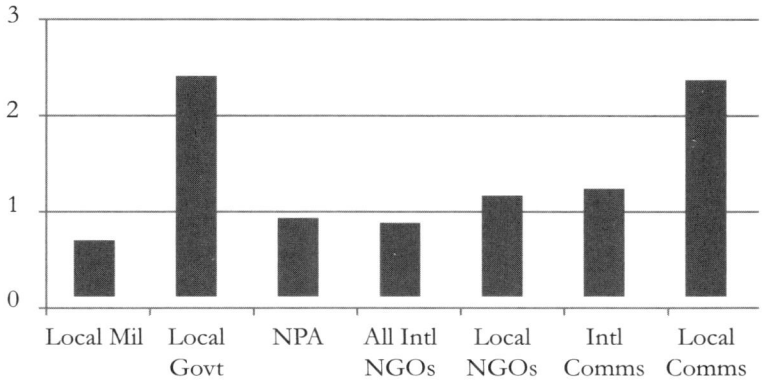

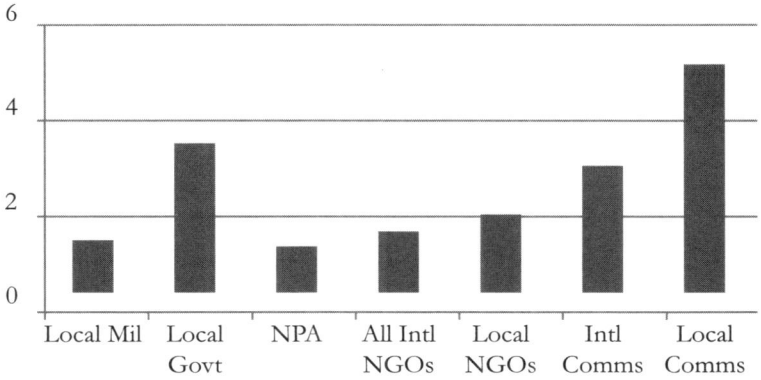

Professional Ethics, the Race to the Bottom and Accountability

NGO personnel in each of the case countries asserted that their better quality and safety record stemmed from their professional ethics. As one manager explained, 'For an organization that is called Danish *Church* Aid, it cannot make certain [unethical] choices.'[43] In contrast with companies, which exist to make a profit for their owners, the international NGOs largely exist because of some sense of social responsibility. For NPA, it is solidarity with the oppressed; for HALO, a traditional notion of charity; for DanChurchAid, it is Christian service to those in need. NGOs argue that these values extend to their work. Unnecessarily risking the lives of deminers, or future users of the land, would contravene international NGOs' very *raison d'être*. Within the commercial sector, there is variation in professional ethics. A representative of ArmorGroup in Bosnia argued that the international commercial companies (which had fewer accidents than local companies in Bosnia) were constrained by their worldwide brand image, which prevented them from taking the risks that local companies did. Likewise, in Afghanistan, RONCO was able to spend much more money on training, and consequently have a better quality and safety record than other commercial companies because its reputation and relatively secure position as the market leader meant it was not forced to cut costs as harshly.[44]

Further, the professional ethics of NGOs may be supported by their secure funding as they are not forced to make sacrifices. In Bosnia, UNMAC argued that NGOs had fewer accidents because NGOs were 'normally not so productivity-orientated as commercial organisations are and can progress at a rate not driven mainly by the financial aspects of an operation.'[45] In contrast, driven by the need to turn a profit at the end of every task, commercial companies often have to speed up the demining process, perhaps leading to greater risk of errors.[46] One UNMACA employee said that for commercial companies to make profit on clearance, 'you have to tell the deminers "go, go, go and go faster!" and then – accidents.'[47] In Bosnia, some companies even paid deminers by square meter cleared, which encouraged deminers to rush their work.[48] In a sense, commercial companies struggle to afford the expensive morals of NGOs. The demining accidents reveal some commercial companies' general callousness in cutting corners. The following quotes from UNMAC inquiries into accidents by companies in Bosnia reveal a scandalous disregard toward Standing Operating Procedures:

> The [Mine Tech] Team Leader acted irresponsibly and the site was marked inadequately.[49]
>
> [I]nappropriate emphasis has been placed on [Mine Tech] field workers to complete a stated minimum square metres each day.[50]

Mines were stored in an unacceptable way and were not destroyed at the end of the working day [by UXB/Amphibia].[51]

[T]he flagrant disregard of such a fundamental and common-sensical procedural regulation as the appropriate minimum distancing between deminers constitutes in the Board's view a gross deficiency in supervision at all levels [by RONCO/UNIPAK].[52]

[There is a] lack of a firm quality assurance and quality control policy in the company [AKD Mungos].[53]

Therefore, as competition increases, the space for professional ethics decreases. The most serious problems occur at the cutthroat competitive margin of the market, where companies make up for lack of reputation or brand in speed and price. In all three cases, the most competitive margin of the market seemed to display element of a 'race to the bottom', where quality and safety was sacrificed for profit. In Bosnia, this was seen among the local commercial companies. In Sudan, there was less variation among quality assurance records, but again it was the groups that had to compete for funding that had the poorer overall record. In Afghanistan it was the less well established commercial companies, competing with RONCO, which had the most quality and safety problems. The competitive system created incentives to seek profit not through an improvement of the quality of demining, but through dangerous corner-cutting practices, such as poor spacing of deminers, increased speed, inappropriate use of machines, poor oversight and improper safety equipment.

An illustration of how bad this can get is the corner-cutting of an American company, Explosive Ordnance Technology, Inc. (EODT) in Afghanistan. EODT was clearing a military area of Kabul International Airport on a NATO contract with the US Air Force. They were new to Afghanistan, and had little in the way of assets – they had to rent personal protective equipment (PPE) from another commercial company. When interviewed, the EODT demining operations manager said EODT was able to clear areas much quicker and more cheaply than Afghan NGOs, even though they paid their deminers as much as 30% more. He attributed their speed to effective use of mechanical technologies and the 'motivation that a privately owned company has got.' He said that the increased speed did not come at a cost to safety, 'because of the simple reason that all of our safety procedures, our standard operating procedures are set up specifically for mechanical clearance operations.'[54]

However, a quality assurance report in the files of the Central AMAC outlined observations of life-endangering deviations from mine action standards. For instance, in August 2006, the AMAC report observed, 'Since the bulldozer machine which was busy in clearance of the task was un-armoured, it was unsuitable for mine clearance program.'[55] Incredibly, three months later, the author took the below picture of an EODT digger excavating a bunker in a battle area

(Figure 24). Note that despite the risk of there being unexploded ordnance buried in the ground being excavated, the digger's cab is unarmored, the cab door is open, and the digger's mechanic is hanging out of the door while the digger is in operation. This put the lives of the driver and mechanic in unnecessary danger.

Figure 24: EODT Battle Clearance at Kabul International Airport, 22 October 2006.

Figure 25: EODT Mineline Clearance at Kabul International Airport, 24 November 2006.

In the same August report, the Kabul AMAC team observed that 'The relevant site supervisor and his colleagues were standing near the machine without PPE [Personal Protective Equipment] and maintaining safety distance.'[56] In November, the author was told by the demining operations manager that the digger in the photograph below was excavating a suspected mine line (Figure 25). Note that the digger, while this time armored, again has its door open, and the machines' spotter, lacking a visor, is not wearing the required PPE.

Despite commercial corner cutting, the commercialization process, as a form of deregulation, can actually lead to lower levels of accountability and supervision. Without effective rule of law and centralized authority, all the three case countries lacked public agencies able to enforce the kind of regulation necessary to prevent a race to the bottom. Maintaining a professional standard was often left to the internal checks on an agency's behavior – whether 'good intentions' or a reliable brand.

In Afghanistan, the commercialization process eroded the regulatory power of UNMACA. For instance, the EODT example illustrates that UNMACA struggled to impose its authority on the company. The author overheard an EODT manager express confusion and incredulity about why UNMACA inspectors must visit the site and certify clearance. While UNMACA could take away their accreditation, it had less power over commercial companies than the local NGOs, since it did not fund them. Many organizations that were contracting commercial demining did not include clauses requiring coordination with and submission of data to UNMACA. Therefore, UNMACA felt the commercial companies were 'not properly reporting to us'[57] and there were 'multiple incidents where an organization will come us and say "we need a clearance certificate for our task that we have completed," … and we will go, "What task?"'[58] Some clients hired their own independent quality assurance personnel to supervise commercial deminers' work. However, this system was decentralized and privatized – reports were not submitted to a central and public repository. Indeed, the limited attempts by the author to obtain access to their reports were unfruitful. This could create 'information asymmetries' as information on quality of work was not available to all potential clients.

UNMACA also had difficulty keeping track of commercial demining on US military bases. Afghans working as UNMACA's quality assurance inspectors found it difficult to gain the security clearance to go on base. While the US military had its own office controlling demining on US bases, coordination and information sharing with UNMACA was patchy. Therefore, despite UNMACA's mandate of ultimate responsibility for all demining on Afghan territory, the Coalition bases had a parallel and stove-piped system. Thus, the commercial companies in Afghanistan were subject to far less oversight than the NGOs. The graph overleaf shows that in 2005, (the only year where complete quality assurance data was available at the time of the fieldwork) quality management investigation teams from the Central Area Mine Action Team did far fewer spot checks on commercial companies than NGOs.

Figure 26: Level of Quality Assurance Surveillance (Inspection Reports per Million Square Meters Cleared) over NGOs and Commercial Companies in the Central Area Mine Action Center, Afghanistan.[59]

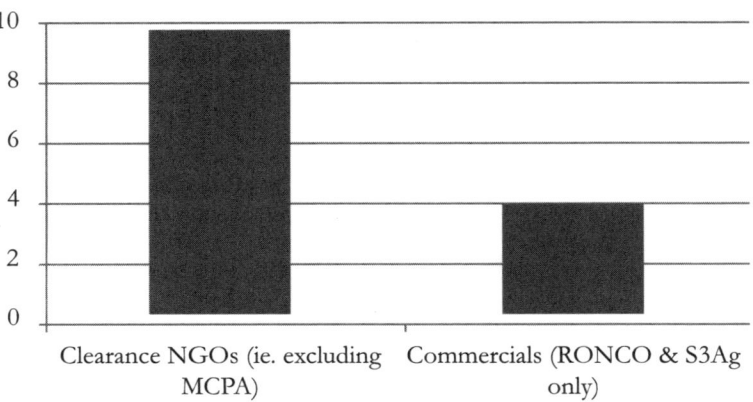

In Bosnia, it is also possible that commercial companies, especially local ones, faced lower standards of accountability. Firstly, they were less able to afford the high levels of internal supervision, in terms of a larger management and administrative staff. In contrast, NPA had regular evaluation visits from staff in Oslo and paid for an external evaluation by the Geneva International Centre for Humanitarian Demining. Secondly, the early corruption of the commercial demining sector suggests that many companies were able to buy their way out of accountability. Finally, through their consultations with local community leaders and beneficiaries, international NGOs like NPA were probably more accountable to the local population. They would be far more likely to notice and acknowledge complaints from local residents.

In Sudan, it is possible that the commercial companies were actually subjected to tighter control from UNMAO than the NGOs, as UNMAO was their donor and contract supervisor, in addition to being their regulator. However, this has also meant that they have been subject to less independent external criticism. The bilateral donors and individual NGOs have commissioned several independent evaluations of their work. In contrast, the UN-led system has only been evaluated internally in a manner that several have described as 'covering the UN's back.'

There is a danger then, that without proper control and oversight, the increased productivity and lower price of commercial demining may come at a human cost – a loss of safety – and a reduction of quality. This is suggested both by quality assurance records and accident data. Analyzing qualitative data from the cases indicates that this reduction in quality may come from the inability of commercial companies, especially those competing at the margin, to afford the 'luxury' of professional ethics.

Modeling the Results

While one must be cautious about creating a generalizable theory from only three case studies, the data presented above is sufficient to begin a process of model-building for others to test elsewhere. The following looks at what the above data might say regarding the impact of competition on demining and agency versus stewardship models of contracting. From the data, it appears the following can be said about the impact of competition and profit-seeking on key demining indicators:

1. **Price**: Increasing competition and profit-seeking lowers the price of clearance per square meter, though eventually a law of diminishing returns applies. One should note however, that this proposition is based on some of the lowest quality data.
2. **Speed**: Increasing competition and profit-seeking leads to an increase in speed (though at a slower rate than the decrease in price), though the law of diminishing returns eventually applies. One should note, however, that the link between competition and speed was not ironclad.
3. **Task Complexity**: By moving away from a monopolistic market, increasing competition and profit-seeking initially makes demining agencies more accountable and thus more likely to target their efforts on more difficult tasks, but eventually the drive to cut costs leads them to seek the 'lower hanging fruit' of easier work. This correlation was fairly strong.
4. **Quality/Safety**: By moving away from a monopolistic market, increasing competition forces demining agencies to be accountable and increase the quality and safety of their work, though as competition and profit-seeking becomes more cutthroat, there is a race to the bottom at the expense of quality. This correlation appeared to be strong across the case studies.

These variables are depicted graphically in Figure 27 overleaf. As indicated on the graphs, there seem to be three points where demining programs can settle, depend on the level of deregulation and competition in the market. The first point, with high prices, low speed, low complexity and low quality/safety, represents a monopolized market, where an agency has no incentive to improve the efficiency or quality of their work. This might be approximated by the poor record of the local government civil protection teams in Bosnia. This is the result of a statist model of demining or of non-competitive cost-plus contracts.

The second point, with fairly high prices and low speed but high levels of task complexity and quality/safety, represents granting to international NGOs. There is more competition than in a monopoly, as the NGOs must apply for their funding, which can always go to an alternative, competing NGO. However, it is a more protected market than commercial tendering. This is represented by the Human Security-Civil Society demining complex, which puts greater premium on the value of preventing accidents and errors than on saving money.

The third point, with lower prices and higher speeds, though lower task complexity and quality/safety, represents the relatively open market of commercial tendering. The higher levels of competition and profit-seeking enforce a level of efficiency not seen in the previous two points, though this comes at the cost of quality and targeting efforts at more difficult demining. This is represented by the Strategic-Commercial demining complex, which assigns a high value to profit, and thus efficiency.

The three points also represent three alternative examples of how increasing competition and profit-seeking creates and destroys incentives and space for morally-driven action – service provision guided by professional ethics. A complete monopoly gives no incentive for having humanitarian intentions – in fact it invites abuse. At the other end of the spectrum, cutthroat competition leaves little room for ethics, as cutting corners is the only way to avoid being priced out of the market by the lowest common moral denominator. However, a system of grants, blending a low-level of competition with enough space to focus on longer-term time horizons, allows agencies to encourage a higher regard for professional ethics. Between grants and cutthroat competition, some companies may be able to trade on a high quality international brand, rather than corner-cutting, to gain moral space.

As outlined in chapter two, when donors contract to commercial demining companies, they operate according to a 'principal-agent' relationship, in which the donor relies on the company's pursuit of profit, and the contract's stipulations, to ensure demining objectives are met. When donors give long-term grants to trusted partner agencies, they operate according to a 'principal-steward' relationship, relying on the implementer's normative and professional commitment to the achievement of the donor's goals. Studies of other public service contracting have also found that the high levels of competition and re-tendering in the 'principal-agent' model can encourage agents to take advantage of information asymmetries or engage in a destructive 'race to the bottom' with competitors. Counter to free-market conventional wisdom, some scholars have found that the 'principal-steward' model can actually hold stewards more accountable than agents and that their ideological commitments can spur them to produce more effectively. In other words, stewardship employs forms of accountability beyond free market competition, such as trust, commitment and the importance of maintaining a trustworthy organizational brand, that maintain high standards of ethics.

Figure 27: The Impact of Competition and Profit-Seeking on Demining

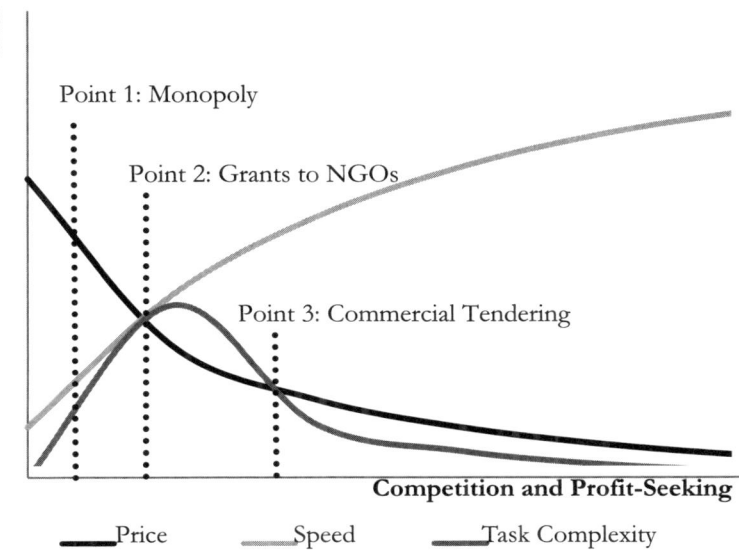

Figure 28: The Impact of Competition and Profit-Seeking on Demining

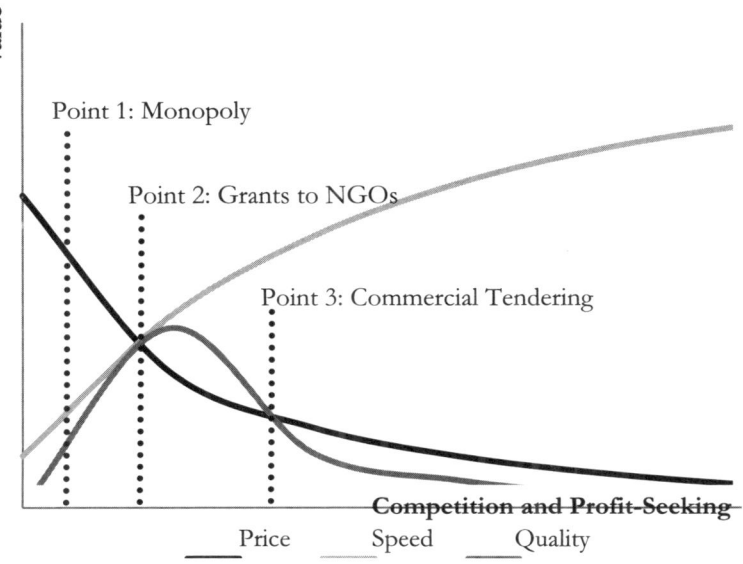

Figure 29: The Impact of Competition on Moral Space

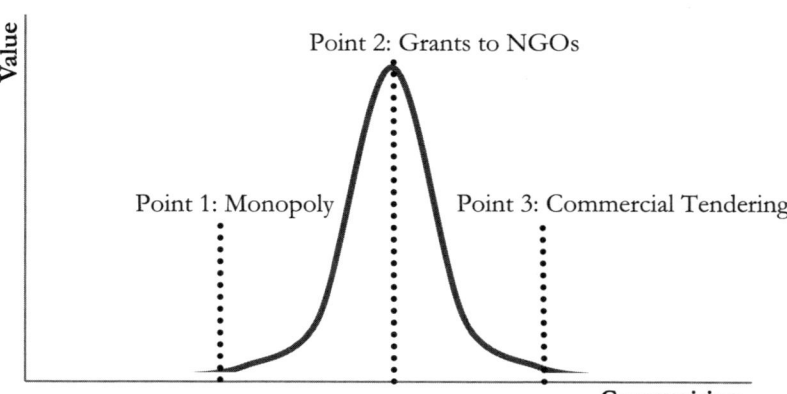

In the real world, of course, examples of contracts and agencies fall somewhere between the poles of 'perfect competition' and 'perfect stewardship.' This may explain the continuum of results seen, for example, in Bosnia, where local NGOs faced less competition than commercial companies but more than international NGOs. As a result, in many indicators they fell between these two other organizational types. Likewise, while an international company may face considerable competition in obtaining tenders, it may nonetheless interpret its reputation and brand as a kind of stewardship placed upon them by valued clients. They are thus less likely to cut corners and 'race to the bottom' (see, for example, their better record in Bosnia compared with local companies) because it would damage their clients' trust.

To put it bluntly, mine action donors are faced with a value judgment of how much quality, safety and task complexity they are willing to sacrifice in order to lower the cost and speed of demining. Increasing competition and profit-seeking, while initially encouraging accountability, eventually shifts agencies' attention away from quality to ensuring profitability. This gives less room for ethical considerations and professional ethics. Whether it is worth risking the lives of deminers in order to more quickly reduce the risk to the lives of civilians in proximity to dangerous areas is a value judgment that needs to be considered carefully by the donor. Certainly the preliminary analysis from Bosnia in 2002 suggests that the trade off, in increased accidents, for saving lives through increased speed is not worth it. Operating according to a 'principal-steward' model may offer the appropriate balance between accountability and trust necessary to encourage efficient and high quality demining.

Conclusion

Whether the correlations drawn in this paper are generalizable will depend on further analysis and study in other cases. However, there appears to be clear indications that the two demining complexes, rooted in their differing understandings of security, constituent organizations and contractual relationships, produced differing outcomes. Due to the pressures of strategic expedience and the commercial bottom line, the Strategic-Commercial Complex produced greater volume of square meters cleared at lower cost. However, this was partly because the tasks prioritized were not always the most difficult nor of the greatest humanitarian impact. Moreover, donors and implementers chose a higher level of risk to the lives of deminers and future land users, reducing the quality and safety of the process. This was incentivized by the high levels of competition and profit-seeking that meant implementers were forced to prioritize the bottom line over humanitarian considerations.

In contrast, the slower speed and higher cost of demining done by the Human Security-Civil Society Complex was in part due to choosing tasks according to humanitarian priorities, yielding more difficult working conditions. However, it also derived from the lower willingness of the donors and implementing agencies to risk the lives of their deminers and future land users. This was incentivized by long-term grants to trusted 'stewards', encouraging implementing agencies to invest time in training; good equipment and working conditions; providing high levels of supervision and oversight; and surveying land to make sure it was of high humanitarian priority. This encouraged demining agencies to prioritize humanitarianism and professional ethics over profitmaking.

The analysis of differing outcomes cannot, however, stop there. The two complexes also had a wider impact on the socio-political context of the beneficiary country. Examining economic policy in Bosnia, Donais wrote, 'in post-conflict situations economic reform efforts should be measured not only in terms of narrow economic impacts, but also – equally if not more importantly – in terms of the impact of such efforts on the broader peace process.'[60] Applied to demining, this will be the subject of the next chapter.

6

IMPACT ON PEACEBUILDING

'The challenge of peacebuilding...lies...in finding ways to dismantle the very structures that prevent such a post-war order from taking root, and which in many ways replicate and reinforce the logic of conflict.'
– Timothy Donais.[1]

The literature on the political economy of aid in conflict[2] – notably by Mary Anderson[3] and Jonathan Goodhand[4] – highlights aid's embeddedness in a web of political, economic and social relationships. Aid agencies' activities invariably impact the local political economy and are interpreted through local socio-cultural lenses. Therefore, aid organizations can inadvertently contribute to conflict by collaborating with vested interests or appearing to take sides. Conversely, their activities can contribute to peacebuilding by strengthening local constituencies for peace or by taking a principled stand for nonviolent politics. This chapter draws conceptually from this literature, showing that demining agencies also have a wider impact on a transitions from war to peace.

This chapter argues that shorter-term profit-seeking objectives of commercial companies supported by the US (and the UN in Sudan) constrained them with the necessities of expediency. They had to collaborate with the powerful political actors, both donors who might be engaged in the conflict, as well as local vested interests. As a result, the Strategic-Commercial Complex probably contributed to military stabilization in the three cases, but was also more likely to become embedded in the illiberal political economy of war. Moreover, by contributing to the privatization of protection, Strategic-Commerical Complexes may have exacerbated the fragmentation of security provision for ordinary people.

In contrast, as stewards of a humanitarian mission, constrained by public expectations of strong professional ethics and afforded the space of long-term time horizons, the international NGOs supported by Norway were able to invest in contributing positively to the peace process. Specifically, this Human Security-Civil Society Complex contributed to integrative and non-violent politics through inclusive hiring practices, resisting the political economy of conflict and advocating for limits on the technologies of violence. Moreover, they helped restore public security by distributing protection from explosive remnants of war by need, rather than ability to pay or strategic rationale. That said, by using international NGOs and creating organizations that were independent from other local institutions, it is possible that the Human Security-Civil Society Complex was less effective at building the capacity of the local state, especially in terms of its ability to respond to violent threats.

The chapter first considers the wider impact of the Strategic-Commercial Complex, looking at issues of state stabilization, commercialization of protection, local capacity and the political economy of war. It then examines the impact of the Human Security-Civil Society Complex, focusing on needs-based distribution of protection, local capacity, inclusivity and advocacy for limits on violent politics.

Wider Impact of the Strategic-Commercial Complex

As outlined in earlier chapters, the motivations and sources of power of the Strategic-Commercial Complex influence its priorities and systems of implementation. The interests of Strategic-Commercial implementers and their use of expediency in guiding their relations with other groups, made them more likely to support the security structures of their donor's intervening forces and those of the host state (if it was in their donor's strategic interests). This may mean that commercial demining companies contribute more than international NGOs to stabilization and the national security of the host state. However, using commercial companies erodes the concept of a broad public provision of security, contributing to the privatization of protection. Moreover, requiring rapid implementation and guided by expediency, implementing bodies were also more likely to collude with the political economy of conflict. Their embeddedness in these illiberal socio-political networks meant they had difficulty resisting capture by vested interests, occasionally facing 'blowback' – unintended consequences – from strengthening chauvinistic or reactionary groups.

Stabilization and Building State Security

In all three cases studies, international authorities and local state institutions instrumentalized demining as a means to support the process of political stabilization and the assertion of state authority. Probably due to its greater comfort working with military institutions, the 'Strategic-Commercial' approach was generally more likely to pay attention to bolstering state security than the 'Human Security-Civil Society' approach.

In Afghanistan, commercial demining cleared land for the development of military and police infrastructure used by NATO forces and the Afghan government. Demining companies also collected and destroyed insecure ammunition, denying weapons to the insurgency. Moreover, the commercial companies hired many former soldiers from the UNDP demobilization process, giving them an alternative source of employment than armed groups. Bob Gannon of RONCO argued that commercial deminers supported the peace process more effectively than demining NGOs, because, by supporting the security structures of the US, NATO and the nascent Afghan state, it was aiding the stabilization of the country:

> If you can get a stable government in a country then it's going to go a long way towards bringing proper peace. And the way to get a stable government, is to make sure the government's got the land, the property it needs to put good governance in place.[5]

In Bosnia, linking USAID infrastructure demining to US Army priorities probably contributed to NATO's stabilization mission. Moreover, the close integration between the commercial companies and local NGOs with ethno-nationalistic structures may initially have contributed to stability in Bosnia. The strategy of the international community in Bosnia incorporated a combination of both conflict and collusion with nationalists. Accommodation – allowing capture of foreign aid, tax evasion and participation in privatization – may have encouraged them to 'buy-in' to the Dayton state structures. In contrast, NPA was initially resistant to state coordination, preferring to work in the areas where it felt need was greatest. Moreover, in interviews with the author, BHMAC officials claimed that NPA, with its deeper pockets, was damaging BHMAC capacity by hiring away some of its most talented people – notably former BHMAC deputy director Darvin Lisica, now NPA's country program manager.

In Sudan, by prioritizing road clearance, commercial demining has helped UN peacekeepers to deploy throughout the South and oversee implementation of the CPA. Likewise, RONCO acted in support of the Joint Military Commission in the Nuba Mountains, enabling them to conduct the mission mandated by the ceasefire. In contrast, international NGOs seem to have had a more combatative relationship with Sudanese state institutions – suspicious of attempts by the northern government and the SPLA to control their activities. Some NGOs tried to develop their own institutions outside the 'para-statal' NGOs, OSIL and JASMAR, independent of the local authorities. This drew criticism from governmental officials who believed donors should channel more assistance through state organs. Indeed, international NGOs have long been criticized for creating 'local substitutes for state administration' and 'contributed unintentionally to the erosion of the authority of the very weak state' in South Sudan.[6] Therefore one might argue that mine action NGOs exacerbate South Sudan's fragmented polity, instead of channeling their assistance through and helping to develop the capacity of state institutions.

While the Strategic-Commercial approach to demining probably does more to assist in stabilization and re-establishment of the security structures of the state, when examining any political or governance system, it is important to consider how protection is distributed among the population. For, as with any service or commodity, it is rare that security is distributed evenly. Generally those with the most access to political power determine who is protected and how. The following sub-sections will examine how systems of security and protection were organized by the Strategic-Commercial Complex.

Commercialization of Protection

Commercial demining contributed to and was influenced by broader processes of commercializing and privatizing protection and security. It was part of a system in which protection from the threat of landmines and other insecurity was provided and distributed through market, rather than public bodies. Tendered demining reinforces the commercialization of security provision because the private security and commercial demining sectors are closely associated.[7] Many companies provide both services and draw on an overlapping workforce of demobilized soldiers. As such, mine action may provide a figleaf for the private security industry – a service more publicly palatable, employing the same type of people, operating in the same environment and providing the contacts and groundwork necessary to launch other operations.[8] While mine action is not as profitable as private security operations, it provides a logical diversification for such companies, providing them with some lower risk business in their portfolio. While private security is a response to real demands for security, it may also contribute to the fragmentation and delegitimation of public control over violence. Mine action can facilitate this process by providing such companies with a stabilizing element of their portfolio and making them seem more socially acceptable.

In Afghanistan, many companies doing commercial demining had other divisions that provide private security services. They argued that the Afghan state was incapable of fielding adequate security for demining operations and that private security work provides employment for people who might otherwise use their military training for more nefarious activities.[9] However, Mohammed Sediq of UNMACA worried that a key to mine action's success in Afghanistan – its 'flag of impartiality and neutrality' – may be eroded by these companies' image.[10] For instance, all DynCorp mine action staff took weapons training, carried a pistol and rifle while traveling and had access to a DynCorp rapid reaction force if they came under attack.[11] DynCorp itself has faced much criticism for its Afghan security operations.[12] ArmorGroup had an entire military style-barracks on Kabul's outskirts and was responsible for guarding the US Embassy.[13] RONCO provided static guards around Kabul and bomb detection dogs to other security companies.[14] The USAID demining coordinator, who, interestingly, was also the coordinator for security for USAID projects, told the author that even if the demining company did not have a private security arm, other private security companies would guard US reconstruction demining sites.[15] In contrast, the HALO Trust compound and convoys were guarded by the Afghan police. While this does associate HALO with the Afghan state, perhaps compromising their neutrality, it does send the message that HALO supports public control over the use of force.

It would not be surprising then, if some Afghans, seeing commercial deminers carrying guns, jump to the conclusion that all demining is related to military operations. Research by Rimli and Schmeidl has shown that ordinary Afghans cannot distinguish private security contractors from other international organizations.[16] This is not a good thing for demining, as private security companies, especially international ones, have a bad reputation among Afghans. Many security companies have contracted militia commanders as 'an expedient and convenient way to obtain armed and trained guards', which Rimli and Schmeidl argue means they are 'essentially paying individuals to protect them, who might be the main source of insecurity in the region to begin with.'[17] Furthermore, private security contractors are less legally accountable than Afghan police or NATO military personnel, who are at least nominally accountable to military code. In late 2007, the Afghan Interior Ministry cracked down on security companies, claiming that they contributed to lawlessness, having been linked to several high profile murders and armed robberies.[18] Indeed, many Afghans say the presence of private security firms actually makes them feel less secure.[19] Thus some have argued that the privatization of security has actually increased insecurity for those who are not able to afford protection and that the commercial mine action contributes to this trend.

This has the potential to erode the popular legitimacy of demining and its long-cultivated neutrality. One Afghan NGO director explained that when providing demining to support the reconstruction of the Kabul to Kandahar road, his NGO was required to have an escort from an American private security contractor. He said that this 'damaged' the NGO's reputation because insurgents 'thought that we are now with the Americans and with the military.'

In both Bosnia and Sudan, the links between commercial mine action and private security have had less impact than in Afghanistan. Generally, the private security companies engaged in demining at least claimed they were not involved in any security operations. However, there have been some less high-profile links. Both RONCO and DynCorp have recruited Bosnian deminers to work in Afghanistan. The Bosnian firm OKTOL actually turned into a private security company when their profits from demining decreased.[20] There have also been reports (the author was not able to gain firm confirmation) that some of the nationalistic networks involved in demining have also been supplying private security personnel to international companies.

In Sudan, Mine Tech's sister company Frontier Medical has been supplying medical supplies to the African Union military force in Darfur, an experience which probably helped Mine Tech win the UN ordnance disposal contract in Darfur.[21] Moreover, the private security industry has lobbied the UN and Western governments to hire them to restore security in Darfur. Even if they are not given such a 'kinetic' role, it is likely that the ongoing expansion of the UN peacekeeping force there will require contractors willing to supply them goods and services. It may thus be possible that private security companies like ArmorGroup see Sudanese mine action as a beachhead into the possibility of more lucrative private security or military contracting work.[22] In contrast, when the HALO Trust discovered that its local NGO partner was making forays into private security provision, by renting out armed escorts, it demanded SLR stop immediately. When the partnership fell apart, HALO alerted mine action donors to SLR's private security work.

Privatizing security provision also affects the distribution of protection among the population. In contrast with the international NGO approach, the 'Strategic-Commercial' approach taken by the US (and by the UN in Sudan) generally prioritized demining that paid some strategic or commercial dividend. Strategic priorities were either directly related to military operations (such as demining military bases) or indirectly strategic, such as opening roads to give troops (and everyone else) access to remote areas. Commercial priorities included preparing the way for commercial infrastructure (such as telecoms antennae or oil pipelines) or to open access to primary commodities such as oil.

In Afghanistan, the US-led commercialization of demining challenged the UNMACA-led need-based system. By setting up a mixed market system, persons or organizations that wanted demining performed immediately no longer had to wait for higher priority tasks to be completed first. If they had the ability, they could pay for a commercial company to demine it immediately. In other words, demining became a purchasable commodity. For instance, the Afghan telecommunications company Roshan unilaterally contracted demining with commercial companies. The potential advantage is that it may introduce flexibility – a two-track system – like the mixed public and private health care system in Britain. Those who have money do not need to wait in line. Since there is a surplus of trained deminers in the market, at this point there is not too much of a danger of the commercial market drawing significant human and physical resources away from the humanitarian demining system. In fact, the commercial capacity has soaked up some of the excess labor created by a UN-sponsored demobilization program that trained ex-soldiers to be deminers.[23]

However, since commercial operators pay more, there is a danger that the better quality people will be drawn away from the humanitarian sector and into the commercial sector – just as the commercial legal sector draws the best lawyers away from public defense offices in the US and Britain. There will be significant incentive for the best deminers to abandon NGOs, redistributing the best talent to the commercial sector. There has already been some bad feeling about this. One local NGO director complained that the commercial companies 'steal the expertise' from the NGOs.[24] UNMACA chief of staff Kerei Ruru admitted that there was 'a wee bit' of 'poaching of talent' by the commercial companies.[25]

In addition to making it more possible to allocate protection from mines through market-based 'ability to pay', the security objectives of US aid have posed a further challenge to the UNMACA/NGO system. By taking money out of the UN-led system, USAID has directed its demining funds to support the US reconstruction efforts. While Afghanistan certainly needs the roads, powerlines, schools and clinics built by these efforts, studies by others showed that many of these massive projects seem to be more for political and symbolic, rather than humanitarian, impact, sometimes affecting the quality of the work.[26] Indeed, studies by UNDP/World Bank and MCPA have shown that the highest economic returns come from focusing demining efforts on irrigation and irrigated agriculture, not the roads that USAID has prioritized.[27]

Moreover, one must problematize the word 'security.' There is no doubt that security is a desperate need in Afghanistan. However, one must question whose security the US is aiming to secure. For instance, a significant portion of US troops and military budget are assigned to the hunt for Bin Laden and Al Qaeda, which affects the national security of the US more than human security of the Afghan people.[28] By shifting the focus of US-funded demining away from the needs-based system of the UN to supporting the Coalition's military objectives, the US is prioritizing its own security over that of the Afghan people.

One sees similar patterns in Bosnia, where the commercial system set up by the US and other donors allowed certain clients with large purchasing power to sidestep the need-based prioritization systems set up by public agencies. For instance, the para-statal energy giant Energoinvest contracted the US company UXB to do several tasks. During the era of World Bank demining funding, the public prioritization of tasks was easily influenced improperly by vested interests linked to the commercial demining companies (for instance prioritizing easy tasks that were more likely to turn a profit) or for ethno-chauvinistic reasons. While the ITF system supported by the US State Department accepted priorities chosen by the national body, BHMAC, USAID funding prioritized preparing the way for macro-infrastructure (mostly energy) and projects selected by the US Army.

Following this pattern, the commercialization of Sudanese demining has allowed clients with deep pockets to sidestep the 'waiting line' of national priorities. For instance, an oil company contracted Zimbabwean demining firm The Development Initiative to support operations in South Sudan.[29] Moreover, UNMAO prioritized road demining, based on the UN's strategic necessity of opening access for UN peacekeeping troops. Therefore, UN-funded commercial demining in Sudan bore similarity to the commercial approaches in Afghanistan and Bosnia. Clearance was prioritized on the basis of strategic rationale and ability to pay.

In short, the 'Strategic-Commercial' approach to demining allocates protection according to strategic rationale or the purchasing power of the client rather than an analysis of where human need is greatest.

Local Capacity Building

Any international agency working in the local environment will obviously need to engage with at least some local professionals or partner organizations to gain local knowledge and contacts. However, if a commercial company (or for that matter, an NGO) is too successful at developing local institutions, in effect it creates its own competition. For example, in one of DSL's (later called ArmorGroup) first contracts in Bosnia they trained local deminers who then created local companies that beat DSL/ArmorGroup in tenders. Therefore, one can expect an institutional reluctance (even if unspoken or unconscious) among international commercial companies to build local capacity 'too effectively.' This may be compounded by a donor tendency to favor companies from their own countries.

Indeed, they may have no incentive to build institutional structures that will persist beyond their contracts. As one Afghan NGO director said, 'They try to do it quick, in order to make money...and then they go.'[30] Similarly, in Sudan the UNMAO head acknowledged that 'commercial companies...don't put up structures that can stay behind.'[31] As a result, there are no viable local commercial operations in Sudan and international companies have relied heavily on expatriates and mechanization, avoiding hiring large numbers of Sudanese. A commercial demining supervisor explained that this was 'because of the timeframe, we needed the job to be done quickly'[32] – hiring and training local Sudanese would take too much time and investment. Distrusting the local population and fearing importation of local political tensions into the workplace, some commercial companies find it easier to 'parachute' in deminers from other countries. In Sudan, several commercial companies imported deminers from Mozambique or Zimbabwe. Likewise, while the State Department provided resources to the local Afghan NGOs, simultaneously USAID and the US military eroded the strength of local NGOs by encouraging the entry of international commercial companies into the country. These companies use far more expatriates than the NGOs, including HALO. Their tasks inside US military bases use almost no Afghans, since few can obtain security clearance.

Nonetheless, there are cases where the interests of strategic or commercial motivations encourage cooperation with local society. Companies may find that collaborating with local subcontractors gives them a commercial advantage, due to connections with local authorities and ability to better recruit and manage local labor. In the early years of Bosnian demining, RONCO dominated the market using subcontractors intertwined with the local political economy. In Afghanistan, both RONCO and DynCorp were actually contracted to build local NGO capacity. Some have argued that certain international companies in Afghanistan have pushed for the creation of local companies, because they see them as potential partners in challenging RONCO's market dominance.

A technocratic analysis of capacity-building would probably stop here. However, a social scientific understanding of capacity has to be rooted in understandings of power. As postmodern theorists have argued, organizations are not simply containers for power that can be easily filled up through capacity building efforts. Rather, power is a relational concept – it is exercised in relation to other people and organizations that may have other interests and objectives. Without a serious discussion of power or the political struggles that accompany the development of any institution, 'capacity' becomes a vacuous concept and a technocratic euphemism. The scholar Michael Shafer's work on state capacity is particularly helpful. By comparing South Korea and Zambia, Shafer sought to 'replace the current tautological and static understanding of state strength in economic policymaking with a dynamic, relational one.' [33] He argued that state capacity and strength

> on a given issue at a given time is relative to the strength of the interest groups the state confronts and of coalitional allies it might invoke. Over time state strength will vary as a function of changes in its capacity in relation to changes in the capacity of its opponents and allies.[34]

As a result, the nature of an organization and the impact it will have are determined in part by the social sources of its power. In the process of choosing allies (such as employees, fixers, government supporters) to help carry out their agenda, organizations compromise their agendas, become embedded in other social networks or even become captured by other interests. Therefore, building the capacity of an organization may inadvertently grant power to particular social networks. While the capacity building literature often speaks as though 'local capacity' and 'local ownership' are inherently good, one has to ask 'building the capacity of whom and of what socio-political structures?' It may be that there are some social institutions, such as criminal or chauvinistic networks, which are better left without capacity.

The next subsection shows that if capacity-building is understood as empowerment of particular social actors, the Strategic-Commercial Complex often built the capacity of networks partially responsible for the country's ongoing instability. Motivated by strategic and commercial interests, Strategic-Commercial Complexes were incentivized to act quickly and produce the results dictated by the donor. Expediency encouraged them to accept and collude with the status quo, borrowing the power of existing social structures. This left them embedded in the local political economy of war, where strength derived from ability to wield money and violence.

Embeddedness in the Political Economy of War

Literature on the political economy of war[35] has highlighted how international aid can interact negatively with violent and illiberal structures led by warlords, profiteers and chauvinistic networks aiming to capture rents.[36] By reinforcing structures responsible for violent politics, aid can actually contribute to conflict.[37] Mine action is not immune from these risks. In fact, it is probably more susceptible, because in addition to pumping large amounts of money into conflict zones, for an aid sector, demining has an uncommon level of interaction with the security structures of the affected country. In order to find out where mines are, demining agencies must persuade armed groups to share military secrets, such as the location of minefields, frontlines and garrisons. They must also persuade these groups to allow removal and destruction of mines, which factions may rather leave in the ground. Finally, and perhaps most importantly, most deminers working for NGOs and commercial companies tend to be former soldiers. One often sees the organizational structure of armed factions reproduced within demining teams, with officers acting as supervisors and footsoldiers becoming deminers. Thus there are often significant connections between demining agencies and the security forces, armed groups and paramilitary networks that developed during the conflict. This carries obvious risks. On one hand, demining agencies face 'capture' by the political (and/or economic) objectives of an armed group. On the other hand, drawing deminers from multiple sides risks importing broader societal conflicts into the organization. If they become too deeply embedded in the political economy of conflict, demining agencies may find their work shaped and directed by interests that have objectives at cross-purposes to their own.

Commercial companies, with their shorter-term time horizons, are drawn to people and social structures, both local and international, that can 'get the job done'. These people are often those who are well-connected, powerful and who learned logistical skills in organizing wartime supply chains. This political 'path of least resistance', accepts social divisions as they are, leaving companies less able to transcend existing political structures. Some have argued that by working with such groups it may be possible to co-opt them into the agenda of the intervening international agencies. By encouraging potential spoilers to engage in demining instead of violence, they may be able to contribute to, rather than threaten, the reconstruction process. For instance, the former vice-president of RONCO said there was an awareness that some of their local subcontractors in Bosnia were 'thugs,' but

> the fact was we were able to get some demining done through those thugs. What do you do at the time? If you want to wait till you can get a missionary-type group to be formed that you can work with, you might be waiting for a long time and the minefields stay out there and nothing happens.[38]

However, observation in the field suggests that costs of this approach, in terms of damage to the political, social and economic foundations of society, are very high. In Afghanistan, US-supported 'reforms' to the sector have eroded the long-cultivated neutrality and separation of demining from armed groups. With no incentive or motivation to resist their donors and clients, demining companies have readily accepted their integration into the US politico-military effort. As a result, US-led commercialization has blurred the lines between civilian and military demining. Instead of using uniformed military personnel to conduct military demining tasks, they are done by the same commercial companies that also bid on civilian tasks. Many of these companies also engage in private security operations. Occasionally one sees a merging of the two kinds of operations when demining companies provide bomb dogs to private security units or support unpopular US-funded forced poppy eradication.[39] Through integrating demining funding into security operations and funneling money through commercial channels, the US has eroded the power of implementing agencies to resist capture by the agenda of a party to the conflict. Unlike the NGOs, who have tried to guard their autonomy by declaring political neutrality, commercial companies are fully embedded in US policy.

In Bosnia, both the US (through RONCO, the US Special Forces and NATO) and Norway (through NPA and NATO) worked to develop the capacity of the entity armed forces. While it could be argued that this may have legitimized the very institutions that were responsible for the war, it may also have provided a catalyst for the integration of the armed forces and a productive role for them. One sees a divergence, however, between US and Norwegian-supported demining in its interaction with the more subterranean paramilitary networks that continued to dominate Bosnian politics even after the peace agreement. In contrast to the relatively independent and multiethnic NPA, US-supported demining, contracted through the World Bank, UXB, RONCO and ITF, became deeply embedded in ethno-nationalist and criminal politics. The US reacted passively to the involvement of criminalized elite in demining until 2003, when the international community began a reinvigorated campaign against war crimes fugitives. Rather than taking a proactive role in reorganizing Bosnian demining in a multiethnic manner, it instead acted reactively, while trying to limit its embarrassment in the media.

As such, contracting demining to commercial entities mirrored broader problems with privatization in Bosnia.[40] In Bosnia, the true 'winners' of the conflict were a class of 'war entrepreneurs,' who continued to subvert and undermine already weak central public institutions. In this, they found common ground with the international community's neo-liberal agenda. They benefited from privatization, deregulation and liberalization, capturing lucrative public assets and reducing government oversight of their activities.[41] The result was a political economy dominated by ethnicized political machines, which distributed the windfalls of international aid through ethnic patron-client networks.[42] Said Donais:

> the way the privatization process has unfolded in Bosnia ... has done more to date to entrench the economic positions of the country's nationalists and reduce the prospects of ethnic reintegration than to establish the foundations for sustained economic growth and recovery.[43]

These patterns of 'illiberal privatization' were also reproduced in Bosnian demining. As one former senior UNMAC official said, 'Local mine action was based firmly on nepotism and the patronage of unscrupulous men in high places.' From 1996 to 2000, each Demining Commissioner functioned as patron and protector of a local demining company representing their ethnic group. The Commissioners pressured the World Bank Project Implementation Units to direct contracts to their favored companies and obstructed investigations into corruption or poor safety practices.[44]

The first Serb Demining Commissioner, Radislav Ilic, was brother-in-law of Radomir Kojic, owner of the main Serb demining company, UNIPAK, and director of Kojic's import-export company. Ilic later set up another demining company, MEDECOM, and a quality assurance company, TERAPROM, which won contracts to be an 'independent' evaluator of both UNIPAK and MEDECOM's work. For his part, Kojic allegedly organized the ethnic cleansing of non-Serbs from Pale,[45] supervised torture centers[46] and commanded a unit shelling Sarajevo.[47] Since the war's end, press reports regularly accused Kojic of illegal and ultra-nationalist activities, most notably supporting the network protecting the now captured Radovan Karadzic, who was one of the most wanted war crimes fugitives.[48] State prosecutors also claimed 'grounded suspicion' that during the war, prior to turning its attention to demining, UNIPAK 'performed operations exclusively in the interest of the Government of Republika Srpska and for the purpose of financing war activities.'[49]

While Kojic has so far avoided conviction for any of the above alleged offenses, he and Ilic were arrested in August 2006 for money laundering, tax evasion, and abuse of office.[50] At the time of writing, it appeared the case had run out of steam; the two were still awaiting trial. However, prosecution evidence suggested that UNIPAK had engaged in tax evasion by transferring income in cash to Ilic's companies MEDECOM and TERAPROM.[51] While the author was on fieldwork in Bosnia, Kojic's house was raided by NATO in an attempt to gather further information on the Karadzic network.[52]

The Croat Demining Commissioner, Berislav Pusic had been the 'Head of the Service for the Exchange of Prisoners and Other Persons' in the breakaway Bosnian Croat statelet, the 'Croatian Republic of Herzeg-Bosna.' Currently on trial for war crimes at the International Criminal Tribunal for the Former Yugoslavia (ICTY), Pusic's 2004 indictment alleges he ordered 'deportation of Bosnian

Muslims' from the so-called 'Hezeg-Bosna' region.[53] Various Financial Police documents alleged Pusic had improperly arranged for DECOP to rent donated equipment, tried to pressure an international firm to take DECOP, instead of another local company, as a subcontractor and generally gave the RONCO-DECOP partnership 'privileged status' over other companies.[54] The Financial Police also accused DECOP of falsely reporting income and evading taxes.[55] It is not clear why these indictments and investigations never led to actual court cases.

Pusic also had close links to the Bosnian Croat demining NGO PROVITA. Among the founding members were his son-in-law and his daughter, Sandi Kozul, who was PROVITA's first director.[56] The Financial Police claimed that Pusic arranged for PROVITA to rent donated equipment to DECOP and divert a portion of the rental fees to a personal slush fund, which he used to make unauthorized loans, donate to philanthropic causes, pay traffic fines and road tolls and purchase gifts and personal affects like clothes, paintings, perfume and a hunting rifle.[57] Several of the items were bought from Biokomerc, a company owned by his brother.[58] He even apparently transferred money to the Service for the Exchange of Prisoners and Other Persons, his war-time employer, that, according to the ICTY, had facilitated ethnic cleansing of Herzegovina.[59]

Although considered by many to be 'better behaved' than the others, Bosniak Commissioner Enes Cengic was the former director of the discredited MPRA. He was a member of the Bosniak nationalist party and held several influential political positions in Sarajevo local government during and after the war.[60] He was also close friends with and a former colleague at MPRA of Adnan Gradasovic. Gradasovic, with his brother Mirzad, was co-founder of the main Bosniak company SI Company (which formed a partnership with another Bosniak Company OKTOL). According to Financial Police files, SI/OKTOL rented US-donated equipment from MPRA at a highly discounted rate, even before receiving demining accreditation or hiring more than one employee.[61] Furthermore, a Financial Police indictment accused SI/OKTOL of evading taxes by failing to properly report profits from its partnership with RONCO.[62] Again, it is not clear why these indictments did not end up in actual court cases.

As narrated in chapter four, the corruption of the demining program deeply concerned many actors in the international community operating in Bosnia. The World Bank initiated an investigation with the collaboration of the Bosnian Financial Police. In October 2000, Wolfgang Petrisch, the High Representative mandated by the peace process to oversee the country's political institutions, fired all three Demining Commissioners. Despite the involvement of these local companies in high profile scandals, US money continued to flow to them through the ITF until 2003, when the international community finally took action against Kojic. The US has continued to fund PROVITA through the ITF to the present day. They also continued to fund MEDECOM, run by Kojic's brother-in-law Ilic, until 2005.[63]

Whether or not the US intentionally supported them, these patronage networks displayed mafia-like behavior, drained public budget and played into the nationalistic logic of Bosnia's conflict. While neo-liberals see privatization and deregulation as encouraging peaceful interdependence and trade, Duffield has shown that in many 'new' conflicts, such policies allow 'local strongmen' to capture the political space and funds necessary to forward a subversive 'illiberal' agenda.[64] Likewise, Bosnian political machines – funded, equipped and trained by the international community – created three 'ethnically clean' companies and an ethnicized political structure to support them, enriching and strengthening nationalist elites. Several key actors in the demining patronage networks were deeply implicated in the nationalist-criminal networks that have paralyzed the peace process. As Donais observed about Bosnia generally,

> giving international blessing to what has essentially been an ethnically divided privatization process appears to be clearly inconsistent with broader international goals of multi-ethnicity and ethnic reintegration.[65]

There has been a tendency in the mine action literature to assume the Bosnian illiberal political economy was a 'problem with the locals.' However, this would be mistaken. Many of the international managers of commercial demining companies also came from military and police backgrounds, several having fought for illiberal regimes. For instance, as a member of the South African police, MECHEM's early program manager in Bosnia, Johannes 'Sakkie' van Zyl, had been involved in covert extrajudicial killings of anti-apartheid activists.[66] Col. Lionel Dyck, head of Minetech, had been an elite officer in both the Rhodesian and Zimbabwean armies and had commanded elite paratroopers operating in Mozambique in support of FRELIMO.[67] As an Australian Army officer David Rowe, early manager of DSL in Bosnia, claimed he had been involved in a covert effort pitting mercenaries against the Viet Cong.[68] It is therefore possible that these people would be more used to 'doing deals' with unsavory characters and feel less uncomfortable about working with groups deeply embedded in Bosnia's political economy of war. Indeed, Dyck told investigative reporter Hugh Griffiths that he had been aware of possible links between Kojic and Karadzic while using UNIPAK as a subcontractor.[69] To be fair, international NGOs have often hired people from similar backgrounds. However, from observations and conversations in the field, it appears international NGO personnel feel more uncomfortable with doing so and try to balance their staff with employees from a more diverse set of backgrounds.

In Sudan, UNMAS has attempted to develop a 'one-country' approach to mine action,[70] in the framework of the Comprehensive Peace Agreement. As a result, UN mine action has continued to contribute to the peace process in a number of ways. Firstly, UNMAS brought together representatives from both sides to develop the mine action sections of the peace treaty at a much faster speed than other parts of the treaty. Participants in these talks say that the experience of working together on mine action issues in the Nuba Mountains built the trust necessary to negotiate in good faith. Secondly, by clearing roads and settlements, mine action agencies opened the country to peacekeepers, aid, commerce and the census-takers who will prepare the way for the CPA-mandated elections. The UNOPS and WFP program have also taken into consideration return issues when determining priority roads.

In reality, however, the UN mine action program, operating in a Strategic-Commercial manner, has done little to transform the constellation of Sudanese politics. The CPA is essentially a pact between two armed groups – the Government of Sudan and the SPLA – guaranteeing hegemony in their respective regions of strength, to the detriment of other groups.[71] Moreover, despite a declared 'one-country approach,' in reality there was little cooperation between the two local authorities beyond macro-policy making.

Given the UN's structural bias towards operating through national institutions, has tended to accept the status quo of the polity. UNMAO works through local structures intimately linked to the political economy of the war. In the north, the UN's local counterpart is the National Mine Action Authority (NMAA), co-chaired by the Sudanese Ministry of Defense. Moreover, the Secretary General of the National Mine Action Committee, who has also represented Sudan at international mine ban conferences,[72] is Ahmed Haroun, the northern government's Humanitarian Affairs Minister who is wanted by the International Criminal Court for crimes against humanity in Darfur.[73] While UNMAO now limits interaction with him, they worked more closely with him in the past, even paying for him to go on a 'capacity building' trip to Jordan (an example of how the technocratic understanding of capacity building is sometimes blind to political sensibilities).[74]

In the south, UNMAO's counterpart is the South Sudan Demining Commission (SSDC) that, like all other South Sudanese government institutions, is dominated by the SPLA. UNMAO's inability to challenge the power of the SPLA-created SSDC was indicative of their slow response to protect HALO's assets in the debacle with its local partner NGO SLR. SLR attempted to misdirect significant resources from the partnership and as relations with HALO deteriorated, SLR engineered HALO's expulsion from South Sudan. They then attempted to grab HALO's assets and bank accounts. The extremely slow action by SSDC, which then hired SLR's former deputy director and rented land from SLR, suggests it was deeply complicit in this scandal and that SLR was protected by prominent figures in the SPLA. While UNMAO was finally able to secure HALO's equipment, at the time of the fieldwork, it lay in limbo at a UN compound. UNMAO never took a strong public stand supporting HALO nor publicly condemned the SSDC's complicity in the affair. WFP has coordinated even more closely with the SPLA, letting them influence which roads should be rehabilitated, even before the peace agreement was signed, despite the obvious benefits to the SPLA's logistical chains.

The companies contracted to UNMAO and WFP had limited policy autonomy and generally worked within the larger framework developed by their client agencies. As a result, the commercial companies have operated within the political status quo without attempting to change or transform it. Further, the presence in Sudan of the demining company MECHEM, should be considered in the context of the long history in the country of its arms manufacturer parent company, DENEL. Human Rights Watch accused DENEL of selling military equipment and services to Sudan at least until 1998.[75] Recently, DENEL made an agreement with the Government of South Sudan to help with the creation of a South Sudanese Air Force.[76]

Because the Strategic-Commercial Complex embeds demining in the political economy of conflict, vested interests may have other objectives or invite reaction from opponents, that can result in damaging unintended consequences both to the program and wider society. Therefore, the interaction of demining with the political economy of war may contribute to blowback – spy jargon for unintended effects of an operation that rebounds against its creator[77] – against donor interests. Advocates of 'engagement' with extremists, terrorists and war criminals must accept the fact that money used to buy support for a policy may be used to subvert it by the very same people. A classic case of blowback is the US and British arming of Saddam Hussein's Iraq, which later enabled it to invade Kuwait and in 2003 create enormous discord in the UN Security Council.

Afghanistan offers perhaps the most dramatic example of blowback. The early demining program developed at the end of the 1980s operated in the context of and supported US covert action. USAID's local counterparts in the Cross-Border Humanitarian Assistance Program were the Pakistani intelligence service and the mujahideen parties. Much of its assistance was distributed directly to resistance factions – a disproportionate amount going to the fundamentalist parties. Said one Afghanistan expert:

> Dependence on Afghan resistance leaders…compromised much of the humanitarian response of the 1980s, giving rise to widespread diversion of resources, and strengthening the new, armed elite at the expense of more traditional structures.[78]

The US is now reaping what it sowed in Afghanistan, say Girardet & Walter,[79] as those jihadist networks funded and trained by the US turn against them.

While it is too early to discern the impact of the recent US-led demining commercialization in Afghanistan, privatizing protection could also contribute to blowback. As mentioned previously, security companies have hired warlords and their militias and legitimized the fragmentation of sovereignty. This legitimation of warlords could lead to problems as the US attempts to stabilize the Afghan state and eradicate the opium trade. Further, the significant public distrust of private security companies is contributing to suspicion of the US effort in Afghanistan, eroding possibilities for local collaboration with US objectives.

As demining began losing its neutral, impartial and nonviolent reputation, it was increasingly targeted by the insurgency. The embeddedness of demining into US security objectives provoked an unintended reaction against deminers. According to UNMACA data, between January and November 2006 more demining personnel died in terrorist attacks than demining accidents. This trend appeared to be increasingly serious. In April 2007 'dozens of Taliban militants' in western Farah province ambushed a RONCO demining team, though it is not clear whether RONCO itself was the intended target.[80] In June 2007, Taliban insurgents in Ghazni province kidnapped 18 deminers from MDC and MCPA and 'threatened to kill them if investigations suggest they are working for U.S.-led forces.'[81] The following month three MDC deminers were shot dead in Kandahar province, leading the organization to suspend its operations in the Kandahar and Helmand region.[82] Unfortunately, if deminers come under increasing attack, it will probably lead to calls for increasing militarization of the process, further reducing the widespread legitimacy of demining agencies.

The Strategic-Commercial approach also generated blowback against the US in Bosnia, as US funds for demining were diverted to nationalistic networks that later hindered US objectives. These networks obstructed the peace process and raised the costs of the NATO peacekeeping mission. Moreover, the alleged connection of Kojic and Ilic's companies to the network supporting Radovan Karadzic was a public relations embarrassment for the US State Department, sending the message that they were either colluding with the nationalists or bungling and uninformed.

In short, therefore, embedding demining in the highly polarized nature of strategic politics and collaborating with violent clandestine structures may have unexpected and highly dangerous side-effects.

Wider Impact of the Human Security-Civil Society Complex

The following section argues that the humanitarian mission of the Human Security-Civil Society Complex led implementing NGOs to attempt to resist or transform the political economy of conflict. Idealistic motivations and long-term grant support, such as from Norway, gave NGOs time to build independent institutions reproducing their cosmopolitan and humanitarian ideals. Autonomous of state organs and the local political economy of war, they were able to make principled stands advocating for a non-violent politics and limits on war. However, where countervailing forces were much stronger, or when NGOs became close to chauvinistic groups, they were less able to implement their humanitarian mission or resist capture by vested agendas. This suggests that NGOs can fall somewhere on the continuum between the ideal of a Human Security-Civil Society Complex and cooption into other complexes. In general, however, international NGOs did appear to be better than commercial companies at focusing on humanitarian objectives and resisting the politics of war.

Needs-Based Distribution of Protection

Unlike the Strategic-Commercial Complex, which distributed protection from landmines according to strategic rationale and/or a market system, the Human Security-Civil Society Complex was more receptive to local community needs and focused on areas of high human impact. In Afghanistan, the UNMACA system, implemented through local NGOs and the HALO Trust, operated like a government public service. Beneficiaries received demining for free and UNMACA claimed prioritization resulted from assessment of need and potential humanitarian and socio-economic benefits of clearance. Thus beneficiaries' ability to pay for the service should not have been reflected in priorities, nor the allocation of demining resources. This system would be beneficial to the poor who were unable to afford to clear minefields threatening their lives and livelihood. UNMACA has drawn criticism that is common of many public services – that priority-setting was non-transparent, occasionally influenced inappropriately and focused more on the processes of reconstructing destroyed infrastructure (which benefits those who were privileged to have infrastructure previously) than providing new infrastructure for the poor (which expands access). However, despite these admittedly serious problems, there still existed elements of fairness in the distribution of demining resources – at least lip service to the allocation by need, not ability to pay.[83] Moreover, the completion of the Afghanistan Landmine Impact Survey (LIS) in late 2005, which categorized communities according to the impact of landmines on public safety and socio-economic development, has allowed for more precise targeting of priorities.[84]

Likewise, in Bosnia, NPA invested significant effort into ensuring that its demining priorities were driven by needs-based concerns. NPA focused on areas where there was both high concentration of mines and high impact upon dense human populations. Hence their efforts concentrated on Sarajevo, Mostar and Brcko. For each individual demining task, NPA carried out a 'Task Impact Assessment,' a simple socio-economic assessment of the impact of the minefield on the surrounding community. This aimed to ensure that the suspected hazardous area posed a genuine threat to people and/or their livelihoods, and that the task was not prioritized simply by vested interests. NPA then followed up post-clearance to evaluate how cleared land was used, in order to better target future efforts on land that would be used productively. Recently, NPA have piloted a new methodology of mine action planning developed by UNICEF and BHMAC called 'Task Assessment and Planning' that helps local communities map their mine problems and prioritize the responses to the problem. This community-based planning takes the views of local people seriously, rather than directing priorities from the central government.

NPA did occasionally do tasks prioritized for reasons other than simply need. However, these were generally intended to further a political agenda of cosmopolitanism and transitional justice. For instance, one of NPA's first projects in 1996 brought a team of its mine dog handlers from Mozambique to provide demining support to forensic scientists opening mass graves for the international war crimes tribunal at The Hague.[85] It also dedicated considerable resources to the complex and difficult process of demining the Sarajevo Jewish cemetery, the second oldest in Europe.[86] The Norwegian Ambassador explained that they wanted to send a 'political and symbolic message' that Sarajevo was an 'interethnic city.'[87] In addition, NPA has integrated its work in Bosnia into broader campaigning in favour of the landmine ban.

In Sudan, international NGOs funded by bilateral contributions have tended to pay better attention to need-based considerations in setting priorities. Landmine Action used its own survey capacity to determine where demining was most needed and would have the most impact on local communities. MAG was implementing the Sudanese Landmine Impact Survey and was targeting its operations according to their findings. They also worked with an NGO consortium in Blue Nile State, trying to integrate mine action into broader community development efforts. NPA used the same Task Impact Assessment methodology as in Bosnia, identifying 'land being denied for agriculture or housing or other local cultural needs.'[88] A DanChurchAid employee argued that being a development organization, with a broad range of expertise in interacting with communities, DanChurchAid understood the local context better than commercial companies: it had 'the capacity to think 360 degrees, to see all the problems related to the population.'[89]

The Human Security-Civil Society Complex was also able to contribute to more traditional security stabilization efforts. For instance, both Handicap International and the HALO Trust (with funds from both the US State Department and Norway) engaged in 'humanitarian disarmament' in Afghanistan – securing and destroying abandoned arms caches.[90] Likewise, NPA's demining of major areas of Sarajevo helped make it a viable capital city again, perhaps contributing to stabilization of the city and the re-establishment of government. In Sudan, Landmine Action and DanChurchAid played important roles in facilitating contacts between the government and SPLA in the Nuba Mountains, as well as the leadership level. This undoubtedly contributed, at least in a small way, to the success of the Nuba Mountains ceasefire. Moreover, several mine action NGOs in Sudan were securing and destroying abandoned caches of mines, ordnance and weapons.[91] This may have helped reduce arms available to armed groups.

In short, the 'Human Security-Civil Society' approach set up systems to distribute protection from landmines according to criteria based on need and fairness. This resulted from commitments to humanitarianism and justice, but, as the next sub-section shows, was also enabled by their independence from vested interests, both of the donor and the local political economy.

Local Capacity Building

Just as with commercial companies, the international NGOs also have some disincentives to build local capacity. NGOs have material interests in institutional survival that may disincentivize creating local competitors and expatriates may be reluctant to 'work themselves out of a job.' Indeed, in both Bosnia and Sudan, international NGOs appeared to have an institutional reluctance to hand over operations to local people. There are also political disincentives to local capacity building. As noted previously, the local people best placed to be demining entrepreneurs are often those linked to the political economy of war. Therefore, while Norway trumpets the importance of 'national ownership' in mine action, in the each of case studies it opted to support international agencies and was nervous about directly funding organizations embedded in the local political economy. Therefore, there was an implicit concern about building the capacity of the 'wrong' local institutions.

That said, in all three cases, Norwegian-supported international NGOs had a strong commitment to 'capacity build from within.'[92] Many NGO professionals argued this was the best way to organize demining programs; blending both the local and international enabled them to bridge both worlds. Hiring local staff saved money, eased navigation of local political complexities and simplified negotiation with vendors. External ties enable NGOs to diversify their funding base internationally, gain from economies of scale and professionalize staff through training, mentoring and development based on experience the organization gained outside the country.

Examples of this model include the HALO Trust in Afghanistan and NPA in Bosnia and Sudan. In all three cases, while ultimately controlled by their headquarters, local offices had considerable local autonomy. Further, while there was a large expatriate presence at the beginning, they slowly trained up local staff to take over management functions. HALO in Afghanistan is run largely by Afghan managers. Expatriates only rotate in on short term contracts to build capacity on specific issues. In Bosnia, NPA's last expatriate position – that of the country program manager – was handed over to a local staff member in 2006. In Sudan, NPA has developed a detailed capacity building plan and intends to draw down expatriates staff by two-thirds in five years as 'talented local candidates [are] educated and promoted.' The plan outlined training courses and mentoring relationships to be offered to local staff in both technical and management issues.[93] Beyond core demining skills, NPA has trained staff in other skills like management and cross-cultural communication.[94] Many commercial operators would not have the time or budget to invest in these 'softer' skills.

That said, most of the other international NGOs in Sudan have worked in partnership with local partner NGOs, largely, it seems, due to pressure from donors and local authorities. This seems to have had less success, perhaps because little capacity and responsibility were really transferred to the local NGOs. Most meaningful decisions in these partnerships continued to be made by expatriates within the international NGOs and really the only authority left to the local NGOs was to stall and obstruct decisions. Paradoxically, it appeared that the NGO programs that did not have local partners often gave local people more authority.

Therefore, given space by long-term grants, implementing partners operating in a Human Security-Civil Society Complex were not as forced by expediency concerns. This gave them time to develop alternative sources of power, internally within the organization, and through alliances with civil society, both globally and locally. As a South Sudanese government official expressed, NGOs 'are not limited by a contract' and 'can stay in Sudan for maybe six or eight years.'[95] Norway consistently supported organizations that, despite initially hiring many expatriates, and continuing to maintain international links, developed into strong, healthy and sustainable local institutions that were at least somewhat independent of the state and the local political economy of war. As a result, the following subsections will show that the Human Security-Civil Society Complex was able to build alternative structures that were more inclusive than many local organizations, could avoid co-option by vested interests and advocated humanitarian limits on war.

Inclusivity

War, by its very nature, is exclusive, for violence excludes the voice of the violated. In every conflict, certain people are excluded from decision-making, access to resources and protection from armed groups. These may be those considered 'the enemy', civilians victimized by armed factions or disempowered groups (such as women, children, minority groups and refugees). Marginalization can be institutionalized and continue in the post-war era, stoking grievances and contributing to further social unrest. It can also be reproduced in the agencies involved in reconstruction, including demining. However, the international NGOs operating in the three cases made more effort than commercial companies to create inclusive workplaces.

In Afghanistan, the UN and USAID-created demining NGOs first recruited refugees from the camps in northern Pakistan. These camps were largely populated by people of Pashtun ethnicity and mujahideen supporters. In the years after the fall of the communist regime, the various resistance groups began fighting each other. The demining agencies worked hard to span the mujahideen parties to avoid becoming partisan. The local demining NGO ATC had employees surrender their party ID cards, replacing them an organizational card and saying 'from today, you are not a party-member, you are an ATC member, you are neutral ... you have to follow the policy of the United Nations.'[96] Today demining continues to be dominated by Pashtuns and former mujahideen. This has also carried over into the commercial companies operating in Afghanistan. The HALO Trust is a key exception, however. Largely because its area of operations has traditionally been Kabul and the north, which are ethnically diverse regions, it has both a multiethnic leadership and workforce. Though it began in communist government territory, it now employs a mix of both former communist and former mujahideen supporters.

One sees similarities with NPA in Bosnia. While most of NPA's operations took place in the Federation, it has always focused on the areas of Bosnia which have the highest levels of mine contamination and on cities with multiethnic populations. NPA has worked on both sides of the inter-entity boundary line, claiming that 'Any effort of reconciliation...must include assistance to both entities.'[97] While most of its demining staff was drawn from people who fought on the Bosnian government side, many of whom are Bosniak, it has been more multiethnic than most other demining organizations, especially the local companies and NGOs. Its current country program manager is Croat. In 2000, it also incorporated a team from the Republika Srpska, transferred from a UNHCR demining project. It is thus one of the very few multiethnic Bosnian-led institutions in the country – no small achievement. In contrast, the US played into the 'Dayton logic' of ethnic parity by supporting the creation of ethnically exclusive demining organizations.

In Sudan, the international NGOs, especially DanChurchAid (supported by both Norway and the US State Department) and Landmine Action (supported briefly by the State Department), were more likely to try to build links between deminers from both sides of the conflict than the commercial companies. Landmine Action initially tried to transcend the conflict by building a politically independent local capacity. However, this was blocked by both the Government of Sudan and the SPLA and eventually became polarized and collapsed. HALO also tried to span the north and south but was similarly blocked. In contrast, the commercial companies have avoided 'peacebuilding' activities, considering them outside the purview of their contracts. That said, some NGOs also focused exclusively on the south, including DDG and the SPLA-friendly NPA. Moreover, while MAG (supported by the State Department) had partners both in the north and the south, there was little attempt to bring them into a unified whole. They were, in effect, two regional subcontractors to MAG. It may be that because the international NGOs were less able to build institutions autonomous from the political economy of conflict in Sudan, they were less able than in Afghanistan and Bosnia to resist countervailing exclusive forces.

However, both NPA and MAG have displayed another kind of inclusiveness, by employing of women deminers in South Sudan. NPA has also deliberately tried to achieve more gender balance in back office positions like finance and logistics.[98] Women have often been excluded in South Sudan and have suffered the effects of the war disproportionately. The SPLA, now the ruling group in the South, is an overwhelmingly patriarchal and militarized institution. By including women in a male-dominated process of restoring security, NPA and MAG are contributing to the development of a more gender inclusive society. The deputy director of the South Sudan Demining Commission said she was 'impressed' to see that 'Ladies were leading the demining process.'[99]

In all three cases, then, with some small exceptions, it appears that international demining NGOs, often those supported by Norway, at least attempted to develop more integrative polities than international companies, local companies and local NGOs. This is probably because international NGOs are themselves often cosmopolitan entities with staff spanning many different countries and cultures. Moreover, they tend to be shaped by cosmopolitan ideals and see the value of contributing to the development of an inclusive polity in the host country. Having a diverse workforce may mean that the NGO is less able to count on the loyalty and protection of particular local groups. However, it also gives them greater independence from local social divisions, and may inoculate them to 'capture' by the political economy of war. This enables them to resist violent structures and create alternatives.

Advocacy for Non-Violent Politics and Limits on War

International NGOs, more than commercial companies, attempted to overcome and occasionally transform the politics of violence encountered in the field. Their ability to do this was strengthened by their longer-term time horizons, a concern for human rights and the building of institutions autonomous from both the local political economy of war and 'belligerent donors' (with interests in the conflict). In Afghanistan, in the early 1990s, local demining NGOs, despite largely being staffed by former mujahideen, quickly learned to mitigate potential conflict by cultivating an image of neutrality and dialogue with all factions. Likewise, the HALO Trust's neutrality enabled it to continue work in Kabul and the north, despite the fall of the communist government. Thus, while dialoging with both the Taliban and Northern Alliance, the NGOs avoided being drawn into the conflict. HALO has continued this policy, trying to clearly distinguish themselves from the NATO military effort. They have especially resisted requests to demine military bases or lend equipment to commercial companies doing military tasks.[100]

The local demining NGOs and the HALO Trust also developed an ideology of demining as 'nonviolent jihad' that earned it widespread legitimacy, including with the Taliban and Northern Alliance. Drawing on a passage from the Qu'ran that states 'if any one saved a life, it would be as if he saved the life of the whole people,'[101] demining was framed as an extension of the struggle for liberation from Soviet occupation, an act of service to the Afghan people and a struggle against 'the enemy of everyone.'[102] This passage was quoted by most of the Afghan UN and NGO demining personnel interviewed by the author. It was also often displayed on banners in NGO offices and at meetings. Deminers killed in demining accidents were considered martyrs. UNMACA Chief of Operations Muhammed Sediq said that while many Afghans became disillusioned with the mujahideen when they began fighting each other, Afghans called the deminers 'the real mujahideen' because their work benefited everyone, and unlike the armed factions, they have stayed connected to local communities.[103] This is an incredibly powerful idea, which largely protected the demining agencies during the 1990s. It also represents a counter-hegemonic discourse, posing an alternative form of service, masculinity and sacrifice than the violent discourses of the armed factions. It also contrasts with the image of masculinity projected by the private security companies, in which the male role is defined by their ability to wield violence.[104] The NGO deminers projected a masculinity of nonviolent protection and of restoring the earth to safety.

In Bosnia, while NPA admitted that at times it had had to live with 'tolerable levels of corruption' within the organization,[105] it was able to avoid the transformation of demining into a large politicized patronage machine. Given time by their donor to develop slowly, NPA was able to do this by building a multiethnic institution that maintained an arm's length from the cutthroat ethnic politics of the country. Likewise, in Sudan, many of the international demining NGOs framed their work in terms of peacebuilding. They attempted, with admittedly limited success, to build links between deminers from different factions and strengthen transformative and demilitarizing processes within the country. MAG and several other NGOs insisted on a strict 'No Guns' policy in their vehicles, even when travelling with SPLA liaison officers. Moreover, NGO mine action has contributed to the return of refugees and internally displaced persons. Both MAG and DDG (which is funded by UNHCR) have focused their efforts on clearing area that would have the maximum impact on returning refugees and displaced persons.

Among the NGOs, Landmine Action made the most explicit connection between demining and peacebuilding. Their 'crosslines' work in the Nuba Mountains attempted to build links between deminers from both government and SPLA backgrounds. Because Landmine Action framed the project as an attempt to work outside the polarized context of the war, the government and SPLA naturally viewed it suspiciously. The northern government blocked the registration of a local NGO affiliated with the project and tried to get Landmine Action to work with the government-affiliated JASMAR instead. When Landmine Action left Sudan, intending to hand over its project to HALO, HALO was blocked from registering in the north. The SPLA also initially blocked Landmine Action's attempts to register a local NGO in the south, though finally allowed it to create Sudan Landmine Response (SLR). While this initially was viewed as a success, it actually meant that powerful interests within the South, well-connected to the SPLA, were able to capture the organization. When HALO took over as SLR's partner, it discovered what it alleged were massive misallocations of resources and nepotism, protected by powerful men in the SPLA.

This shows that while international NGOs were more motivated to transform the political economy of conflict within Sudan, their capacity to actually do so was fairly limited. In the absence of a broader political transformation, the ability of NGOs to contribute to peace is obviously limited – they face powerful countercurrents against their liberal visions.[106] Indeed, NPA has made no attempt to build links between north and south, preferring to focus its attention on the south and remain politically close to the SPLA. Moreover, the international NGOs working with local NGOs, such as JASMAR and OSIL (which are more like parastatals linked to the northern government and SPLA respectively), were rarely able to integrate their northern and southern programs into a meaningful whole. So while MAG and DCA work 'crosslines', their local partners rarely do. Furthermore, the structure of these partnerships probably encouraged rent-seeking. While the international NGO provided funding and had some influence over staffing within the local NGO, the local partner still maintained better relations with local authorities. This enabled local NGOs them to influence the outcomes of disputes between them and their international partners. This meant the international NGOs had less control over staffing and resource allocation than if they had hired their own deminers.

That said, the work of Landmine Action and DCA in the Nuba Mountains have been widely praised for the humble, but not insignificant impact they had on facilitating links between the government and SPLA between the Nuba Mountains Ceasefire and the CPA. Mine action analysts Rebecca Roberts and Mads Frilander said, 'Professional working relationships have been developed at the national, intermediate and local levels, and in some cases personal friendships have followed.'[107] Even Sudanese who later clashed with Landmine Action admit that involvement in the Nuba Mountains crosslines work made them more trustful of their counterparts when negotiating the mine action sections of the CPA.

Another way international NGOs contributed to the demilitarization of politics was through advocating humanitarian limits to technologies of violence. Many international demining NGOs participated in the mine and cluster munitions ban movements. In contrast, commercial companies in all three cases limited lobbying efforts to advocating further funding and facilitation of operations. For example, many international NGOs have been critical of the US refusal to join the landmine and cluster munitions bans, whereas commercial companies' public relations efforts highlighted their support for the US 'War on Terror.'[108]

In Afghanistan, while the local NGOs were more like UN contractors than traditional notions of an independent 'civil society,' they also played an advocacy role – highlighting humanitarian issues and recommending policy. The Afghan NGOs together formed the Afghan Campaign to Ban Landmines (ACBL), which successfully pushed Afghanistan to sign the Mine Ban Treaty and has raised objections to US mine ban policy.[109] Commercial companies are far less likely to question US policy in Afghanistan and have much less organizational autonomy from the US government.[110] When the author asked RONCO's president to comment on their programs in Afghanistan he said, 'I suggest you talk to the State Department because I don't want to say something that might jeopardize the position of the company.'[111] Likewise, DynCorp's mine action coordinator told the author, 'State Department is paying DynCorp to represent them on the WRA [State Department Office for Weapons Removal and Abatement] project so our logo says WRA on it, our letterhead says WRA and we are listed here in country as WRA and we are accredited as WRA Afghanistan.'[112] Thus commercialization of mine action could function as a way to neutralize advocacy and opposition from civil society actors involved in mine action.

In Bosnia, NPA and other mine action NGOs have leveraged their field experience to assist in global advocacy efforts. With the US NGO Landmine Survivors Network, NPA helped to organize Princess Diana's visit to Bosnia (just two weeks before she died in August 1997) publicizing landmine issues. Photographs from this visit played a significant role in the ban campaign. More recently, NPA lobbied the Bosnian government to support the Norwegian cluster munitions ban initiative.[113] The US-supported commercial companies and local NGOs have been almost invisible in these campaigns, focusing solely on their operations.

Finally, mine action NGOs may have had small but important effects on the behavior of the Sudanese government's military and possibly the SPLA. The significance of the Sudan Campaign to Ban Landmines' (made up of local northern Sudanese NGOs) successful lobbying of the Sudanese government to sign and honor the landmine ban should also not be downplayed. The Sudanese government has a history of suspicion toward NGOs and the fact that SCBL persuaded the government to cede the right to use a weapon and allow non-state actors to engage in gathering information about a security threat (when NGOs assess the mine situation), should be considered a small victory for civil society. As one evaluation of SCBL noted, 'SCBL managed to transform the image of…landmines from a military and security issue to its humanitarian dimension.'[114] Indeed, exposure to mine action and a local political constituency committed to the landmine ban may have been why the Sudanese government has largely avoided the use of mines in Darfur.

Conclusion

Instead of re-establishing and strengthening the power structures most associated with the politics of violence, those interested in peacebuilding must seek to resist them, establishing in their place institutions that replace relationships of coercion with those of trust and nonviolence. A demining organization's ability to play a positive role in this process is related to its motivations, interests and the nature of the networks of power relations in which it is embedded. An organization's capacity to act is both enabled and circumscribed by its relationships to other groups in society. When it is deeply embedded in the networks involved in the conflict, it is unable to resist or transform them into something new.

Therefore, implementing agencies operating in a Strategic-Commercial mode are less likely to have a positive impact on the political processes of peacebuilding, though they may contribute to the donor's strategic interests. This is because their donor may have interests in the conflict (such as in the US in Afghanistan). Even when this is not the case, a modus operandi of expedience and pressure to achieve results quickly, creates strong incentives to cooperate with those currently in power, who can 'get the job done.' Though it may contribute to the strengthening of state security organs, this means commercial demining is also more likely to strength the fragmentation and privatization of security, granted only to the few with political and economic power. As a result, they may generate unexpected blowback against the very strategic interests the donor aimed to protect.

In contrast, the humanitarian mission, lack of donor strategic interest and the flexibility of funding of implementers operating in a Human Security-Complex gives them time and opportunity to build new structures that are inclusive and not captive to the status quo. This allows them to resist the politics of violence, advocate for limits on the technologies of war and set up systems that distribute protection according to need.

The complexity of real world examples displayed in this chapter indicate that implementing agencies may fall somewhere on the continuum between these two poles. However, it appears that the closer they came to the ideal of the Human Security-Civil Society Complex, the more effective they were at having a positive impact on the long-term peace and reconstruction processes.

CONCLUSION AND REFLECTIONS

> *'To live in a secure world, a new partnership for human security must be forged.... Governments and civil society must work together to advance human security as a viable alternative to militarism and war.'*
> – Jody Williams.[1]

Political life in the twenty-first century is marked by a shift from nationalized relationships of government command-and-control to globalized governance by coalitions, partnerships and contractual relationships between diverse sets of public and private actors. The highly structured bureaucracy has been augmented and/or replaced by hollowed-out, outsourced and privatized networks. As a result, the statist manner in which scholars traditionally conceived security is inadequate for an era of globalization and transnational security threats. In fact, it is unlikely that international politics was ever as straightforward as a simple game of utility-maximizing states. To correct this misguided assumption, the study of politics, as Mark Duffield has argued, must be rooted in an examination of how the messy mix of state and non-state actors organize themselves into networks to address security challenges. Nowhere is this more clearly demonstrated than in the conflicted 'frontiers' of the international system. Countries such as Afghanistan, Bosnia and Sudan are ruled not by the classic Weberian state, but rather by complexes – dizzying arrays of multilateral agencies, bilateral donors, foreign troops, NGOs, commercial companies, warlords, armed factions and criminal networks. These actors come together in shifting relationships of conflict, collusion and compromise.

However, to describe these networks as complexes does not mean that one cannot discern patterns and logic to the complexity. To do this, one must get in close – examine the complexes at the macro and micro level, understanding the motivations, institutional structures and actions of each of the actors and the manner in which they interact. This book has attempted this task by looking in detail at one particular sector of international intervention in conflicted regions – the mitigation and neutralization of the threat of landmines and other explosive remnants of war.

While Mark Duffield has a tendency to conflate these new networks into one 'emerging system of global liberal governance,'[2] this can actually prevent understanding. There is not one emerging system, but many, driven together into constellations by shared interests or norms. The author's research discovered that there were actually at least two forms of complexes governing the clearance of mines and UXO, incorporating different sets of actors, operating in different ways and producing different outcomes, both in terms of the demining process and the impact on the surrounding context:

1. In **Strategic-Commercial Complexes**, mines and cluster munitions were seen as legitimate weapons in a great power arsenal. Mine clearance was shaped by the securitized foreign policy of a governmental donor with a high level of strategic interest in the region and implemented largely by private security companies and military units. The complex was held together by a sense of national interest and the profit motive of contracted agents. It prioritized demining tasks that furthered donor's particularist interests and emphasized speed and cost-efficiency over quality and safety. Its tendency toward expediency meant that it was more likely to compromise, even collude, with the political economy of conflict. As it was a network that included private security companies, it may also have contributed to the privatization and fragmentation of the public monopoly on the use of force.

2. In **Human Security-Civil Society Complexes**, middle powers saw mines and UXO as a threat to the global public good that should be regulated, proscribed and neutralized. Clearance programs were shaped by notions of humanitarianism and implemented in partnership with UN aid agencies and international NGOs. The complex was held together by shared norms, missions and values. It prioritized demining tasks with high human impact (in terms of causalities, refugee return and livelihood opportunities) and emphasized targeting, quality and safety over speed and cost-efficiency. Due to its longer-term time horizons and normative commitments, it tried to transform the political economy of conflict by building inclusive institutions and advocating for limits on the politics of violence.

These main findings are summarized in the table below (Table 12). This concluding chapter reflects broadly on the roles of ideals in mine action policy, provides recommendations regarding the implications of this research for mine action policy and possible directions for future research. Finally, it closes with a few reflections on security in a post-statist world.

Table 12: Two 'Emerging Political Complexes' of Demining: Summary of Key Findings

	Strategic-Commercial Complex	Human Security-Civil Society Complex
Example Donor	- United States	- Norway
Understanding of Interest	- National interest and the profit motive	- Global public good
Implementing Agencies	- Commercial companies - Private security companies - Military demining units	- International NGOs - Local NGOs - UN Humanitarian Agencies
Governance of Internal Relationships	- Principal-Agent Model	- Principal-Steward Model
Attitude to International Laws on Mines and UXO	- Prefer as few restrictions as possible	- Campaign for strict regulation
Price of Demining	- Low	- High
Speed of Demining	- High	- Low
Complexity of Demining Tasks	- Low	- High
Quality of Demining	- Low	- High
Safety of Demining	- Low	- High
Priority Tasks	- Military Bases - Insecure Weapons Caches - Access Roads Important to Military - Large-Scale Infrastructure Projects	- Community-Based Priorities - Areas of High Human Impact - Focus on Marginalized Populations
Impact on Peacebuilding	- Strengthens political economy of war - Contributes to privatization of force	- Builds links across social cleavages - Builds institutions independent of the political economy of war - Advocates for limits on the politics of violence

Agency and Structure: The Roles of Interests, Ideals and Institutions

Some of this book's dominant underlying themes have been the ways in which interests, ideals and institutions interact to create systems of demining and impact the surrounding context. Motivations (interests and ideals) make a considerable difference. They shape the way in which agents interacts with others around them and order their internal organizational incentives and structures. Nevertheless, the systems and institutions that structure and guide the behavior of these agents are also of crucial importance. Institutions (such as negotiating fora or donor contracts) shape the extent to which agents are incentivized to act according to ideals, self-interest or the public good. Therefore, this study has found that both agency and structure matter in the creation of effective demining programs. Further, agents actively engage with, shape and change structures, in order to constrain and influence the behavior of other agents in the system (such as the shifting the locus of negotiations from the CCW to the Ottawa Process).

At the global level, it has been norm-motivated entrepreneurs (such as NGOs, churches and middle powers) not the more strategically-driven great powers or profit-driven companies that have driven the development of a global regime governing mines, cluster munitions and other explosive detritus of war. However, the structure of negotiating fora used by states to develop regimes governing the use of mines and cluster munitions seem to matter as much as the motivations of the entities themselves. The Convention on Conventional Weapons, with its consensus process, effectively encouraged states to adopt the lowest common denominator of the most self-interested state. By contrast, the 'opt-in' system of the Ottawa and Oslo processes encouraged states to take maximalist positions; the inclusion of NGOs in the discussions further pushed them to do so, in the face of potential embarrassment. Nevertheless, these new structures have been intentionally created through the active collaboration of agents who felt too constrained by the traditional arms control game.

At the donor level, Norway has an idealist foreign policy, but as chapter four showed, this may be as much about its place in the international system as its internal motivations. While it has consistently framed its foreign policy in terms of normative commitments, as a small state, Norway may benefit from ensuring that great powers (and other agents threatening their vital interests) play by the rules of law-governed world. Thus its advocacy for humanitarianism and international law may be more about state survival than high-minded ideals. Likewise, there is no shortage in the US of idealist churches, NGOs, peace activists and concerned citizens. However, as Egeland has argued, the US government is constrained by transnational strategic and economic interests, which often prevent it from taking the moral high ground. It is likely that in order to change US mine action policy, it will require effort by idealists to shift and reform the structures of US foreign policymaking, in order to make such structures more amenable to idealists' concerns.

At the level of implementation, international NGOs were more likely than commercial companies to be expected (by donors, employees and the public) to live up to humanitarian ideals and a concern for the global public good. This pushed them to adopt high expectations of quality, safety and conduct. These motives differed across different agencies. HALO Trust's conception of demining as an act of charity is a more traditional view of humanitarianism than NPA's left-leaning conception of demining as an act of solidarity with the oppressed. Both of these secular institutions diverge somewhat from DanChurchAid's religiously driven understanding of Christian service. However, most of these international NGOs did hold key ideals in common. Firstly, they were driven by considerations of justice – that is, demining should be distributed by need, rather than ability to pay. Secondly, they seemed to have a stronger professional ethic than the commercial companies. These humanitarian intentions did seem to have a positive impact on demining performance. They increased the willingness of agencies to spend money on better targeting of resources and increased quality and safety. Moreover, many NGOs were motivated by a vision of transforming the conflict – thus they were nervous about interacting with violent socio-political structures and tried to build institutions that bridged or transformed social divides and modes of exclusion. Since these objectives were also integral to Norway's foreign policy, NGOs were given freedom to pursue aims that were not purely about demining.

However, these humanitarian motivations did not exist in an institutional vacuum. The long-term granting mechanisms employed by Norway, with little competition for funding and multi-year funding almost guaranteed, allowed international NGOs to operate with longer-term time horizons. While these NGOs obviously had some material interests (competing for additional funding, ensuring decent salaries for staff, etc.), their core grants from Norway allowed them to operate outside the ordinary constraints of a material economy – decisions could be made based on what was considered the 'right thing' rather than the cheapest option. Therefore, there was less need to sacrifice idealistic motivations. Freed from the need to make a profit, NGOs could sacrifice some elements of cost-effectiveness in exchange for taking time to ensure work was well targeted, of top quality and met high safety standards. In other words, the institutional structure set up by donors in negotiation with implementers opens or closes space for a demining agency to act more morally. It appears that granting demining funding to international NGOs (as Norway usually did) gave them more ethical space than commercial tendering (preferred by the US). Whether or not the agency uses this space productively or squanders it, will depend on the internal drivers of the particular organization.

In contrast, the implementing partners in the Strategic-Commercial Complex were constrained by competition, the strategic interests of their donors and the short time horizons of their contracts. Thus implementers' only interest in influencing donors would be to encourage them to make the process more profitable. Likewise, the discipline of competition and profit-seeking mean that quality is less important than cost-effectiveness – especially at the margin of the market. With regard to the surrounding socio-political context, attempting to engage in transformative activities would be seen as a distraction from the priority of completing the terms of the contract. Therefore, accepting and working with the socio-political status quo is a natural outcome. If the contract is for demining, there is no money available for peacemaking, extra staff training or advocacy. As a result, even when the donors that tend to operate in a Strategic-Commercial mode give money to NGOs, they rarely give them the same kind of room to maneuver as the grants from Norway. For instance, the US grants to NGOs in Sudan were on an annual basis and NGOs competed with each other for funding. Likewise, UN tenders awarded to NGOs in Sudan followed the same contracts as those with commercial companies. This competition for resources left little room for broader idealist visions, and implementers often simply transmitted the strategic interests of their donors.

In short, in protecting populations from mines and UXO, the idealistic or self-interested motivations of organizations matter. Indeed, this research has found that cosmopolitan and humanistic ideals can play crucial social and political functions. At least for beneficiaries, mine action, post-conflict reconstruction and foreign aid are better instruments of social, political and economic change if they are shaped and driven by a commitment to greater human welfare. However, the institutional structures in which organizations are embedded also shape the way they operate, though agents within the system are able to shape these structures through negotiation and exercise of power. Therefore, funding for demining should be carefully organized to bring out the best in organizations, rather than forcing them into a prisoner's dilemma where they concentrate more on competing with each other than helping people. While self-interest – the protection of life and livelihood – is legitimate and can be a great engine of creativity and prosperity, it must be constrained and channeled by legitimate norms.

Toward Mine Action with a Human Face

After a decade of neoliberal 'structural adjustment' programs enforced by the IMF and World Bank in the name of development, UNICEF released a report in 1987 calling for 'Adjustment with a Human Face.'[3] The authors argued that development had lost its humanistic vision and failed to focus on the most vulnerable. Instead, structural adjustment had myopically focused on a small set of economic indicators (such as GDP growth, prices, currency reserves and government spending) that did not adequately measure the state of wellbeing in a country. Human welfare, they believed, had to be at the center of any attempt to transform and develop a country.[4]

Mine action also needs to rediscover its 'human face', to remember, as UNMAS has said, that 'It is not so much about mines as it is about people....'[5] This study has shown that mine action is in danger of becoming distracted by commercial and strategic priorities. This is especially true since the advent of the 'War on Terror' and the continuing trend of consolidating the commercial mine action sector into the private security sector. Mine action is becoming a commodity to buy or an activity in support of counterinsurgency. To counteract these trends, we need a 'doctrine' of mine action that puts human welfare at its center.

Ten years ago, a group of NGOs tried to develop a framework to guide mine action. The resulting document, 'Mine Action Programmes from a Development-oriented Point of View', sometimes referred to as 'The Bad Honnef Framework', argued that mine action must be guided by basic principles of participation, co-operation, coherence, sustainability and solidarity.[6] Unfortunately, its influence has been relatively limited; an online search of the *Journal of Mine Action* for the term 'Bad Honnef' produces only two articles, in comparison with 23 for 'national security' and 190 for 'commercial.'[7] 'Mine Action with a Human Face' must rediscover and reinvigorate this document, but also recognize its problems. While rooted in the civil society and participatory development schools of thought, it was somewhat technocratic and apolitical. It lacked a clear recognition of the political economy of conflict with which demining inevitably must interact.

This research implies a return to 'Mine Action with a Human Face', building on the 'Bad Honnef Framework' and guided by the following six principles:

1. Doing no harm
2. Moving beyond technocracy
3. Protecting the vulnerable
4. Participation
5. Stewardship
6. Building peace

The following will outline each of these principles in more detail and explain how they relate to mine action.

1. Doing No Harm

This book, confirming the findings of the political economy of aid in conflict literature, has shown how demining interacts with the politics of the post-war context. Too often, demining agencies have been unaware or unperturbed by way their programs are captured by local vested interests or strengthen networks that are opposed to peace. Demining agencies must become more aware of how their funding, priorities, recruitment, information gathering and management practices can exacerbate conflict or contribute to peace.[8] This will require demining agencies to recruit people from outside its traditional ex-military circles – development professionals, political scientists, anthropologists, sociologists, management specialists and economists – who are able to analyze social and organizational issues from different perspectives.

Particularly important are considerations of whether to accept money from 'belligerent donors' (donors that are involved in the conflict), whether to use demining units of the military forces which were responsible for the conflict, how to vet potential deminers who have links to war crimes and atrocities and how to prevent the fragmentation of the public monopoly of force when using private security companies. While this leads to some difficult choices, doing no harm must always be the guiding aspiration.

This also applies to more technical matters. Demining agencies must put safety first in their clearance efforts, ensuring their deminers are well-trained, managed and equipped to do their dangerous work. They must also open channels of communication to the local communities surrounding their task to raise awareness of the mine problem and explain what is and what is not being cleared. Donors have a responsibility to ensure that their contracting systems do not incentivize corner-cutting and a 'race to the bottom.' Stable and long-term grants to humanitarian and professional institutions encourage investment in safety and quality.

2. Moving beyond Technocracy

Much of the literature on mine action, especially in organizational and policy documents, tends to be very technocratic, stripping demining of the political, economic, social and cultural layers that are so crucial to understanding how it really works. While this study has not really addressed this issue directly, it has attempted to move mine action discourse away from technocracy. For while obviously necessary, the knowledge of how to defuse a landmine, run a minefield database or manage a team of deminers is not sufficient to navigate the complex social milieu of a post-war country. Indeed, my research has confirmed Timothy Donais' findings, that

> the application of technical solutions to what are inherently political problems risks being futile at best and, at worst, reinforcing and perpetuating the very conditions that produced the conflict in the first place.[9]

Minefields are political, economic and social problems, not only technical ones. They thus require political, economic and social solutions. This means mine action agencies and scholars must dedicate more resources to researching the ways in which their work interacts with the political economy of conflict and the social and cultural tensions that may hinder their work.

3. Protecting the Vulnerable

The Bad Honnef Framework declares that 'The needs and aspirations of people affected by mines are the starting point for mine action programmes.'[10] However, as this book has shown, commercialization of clearance can erode this humanitarian principle, allocating demining according to ability to pay, rather than need. Moreover, the integration of demining into the strategic objectives of counterinsurgency campaigns (such as in Afghanistan and Iraq) further corrodes mine action's 'humanitarian space.' Putting military concerns first also allows countries to claim the right to use mines and cluster munitions to devastating effect upon civilians (as seen in Lebanon in 2006 and Georgia in 2008).

Reconceiving security as the protection of those most vulnerable, rather than those most privileged by nation or class, mine action must focus on securing the lives and livelihoods of those most affected by mine and UXO contamination. At the same time, mine action agencies, the UN and progressive states must continue to strengthen and deepen international norms protecting non-combatants from both the short and long term effects of mines and cluster munitions.

4. Participation

Protecting the vulnerable cannot be an exercise in paternalism, for 'As much as any human being, mine affected people and communities have the right to shape their own lives and to participate in political and economic decision making which concerns their interests.'[11] Therefore, 'participation' must be a guiding principle of mine action programs. This means demining agencies should use community-based mapping and priority-setting methods (such as those successfully pioneered in Bosnia) employ local people as deminers and build local capacities to manage the mine and UXO problem.

5. Stewardship

This research project has demonstrated how the 'principal-agent' model of competitive tendering does not always produce the results it promises and is clearly not the panacea that it is sometimes touted to be by the US, commercial companies and some staff within UNMAS and UNOPS. Instead of increasing quality, there is a danger that it will encourage a 'race to the bottom', incentivizing corner-cutting, poor treatment of labor and unsafe practices. Some may argue, as Banks does, that in order to save the lives of the public, donors must sacrifice some of the safety and quality of the demining process in return for volume and that the best way to do this is a commercial tendering process.[12] Though this argument is seductive, it is dangerously so. Firstly, poor quality demining may result in the need to revisit and re-clear tasks done by agencies that have failed to meet adequate quality standards. Demining relies heavily on trust in the process – people who distrust the quality of clearance they will act as if a demined area is still mined.[13] Thus money wasted on shoddy clearance that has to be repeated or is never trusted by the potential users of the land may not be accounted for in price considerations. Secondly, in the politically fragmented and lawless context of a conflict or post-conflict country it may be very difficult to enforce the kind of regulation necessary to prevent a race to the bottom. Finally, many of the international NGOs pay more attention to targeting their demining on areas that have the highest human impact or make the most difference to marginalized populations. Therefore, while the commercial companies may be clearing a higher volume, if their efforts are not carefully targeted by the donor, it may have less impact on reducing risk to the population.

Some have argued that the solution to these problems is better-worded contracts and more effective contract enforcement. Certainly these would not be bad things and might, to a point, lead to improvements in performance. However, the demining process, especially in a conflicted context has many different unknown variables. For example, it is impossible to know precisely how many mines are in a minefield prior to clearance, it is difficult to know the extent politics might influences one's work and one cannot predict precisely how much time is needed to safely clear an area. Therefore, there is a little chance that a contract could stipulate all things a contractor should and should not do. Moreover, it is difficult to know whether the contractor will follow the stipulations anyway, if they are working in a context of poor rule of law. Rather, what is needed are implementers that have a strong internal ethical sense of what is appropriate behavior, so they can be trusted to work in the lawless environments of post-conflict zones.

Therefore, in place of commercial tendering, mine action donors should be guided by a 'principal-steward' model, in which they build long-term partnerships with trusted organizations that show dedication, professionalism and resolve. The 'good intentions' of international NGOs, especially those supported by Norway, (and the necessity of maintaining a reputable brand among some companies) provide an additional mechanism regulating the risk of unethical behavior in the lawless environment of a post-conflict country. By granting long-term funding, donors, in effect, create a property right for the implementer, encouraging them to invest in high standards. Thus donors should encourage and enable their implementers to avoid cutting costs on training, equipment, oversight, insurance and health care for deminers. Only through investing in people and organizations can one expect them to perform to their full capacity.

6. Building Peace

Ultimately, mine action is about restoring confidence that people are safe from violent threats of harm in their daily lives. It is about removing barriers to safe freedom of movement and exchange. Thus, it is an integral part of a society's recovery from war. As the Bad Honnef Framework asserted, mine action programs should 'support peace-building including reconstruction and development of the community and aim at enhancing the socio-economic and cultural infrastructure.'[14] Mine action must work to encourage the return of displaced people, kickstart the legitimate economy and restore cultural symbols of integration such as museums, monuments and centers of learning.

Beyond this, however, mine action organizations should embody the peace envisioned. Espousing a multiethnic, law-governed society while supporting monoethnic demining agencies with links to organized crime, as some donors did in Bosnia, seems counterproductive. Mine action agencies should be inclusive, incorporating people across social divisions of ethnicity, class or political affiliation. It must involve people from non-military backgrounds, to encourage a diversity of perspectives. They should also try to resist patriarchal patterns that entrench violence, by employing women deminers where appropriate and offering alternative constructions of masculinity based on service and saving lives.

Moreover, through campaigning for regulations on the use of mines, cluster munitions and other technologies, mine action organizations should advocate for limits on the politics of violence. They should further embody non-violent politics by limiting links to private military contracting and military forces and, where possible, prevent employees from bearing arms.

A Future Research Agenda

The research in this book may raise as many questions as it answers. Chief among them is how transferable and portable the conclusions are to other settings. Therefore, future researchers may wish to examine the findings presented here in other contexts. For instance, it would be interesting to examine the nature and impact of the two governance complexes in other settings and sectors. It would also be helpful for researchers at other cases studies to determine the impact of different models of contracting on the demining process. Another possible investigatory path might be to question why certain lessons (such as about the problems with commercial contracting or the inadequacy of technocratic blinders) have not been learned by mine action donors. It would be worth asking whether there are institutional incentives that actively prevent such lessons being taken seriously.

An important direction worth following is the in-depth micro-level study of intervention in societies at conflict. Not enough is known about the political and social impact of enclavization, fortification and privatizing protection. The mine action sector in particular could benefit from further social scientific study of its operations and impact. Finally, foreign aid generally, and mine action in particular, constantly struggles to navigate the complexities of conflict. How can contractors be held accountable in rule of law's twilight? How much should one compromise with the powerful, yet illiberal clandestine structures that govern social, political and economic life? How can one keep aid workers or deminers safe from attack by unsympathetic militant factions? How can one build inclusive, non-violent institutions in the midst of violent societies? How can one build the capacity of the 'right' local institutions? All these questions are of vital importance when managing assistance in complex emergencies and post-conflict zones. Yet aid workers often lack adequate answers to them. Academics will not have all the answers either, but they can and should use their research abilities, analytical skills and knowledge to assist those who must make and implement aid policy.

Final Thoughts: Security in a Post-Statist World

In closing with final, more personal, reflections on this research, I ask the reader to indulge a reversion to first person. I write this ending section from an apartment in Nairobi, Kenya. It has bars on the windows and is surrounded by high walls. A private guard stands at the gate. I have spent much of this book pointing out the corrosive impact of such 'hard security' and privatized protection. However, living in a place where insecurity and the lack of adequate state protection is a stark reality, I have developed some sympathy and insight into the very real security dilemmas people face in their daily lives. Security can either be sought by fortifying private and privileged space or through building general good will through persuasion, good offices and improving the lot of the general population. Unfortunately the latter, while more desirable, appears very difficult in the short-term. Thus those who respond to insecurity with militarization and fortification are neither irrational nor necessarily mal-intentioned. However, there is a definite cost to this choice. Driving around Nairobi's up-scale Muthaiga neighborhood, one sees large homes with well-tended gardens, surrounded by razor wire, alarm systems and private guards. A short drive away, the people living in the slums of Kiberra, Mathare and Mukuru live lives exposed to far greater risk – police harassment, gang violence and widespread criminality. Privatized protection, while a rational response to insecurity, leaves out the vulnerable and no doubt increases resentment among those at the bottom of society for those at the top.

On the international scene, we have seen an analogous resurgence of these 'hard' approaches to security. Within the US, the trauma of 9/11 shocked the American people and displayed their vulnerability to attack. It provoked a belligerent and militant response from the US government, which has waged war in two countries (and conducted aggressive covert operations in many more) through a complex network of public and private actors. For many conservatives, the Clinton years had been marked by negligence in the face of the tragic reality of global instability and a misplaced faith in the ability of treaties and alliances to protect American democracy and prosperity. Only through military strength and fortifying borders (such as building a fence along the Rio Grande and limiting immigration) could America become safe again. The rise of US unilateralism during the Bush Administration seemed to provoke, or at least occurred in parallel with, a resurgence of great power geopolitics in other parts of the world. The Russian government's fear of NATO expansion and democratic 'revolutions' in several former Soviet states has prompted a comeback of militant Russian nationalism. The assassination of Litvinenko, use of natural gas as a political tool, power-plays in Kosovo and invasion of Georgia all point to Russia's renewed commitment to a realist foreign policy aimed at restoring its great power status. Similar trends can be observed in the foreign policies of China, India, Pakistan and Israel. Through deploying military and private actors (such as arms traders, oil firms, paramilitaries and private security firms) they are penetrating and governing the borderlands through projecting their 'hard' – military and economic – power.

Realists will tell us that such a reversion to great power militarism is entirely rational, perhaps inevitable, and a necessary result of the need for states to maintain national security. However, there are very real costs to this 'hard' militarist approach. In the international arena, the resurgence of great power unilateralism is creating a new multi-polar security dilemma, which arguably makes the global situation more insecure. The competition between the US (with NATO) and Russia may come at the expense of insecurity in smaller states, like Georgia and Kosovo, which become sites of proxy maneuvering.

Great power competition also reduces the possibility that great powers will one day give up mines and cluster munitions, as they feel the necessity of maintaining all military options – hence the Bush Administration's hardening of its line on mines. Militarism can also then creep into demining programs; in Afghanistan the US conceived of them as an element of counter-insurgency and contracted private security firms for implementation. Such militaristic-commercial approaches fail to provide protection from mines to those who need it the most and can result in sloppy work, fragmentation of use of force and a strengthening of the political economy of war.

While there can often be an overwhelming sense that securitization and private protection are the inevitable and tragic sole solutions to insecurity, this study has demonstrated that another option is possible, plausible and perhaps, in the long term, more able to sustain peace and reconstruction. There is then, perhaps, a 'realism' in the 'idealist' approach – a recognition that peace cannot be built simply through walling off the privileged and managing the 'rest' through counterinsurgency and privatized governance.

Peacebuilding requires the development of governance systems that provide protection to all people, particularly those most vulnerable. It requires the construction of institutions that provide inclusive and non-coercive alternatives to the political economy of war. And it requires resistance to the politics and discourses of violence and exclusion. My research has shown that a coalition of middle powers, UN humanitarian agencies, the ICRC and NGOs have managed to achieve such things in their attempts to govern, mitigate and neutralize the security threat of mines and unexploded ordnance. They have created new international norms, monitoring systems, clearance operations and awareness programs that have targeted efforts at those most vulnerable. They have constructed governance complexes in the global peripheries that are effectively reducing the threat of mines and UXO through humanitarianism, persuasion and international law. As Goose, Wareham and Williams argued, the Mine Ban Treaty

> proved that it is possible for small and medium size countries, acting in concert with civil society, to provide global leadership and achieve major diplomatic results, even in the face of opposition from bigger powers.[15]

Therefore, in spite of resurgent great power militarism, middle powers still have room in the international system to reduce insecurity through 'soft power.' The 'hard power' option is not really open to them – they have neither the money nor the troops to project military power into the insecure borderlands in any significant way. However, through joining forces with a network of multilateral and non-profit agencies, they have innovated means to project soft power, penetrate the global peripheries and govern insecurity in other ways. While the campaign against landmines is perhaps the most prominent of these efforts, these Human Security-Civil Society Complexes have also successfully innovated new norms regarding child soldiers, disability rights, the International Criminal Court and cluster munitions. While they have not been very successful in obtaining great power support for such new international treaties, by creating new customary norms, they have nonetheless been able to constrain the behavior of the great powers. Indeed, few countries now use mines. The US has had many opportunities to use them in Iraq and Afghanistan and has avoided doing so. Moreover, the US has felt obligated to provide significant support to clearance and mitigation programs.

Some might argue that while creating Human Security-Civil Society Complexes may be an appropriate course of action for middle powers, it is unrealistic to expect great powers to follow a similar approach. Indeed, they seem trapped in a security dilemma that requires them to project strength, maintain military dominance and avoid entanglements in limiting treaties. However, perhaps because of the idealist strain in its foreign policy and its democratic politics, this research has shown that the US can adopt a Human Security-Civil Society approach in places where its strategic interests are less pressing. In Afghanistan during the Taliban era, the US supported the UN-sponsored NGO program. In Sudan, following the Comprehensive Peace Agreement, the US provided grants to international NGOs.

Thus, while it may be too much to ask for the US to abandon militaristic approaches to security in Iraq and Afghanistan, it may be possible for the US to take a Human Security-Civil Society approach in times and places where the political space exists. This will give the US an opportunity to implement policies that do justice to the idealist strains in American thinking about foreign relations. It could also build a reserve of good will toward the US in the global peripheries and contribute to the governance of insecurity that is ungovernable simply through military tactics and counterinsurgency. I hope that the ascendency of the Obama Administration will increase the size of this political space to act according to the global public interest, not just the national interest.

Similarly, other, non-state, actors in insecure places – individuals, communities, NGOs, companies – may be driven by sheer necessity to barricade themselves behind walls, fortifying private space and withdrawing into privatized forms of governance, safe from the 'red zones' that surround them. However, we should not delude ourselves that this will ultimately create peace and security for entire populations. If peace is what we want, fortification and private protection can only ever be a temporary measure. When and where space opens for alternative approaches, we must seize the moment. We must come out of our enclaves, dismantle our walls, diffuse security out from the core into the peripheries, provide protection and assistance to the vulnerable, govern human conflict with the rule of law and dismantle the structures of violence that are ultimately the cause of human insecurity.

PHOTOGRAPHS

Russian PMN-2 Antipersonnel Mine, near Kabul Airport, Afghanistan.

Assorted Cluster Munitions, OMAR Mine Museum, Kabul, Afghanistan.

RONCO Deminers with Large-Loop Metal Detector, Kabul, Afghanistan.

Demining Dog Training, Mine Dog Center, Kabul, Afghanistan.

HALO Trust Deminer, Shomali Valley, Afghanistan

Norwegian People's Aid Demining Machine,
Brcko District, Bosnia and Herzegovina.

UNIPAK Demining Site, near Arizona Market,
Brcko District, Bosnia and Herzegovina

Norwegian People's Aid Deminer,
Outside Sarajevo, Bosnia and Herzegovina.

Sign by Roadside, Outskirts of Juba, Sudan.

Unrecovered HALO Trust Vehicle in SLR Compound, Yei, Sudan.

NOTES

Introduction

1. William W. Boyer. (March 1990) 'Political Science and the 21st Century: From Government to Governance.' *PS: Political Science and Politics.* 23(1). p. 50.
2. International Campaign to Ban Landmines (ICBL). (2006) 'What's the Problem?' <http://www.icbl.org/problem/what>.
3. In this book, I will occasionally use 'demining,' 'mine clearance' and 'mine action' interchangeably. However, I focus largely on demining rather than other programs aimed at mitigating the impact of mines, such as awareness and survivor assistance. For background on mine action, see: Geneva International Centre for Humanitarian Demining (GICHD). (July 2003) *A Guide to Mine Action.* Geneva, GICHD; Stuart Maslen. (2004) *Mine Action After Diana: Progress in the Struggle Against Landmines.* London, Landmine Action/Pluto Press.
4. *No Man's Land.* (2001) Directed by Danis Tanovic [Film]. United States, MGM.
5. Details on landmines and their use can me found in: Mike Croll. (1998) *The History of Landmines.* Barnesly, Leo Cooper; Rae McGrath. (2000) *Landmines and Unexploded Ordnance: A Resource Book.* London, Pluto Press; Department of the Army. (1966) *Field Manual FM 20-32, Landmine Warfare.* Washington DC, Department of the Army; Department of the Army. (December 1985) *Field Manual FM 20-32, Mine/Countermine Operations.* Washington DC, Department of the Army.
6. Landmine Action. (2005) *Explosive Remnants of War and Mines Other Than Anti-Personnel Mines: Global Survey 2003–2004.* London, Landmine Action.
7. Landmine Monitor. (2009) *Banning Cluster Munitions: Government Policy and Practice.* Ottawa, Mines Action Canada.

8 ICBL. (2008) "Mine Action." *Landmine Monitor Report 2008*. <http://lm.icbl.org/index.php/publications/display?url=lm/2008/es/mine_action.html>.

9 ICBL. (2008) "Casualties." *Landmine Monitor Report 2008*. <http://lm.icbl.org/index.php/publications/display?url=lm/2008/es/landmine_casualties_and_survivor_assistance.html>.

10 For more information on mine and UXO clearance techniques, see: GICHD. *A Guide to Mine Action*; Maslen. *Mine Action After Diana*; McGrath. *Landmines*.

11 Mark Duffield. (2001) *Global Governance and the New Wars: The Merging of Development and Security*. London, Zed Books. p. 2.

12 Ibid. p. 2.

13 Jan Egeland. (1988) *Impotent Superpower—Potent Small State: Potentials and Limitations of Human Rights Objectives in the Foreign Polices of the United States and Norway*. Oslo, Norwegian University Press.

14 Clifford Geertz. (1973) *The Interpretation of Cultures*. New York, Basic Books.

15 Lorraine Bayard de Volo & Edward Schatz. (2004) "From the Inside Out: Ethnographic Methods in Political Research." *PS: Political Science & Politics*. 37. p. 267.

16 Raymond Apthorpe. (2002) 'Whose Development? An ethnography of aid, Emma Crewe and Elizabeth Harrison.' <http://www.development-ethics.org/document.asp?did=1082>.

1. A Political History of Mine Action

1 Lydia Monin & Andrew Gallimore. (2002) *The Devil's Gardens: A History of Landmines*. London, Pimlico. p. xvi.

2 Olivier Razac. (2002) *Barbed Wire: A Political History*. Jonathan Kneight (Trans.). New York: The New Press.

3 John Ellis. (1986) *The Social History of the Machine Gun*. Baltimore, Johns Hopkins University Press.

4 Mike Croll. (1998) *The History of Landmines*. Barnesly, Leo Cooper. pp. 1-24.

5 Croll. *History of Landmines*. pp. 25-36; Monin & Gallimore. *Devil's Gardens*. pp 39-49.

6 Croll. *History of Landmines*. pp. 37-83.

7 Monin & Gallimore. *Devil's Gardens*. pp 49-63.

8 Arthur H. Westing. (1985) 'Explosive Remnants of War: An Overview.' *Explosive Remnants of War: Mitigating the Environmental Effects*. Arthur H. Westing (Ed.). London, Taylor & Francis. p. 6; Boguslaw A. Molski & Jan Pajak. (1985) 'Explosive Remnants of World War II in Poland.' In: Westing (Ed.). *Explosive Remnants of War*. pp. 17-32; Khairi Sgaier. (1985) 'Explosive Remnants of World War II in Libya: impact on agricultural development.' In: Westing (Ed.). *Explosive Remnants of War*. pp. 33-37; Margaret Buse. (June 2000) 'WWII Ordnance Still Haunts Europe and the Asia Pacific Rim.' *Journal of Mine Action*. 4(2). <http://maic.jmu.edu/journal/4.2/Features/ww2/ww2.htm>.

9 Westing. 'Explosive Remnants of War.' p. 6; Molski & Pajak. 'Explosive Remnants of World War II in Poland.' pp. 17-32; Croll. *History of Landmines*. pp. 84-95; M.J. Jappy. (2001) *Danger UXB: The Remarkable Story of the Disposal of Unexploded Bombs during the Second World War*. London, Channel 4 Books.

10 Donovan Webster. (1998) *Aftermath: The Remnants of War*. New York, Vintage Books. pp. 11-80.

11 Geneva International Centre for Humanitarian Demining (GICHD). (August 2005) *A Study of Manual Mine Clearance: 1. History, Summary and Conclusions of a Study of Manual Mine Clearance*. Geneva, GICHD. pp. 17-21.
12 (2004) '1899 Hague Declaration 3 Concerning Expanding Bullets.' *Documents on the Laws of War*. Adam Roberts & Richard Guelff (Eds.). 3rd Ed. Oxford, Oxford University Press. pp. 63-65.
13 (2004) '1899 Hague Declaration 2 Concerning Asphyxiating Gases.' In: Roberts & Guelff (Eds.). *Documents on the Laws of War*. pp. 59-61.
14 (2004) '1925 Geneva Protocol for the Prohibition of the Use of Asphyxiating, Poisonous or Other Gases, and of Bacteriological Methods of Warfare.' In: Roberts & Guelff (Eds.). *Documents on the Laws of War*. pp. 155-167.
15 (2004) 'Geneva Convention Relative to the Treatment of Prisoners of War or August 12, 1949.' *Documents on the Laws of War*. In: Roberts & Guelff (Eds.). *Documents on the Laws of War*. Section III, Article 52. p. 264.
16 International Committee of the Red Cross (ICRC). (2000) 'The ICRC's draft rules to protect civilian populations 1955-1956.' *The Banning of Anti-Personnel Landmines: The Legal Contribution of the International Committee of the Red Cross 1955-1999*. Louis Maresca & Stuart Maslen (Eds.). Cambridge, Cambridge University Press. pp. 15-18; Stuart Maslen. (2001) *Anti-Personnel Mines under Humanitarian Law: A View from the Vanishing Point*. Antwerp, Intersentia. p. 26.
17 Eric Prokosch. (1995) *The Technology of Killing: A Military and Political History of Antipersonnel Weapons*. London, Zed Book. p. 8.
18 Ibid. pp 33-39.
19 Ibid. pp 81-125.
20 Quoted in Ibid. p. 53.
21 Earl S. Martin & Murray Hiebert. (1985) 'Explosive remnants of the Second Indochina War in Viet Nam and Laos.' In: Westing (Ed.). *Explosive Remnants of War*. pp. 39-40.
22 Rae McGrath. (2000) *Cluster Bombs: The Military Effectiveness and Impact on Civilians of Cluster Munitions*. London, Landmine Action.
<http://www.stopclustermunitions.org//files/Cluster_Bombs%20LMA.pdf>. p. 30. For further details on US air war see: Webster. *Aftermath*. pp. 160-217.
23 McGrath. *Cluster Bombs*. p. 30.
24 Titus Peachey & Virgil Wiebe. (November 2000) 'Appendix 1: Laos.' *Mennonite Central Committee*. <http://www.mcc.org/clusterbomb/report/laos_appendix.html>; Center for International Studies. (1971) *The Air War in Indochina: A Report by the Center for International Studies*. Ithaca, Cornell University; Jim Monan. (1998) *Curse of the Bombies: A Case Study of Saravan Province, Laos*. Hong Kong, Oxfam.
25 Bruce Shoemaker. (March 1994) 'Legacy of the Secret War.' *Mennonite Central Committee*. <http://www.mcc.org/clusterbomb/secret_war>.
26 International Campaign to Ban Landmines (ICBL). (2007) 'Vietnam.' *Landmine Monitor 2007*. <http://www.icbl.org/lm/2007/vietnam>; International Campaign to Ban Landmines (ICBL). (2007) 'Laos.' *Landmine Monitor 2007*.
<http://www.icbl.org/lm/2007/lao>; International Campaign to Ban Landmines (ICBL). (2007) 'Cambodia.' *Landmine Monitor 2007*.
<http://www.icbl.org/lm/2007/cambodia>.
27 Croll. *History of Landmines*. pp. 102-108.
28 ICBL. 'Vietnam.' See also: Vietnam Veterans of America Foundation (VVAF). (2005) *Report on Vietnam Unexploded Ordnance and Landmine Impact Assessment and Technical Survey*. Hanoi, VVAF.

[29] International Campaign to Ban Landmines (ICBL). (1999) 'Lao PDR.' *Landmine Monitor 1999*. <http://www.icbl.org/lm/1999/lao>.
[30] Shoemaker. 'Legacy of the Secret War.'
[31] Robert E. Low. (1985) 'Explosive remnants of war: detection through the use of dogs.' In: Westing (Ed.). *Explosive Remnants of War.* pp. 73-74.
[32] This is mentioned in general terms in several US State Department diplomatic cables, e.g.: US Embassy Vientiane. (1974) 'First Call on Foreign Minister.' Declassified Secret Cable to SECSTATE WASHDC. 1974VIENT02960. <http://aad.archives.gov/aad/>. Section 2, p. 1.
[33] Philip Long. (September/October 2005) 'Ordnance Dangers.' *Archeology*. p. 9; Philip Long. (15 August 2006) Telephone interview with the author.
[34] W. Dale Nelson. (9 March 1982) 'Reports Deal Possible on MIA Accounting.' *Associated Press*; Shoemaker. 'Legacy of the Secret War.'
[35] Shoemaker. 'Legacy of the Secret War.'
[36] Arthur H. Westing et al. (1985) 'Explosive remnants of conventional war: a report to UNEP.' In: Westing (Ed.). *Explosive Remnants of War.* pp. 117-136.
[37] Shoemaker. 'Legacy of the Secret War.'
[38] Ibid.
[39] Bob Eaton. (2 August 2006) Personal interview with the author in Takoma Park, Maryland. Titus Peachey. (15 August 2006) Telephone interview with the author.
[40] Handicap International. (n.d.) 'History.' <http://www.handicap-international.org.uk/page_12.php>; Aid Watch. (2003) 'Handicap International.' <http://www.observatoire-humanitaire.org/fusion.php?l=GB&id=21>.
[41] Shoemaker. 'Legacy of the Secret War.' USAID did provide a $600 Ocean Freight Reimbursement for some of the shovels, representing the 'only U.S. governmental aid for UXO work in Laos between 1975 and 1991.' See also: Barbara Crossette. (25 November 1987) '4 Aid Laos Against Bombis and Other Horrors.' *The New York Times*. p. A9.
[42] Shoemaker. 'Legacy of the Secret War.'
[43] Prokosch. *Technology of Killing*. p. 87-89.
[44] Ibid. p. 93-94.
[45] Fred Branfman (Ed.). (1972) *Voices from the Plain of Jars: Life under an Air War*. New York, Harper Colophon.
[46] Shoemaker. 'Legacy of the Secret War.'
[47] Prokosch. *Technology of Killing*. p. 135-145.
[48] Shoemaker. 'Legacy of the Secret War.'
[49] Prokosch. *Technology of Killing*. p. 98.
[50] North Vietnamese delegate to the Lucerne Conference of Government Experts, quoted in Prokosch, *Technology of Killing*, p. 155.
[51] Torgil Wulff. (1973) *Conventional Weapons, Their Deployment and Effects from a Humanitarian Aspect: Recommendations for the Modernization of International Law: A Swedish Working Group Study*. Stockholm, Swedish Ministry of Foreign Affairs.
[52] Jean-Philippe Lavoyer & Louis Maresca. (1999) 'The Role of the ICRC in the Development of International Humanitarian Law.' *International Negotiation*. 4. pp. 501-525. See also: Maresca & Maslen (Eds.). *Banning of Anti-Personnel Landmines*. pp. 19-393.
[53] Prokosch. *Technology of Killing*. p. 160.
[54] (2004) '1977 Geneva Protocol I Additional to the Geneva Conventions of 12 August 1949, and Relating to the Protection of Victims of International Armed Conflicts.' In: Roberts & Guelff (Eds.). *Documents on the Laws of War*. Part III, Section I, Article 35. p. 442.

55 Prokosch. *Technology of Killing.* p. 155.
56 Frits Karlshoven & Liesbeth Zegveld. (March 2001) *Constraints on the Waging of War: An Introduction to International Humanitarian Law.* Geneva, ICRC; Geoffrey Best. (1980) *Humanity in Warfare: The Modern History of the International Law of Armed Conflicts.* New York, Columbia University Press.
57 J. Ashley Roach. (1984) 'Certain Conventional Weapons Convention: Arms Control or Humanitarian Law?' *Military Law Review.* 105. pp. 3-14.
58 Alva Myrdal. (1977) *The Game of Disarmament: How the United States and Russia Run the Arms Race.* Manchester, Manchester University Press.
59 For more information on the CCW, see: Karlshoven & Zegveld. *Constraints on the Waging of War.* pp. 155-168.
60 Ibid. Protocol II, Article 3. p. 529.
61 Ibid. Protocol II, Article 5. p. 529-530.
62 Ibid. Protocol II, Article 7.
63 Prokosch. *Technology of Killing.* p. 162.
64 Roach. 'Certain Conventional Weapons Convention.' p. 14.
65 Steffen Kongstad. (3 September 2007) Personal interview with author in Oslo.
66 Stephen S. Rosenfeld. (1986) 'The Guns of July.' *Foreign Affairs.* 64(4); Robert H. Johnson. (1988) 'Misguided Morality: Ethics and the Reagan Doctrine.' *Political Science Quarterly.* 103(3); Mark P. Lagon. (1992) 'The International System and the Reagan Doctrine: Can Realism Explain Aid to 'Freedom Fighters.'' *British Journal of Political Science.* 22(1); James M. Scott. (1996) *Deciding to Intervene: The Reagan Doctrine and American Foreign Policy.* Durham, North Carolina, Duke University Press.
67 Croll. *History of Landmines.*
68 International Campaign to Ban Landmines (ICBL). (2006) 'What's The Problem?' <http://www.icbl.org/problem/what>; Monin & Gallimore. *Devil's Gardens.* pp. 91-178.
69 Mary Kaldor. (1999) *New and Old Wars: Organized Violence in a Global Era.* Cambridge, Polity Press; Herfried Münkler. (2005) *The New Wars.* Patrick Camiller. (Trans). Cambridge, Polity Press; Martin van Creveld. (1991) *The Transformation of War.* New York, The Free Press;
Mark Duffield. (2001) *Global Governance and the New Wars: The Merging of Development and Security.* London, Zed Books.
70 Shoemaker. 'Legacy of the Secret War.'
71 International Campaign to Ban Landmines (ICBL). (June 1999) 'Mine Action Programmes from a development-oriented point of view ('The Bad Honnef Framework').' <http://www.apminebanconvention.org/index.php?id=424>.
72 Human Rights Watch (HRW) & Physicians for Human Rights (PHR). (1991) *Land Mines in Cambodia: The Coward's War.* New York, HRW.
73 Maslen. *Anti-Personnel Mines under Humanitarian Law.* p. 15.
74 International Committee of the Red Cross (ICRC). (1996) *Anti-personnel Mines: Friend or Foe?* Geneva, ICRC.
75 Lavoyer & Maresca. 'The Role of the ICRC.' p. 516. See also: Maresca & Maslen (Eds.). *Banning of Anti-Personnel Landmines.* pp. 127-652.
76 Monin & Gallimore. *Devil's Gardens.* pp. 13-16.
77 See also: Anonymous. (26 September 1997) 'Fields of death: only days before her death, Princess Diana pleaded for an end to the use of land mines.' *Current Events.* <http://www.encyclopedia.com/doc/1G1-19849595.html>.
78 State signatories list from: International Committee to Ban Landmines (ICBL). (15 August 2007) 'States Parties.' <http://www.icbl.org/treaty/members>. Population statistics from: Population Reference Bureau. (2007) 'World Population Data Sheet.'

<http://www.prb.org/Datafinder/Topic/Bar.aspx?sort=v&order=d&variable=1>. Military expenditure statistics from: Nation Master. (2005) 'Military Expenditures Dollar Figure (Most Recent) By Country.' <http://www.nationmaster.com/graph/mil_exp_dol_fig-military-expenditures-dollar-figure>.

[79] Leon V. Sigal. (2006) *Negotiating Minefields: The Landmines Ban in American Politics*. New York, Routledge. pp. 89-102, 155-174.

[80] Monin & Gallimore. *Devil's Gardens*. p. 6.

[81] Kenneth R. Rutherford. (2000) 'The Evolving Arms Control Agenda: Implications of the Role of NGOs in Banning Antipersonnel Landmines.' *World Politics*. 53(1).

[82] Rosy Cave. (2006) 'Disarmament as Humanitarian Action? Comparing Negotiations on Anti-Personnel Mines and Explosive Remnants of War.' *Disarmament as Humanitarian Action: From Perspective to Practice*. Geneva, UNIDIR.

[83] Prokosch. *Technology of Killing*. p. 185.

[84] Sigal. *Negotiating Minefields*; Monin & Gallimore. *Devil's Gardens*. pp. 20-27.

[85] Myrdal. *Game of Disarmament*.

[86] Alexander Kmentt. (2008) 'A Beacon of Light: The Mine Ban Treaty Since 1997.' *Banning Landmines: Disarmament, Citizen Diplomacy, and Human Security*. Jody Williams, Stephen D. Goose & Mary Wareham (Eds.). Lanham, Rowman & Littlefield. pp. 26-28.

[87] ICBL. 'States Parties.'

[88] Geneva Call. (4 October 2001) 'Deed of Commitment under Geneva Call for Adherence to a Total Ban on Anti-Personnel Mines and for Cooperation in Mine Action.' <http://www.genevacall.org/about/testi-mission/gc-04oct01-deed.htm>; Yeshua Moser-Puangsuwan. (2008) 'Outside the Treaty Not the Norm: Nonstate Armed Groups and the Landmine Ban.' In: Williams, Goose & Wareham. *Banning Landmines*.

[89] (2004) '1997 Ottawa Convention on the Prohibition of the Use, Stockpiling, Production and Transfer of Anti-Personnel Mines and on their Destruction.' *Documents on the Laws of War*. Adam Roberts & Richard Guelff (Eds.). 3rd Ed. Oxford, Oxford University Press. pp. 645-666.

[90] Stuart Maslen. (2004) *Mine Action after Diana: Progress in the Struggle against Landmines*. London, Pluto Press. p. 7.

[91] Ibid. pp. 24-29.

[92] Andrew Latham. (October 2000) 'Global Cultural Change and the Transnational Campaign to Ban Antipersonnel Landmines: A Research Agenda.' YCISS Occasional Paper Number 62. <http://www.yorku.ca/yciss/publications/OP62-Latham.pdf>.

[93] International Campaign to Ban Landmines (ICBL). (2007) *Landmine Monitor*. <http://www.icbl.org/lm>.

[94] Steven Metz. (April 2000) *Armed Conflict in the 21st Century: The Information Revolution and Post-Modern Warfare*. Carlisle, Penn., Strategic Studies Institute. <http://www.strategicstudiesinstitute.army.mil/pubs/display.cfm?pubID=226>.

[95] Jean Baudrillard. (1995) *The Gulf War Did Not Take Place*. Paul Patton (Trans.). Bloomington, Indiana University Press; Chris Hables Gray. (1998) *Postmodern War: The New Politics of Conflict*. New York, The Guilford Press; Wesley K. Clark. (2002) *Waging Modern War: Bosnia, Kosovo, and the Future of Combat*. New York, PublicAffairs.

[96] Virgil Wiebe. (Fall 2000) 'Footprints of Death: Cluster Bombs as Indiscriminate Weapons under International Humanitarian Law.' *Michigan Journal of International Law*. 22(1). pp. 142-151; Mennonite Central Committee. (2000) 'Cluster Munitions Use by Russian Federation Forces in Chechnya.' <http://mcc.org/clusterbombs/resources/research/death/chapter3.html>.

97 Human Rights Watch (HRW). (June 1999) 'Ticking Time Bombs: NATO's Use of Cluster Munitions in Yugoslavia.' <http://www.hrw.org/reports/1999/nato2/>; Virgil Wiebe. (Fall 2000) 'Footprints of Death: Cluster Bombs as Indiscriminate Weapons under International Humanitarian Law.' *Michigan Journal of International Law*. 22(1). pp. 126-143.
98 Human Rights Watch (HRW). (December 2002) 'Fatally Flawed: Cluster Bombs and Their Use by the United States in Afghanistan.' <http://www.hrw.org/reports/2002/us-afghanistan/>.
99 Human Rights Watch (HRW). (2003) *Off Target: The Conduct of the War and Civilian Casualties in Iraq*. New York, HRW. <http://www.stopclustermunitions.org//files/HRW%20Off%20Target.pdf>.
100 Landmine Action. (2006) *Foreseeable Harm: The use and impact of cluster munitions in Lebanon: 2006*. London, Landmine Action. <http://www.stopclustermunitions.org/files/Foreseeable%20harm.pdf>.
101 Human Rights Watch (HRW). (15 August 2008) 'Georgia: Russian Cluster Bombs Kill Civilians.' <http://www.hrw.org/english/docs/2008/08/14/georgi19625.htm>.
102 McGrath. *Cluster Bombs*. p. 33-35.
103 McGrath. *Cluster Bombs*. p. 36; Monin & Gallimore. *Devil's Gardens*. pp. 183-192.
104 HRW. *Off Target*. pp. 56, 80.
105 Ibid. pp. 104-110.
106 Database query on 16 January 2008: Iraq Coalition Casualty Count. (2008) 'Iraq Coalition Casualty Count.' <http://icasualties.org/oif/>.
107 Christopher Spearin. (November 2001) 'Ends and Means: Assessing the Humanitarian Impact of Commercialised Security on the Ottawa Convention Banning Anti-Personnel Mines.' YCISS Occasional Paper Number 69. <http://www.yorku.ca/yciss/publications/OP69-Spearin.pdf>.
108 Paul Donovan. (September 1997) 'Making a Killing.' *New Internationalist*. pp. 14-15; R. Block & L. Doyle. (6 June 1994) 'UN aid goes to landmine makers; Awarding of clearance contracts to producers means they are being paid twice.' *The Independent* (London). p. 1.
109 Croll. *History of Landmines*. p. 132.
110 Donovan. 'Making a Killing.' pp. 14-15; International Committee to Ban Landmines (ICBL). (1999) 'Kuwait.' <http://www.icbl.org/lm/1999/kuwait>; Webster. *Aftermath*. pp. 218-252.
111 Adriaan Basson & Lynley Donnelly. (5 April 2008) 'SA opposes cluster-bomb ban.' *Mail & Guardian*. <http://www.mg.co.za/articlePage.aspx?articleid=336297&area=/insight/insight__national/>.
112 US Department of State. (27 February 2004) 'New United States Policy on Landmines: Reducing Humanitarian Risk and Saving Lives of United States Soldiers.' <http://www.state.gov/t/pm/rls/fs/30044.htm>.
113 BBC. (26 December 2006) 'Pakistan plans border minefield.' <http://news.bbc.co.uk/2/hi/south_asia/6210057.stm>.
114 (28 November 2003) 'Protocol on Explosive Remnants of War (Protocol V to the 1980 Convention).' <http://www.mineaction.org/docs/1850_.asp>. Article 4. For further details see: Cave. 'Disarmament as Humanitarian Action?'
115 Landmine Action. *Foreseeable Harm*. p. 3; Matthew Bolton. (9 September 2006) 'Shooting ends, but bombs' danger lingers for decades.' *The Examiner* (Independence, Missouri, USA). <http://examiner.net/stories/090906/mat_090906020.shtml>; Reuters. (26 September 2006) 'UN: It will take over a year to clear Lebanon of cluster bombs.' *Haaretz*. <http://www.haaretz.com/hasen/spages/767528.html>.

116 Scott Peterson. (7 February 2007) 'Cluster bombs: a war's perilous aftermath.' *The Christian Science Monitor.* <http://www.csmonitor.com/2007/0207/p01s01-wome.html>.
117 Cluster Munition Coalition. (March 2006) 'Report on CCW March 2006.' <http://www.stopclustermunitions.org/files/Report%20on%20CCW%20March%202020 06.doc>.
118 Jonas Gahr Støre. (20 November 2006) 'Norway takes the initiative for a ban on cluster munitions.' <http://www.norway.org/policy/news/ban+cluster+munitions.htm>.
119 (18 March 2009) 'Convention on Cluster Munitions.' <http://www.clusterconvention.org>.
120 Matthew Bolton. (23 June 2007) 'Important step forward not enough for United States.' *The Examiner* (Independence, Missouri, USA). <http://examiner.net/stories/062307/new_179584763.shtml>.
121 States parties list from: Cluster Munition Coalition. (30 May 2008) '107 Countries on Board.' <http://www.stopclustermunitions.org/the-solution/the-treaty/?id=84>. Population statistics from: Population Reference Bureau. 'World Population Data Sheet.' Military expenditure statistics from: Nation Master. 'Military Expenditures.'
122 John L. Wilkinson. (December 2002) 'Demining During Operation Enduring Freedom in Afghanistan.' *Journal of Mine Action.* 6(3). <http://maic.jmu.edu/JOURNAL/6.3/notes/wilkinson/wilkinson.htm>.
123 Mark Duffield. (1999) 'Globalization and War Economies: Promoting Order or the Return of History?' *Fletcher Forum on World Affairs.* 23(2). p. 31.
124 Reportage by Dan Rather has found evidence suggesting there was a deliberate attempt by Israel 'to create de facto minefields' between Israel and Lebanon. Dan Rather. (2008) 'Bombs Left Behind.' *Dan Rather Reports.* <http://www.hd.net/drr306.html>.

2. The New Complexes Governing Insecurity

1 Lord Curzon of Kedleston. (1907) 'Frontiers.' 1907 Romanes Lecture. <http://www-ibru.dur.ac.uk/resources/docs/curzon1.html>.
2 Mark Duffield. (2001) 'Governing the Borderlands: Decoding the Power of Aid.' *Disasters.* 25(4).
3 Eyal Weizman. (2007) *Hollow Land: Israel's Architecture of Occupation.* London, Verso. pp. 1-6.
4 Ibid. p. 5.
5 Ibid. p. 165.
6 Duffield. 'Governing the Borderlands.' p. 308.
7 Mark Duffield. (2001) *Global Governance and the New Wars: The Merging of Development and Security.* London, Zed Books. p. 2.
8 Weizman. *Hollow Land.* p. 7.
9 Charles S. Maier. (2000) 'Consigning the Twentieth Century to History: Alternative Narratives for the Modern Era.' *The American Historical Review.* 105(3). <http://www.historycooperative.org/journals/ahr/105.3/ah000807.html>.
10 cf. Lord Curzon. 'Frontiers.'
11 cf. John H. Herz. (1951) *Political Realism and Political Idealism.* Chicago, University of Chicago Press.
12 Mary Kaldor. (1999). *New and Old Wars: Organized Violence in a Global Era.* Cambridge, Polity Press. pp. 13-30. For a classical description of this kind of war, see: Carl von Clausewitz. (1997) *On War.* J.J. Graham (Trans.). Ware, Hertfordshire, Wordsworth Editions Ltd.

[13] I include in this category neo-realists like Kenneth Waltz, John Mearsheimer and Hans Morgenthau who use the realist preoccupation with power politics as the foundation for a 'scientific' understanding of international relations.

[14] Sun Tzu (1998) *The Art of War*. Ware, Hertfordshire, Wordsworth Editions Ltd.

[15] Niccolo Machiavelli. (1997) *The Prince*. Ware, Hertfordshire, Wordsworth Editions Ltd.

[16] von Clausewitz. *On War*. p. 6.

[17] Condoleezza Rice. (January/February 2000) 'Campaign 2000: Promoting the National Interest.' *Foreign Affairs*. <http://www.foreignaffairs.org/20000101faessay5/condoleezza-rice/campaign-2000-promoting-the-national-interest.html >.

[18] Ibid.

[19] Jack S. Levy. (1983) *War in the Modern Great Power System, 1495-1975*. Lexington, Kentucky, University Press of Kentucky.

[20] e.g. Danilo Zolo. (1998) 'Hans Kelsen: International Peace through International Law.' *European Journal of International Law*. 9(2). pp. 306-324; John R. Oneal & Bruce Russett. (1999) 'The Kantian Peace: The Pacific Benefits of Democracy, Interdependence, and International Organizations, 1885-1992.' *World Politics*. 52(1). pp. 1-37.

[21] For modern idealism's roots see: Immanuel Kant. (1983) 'To Perpetual Peace: A Philosophical Sketch.' *Perpetual Peace and Other Essays*. Ted Humphrey (Trans.). Indianapolis, Hackett Publishing Company. pp. 106-143.

[22] Examples of scholarly works by modern idealists include: Charles A. Kupchan & Clifford A. Kupchan. (Summer 1995) 'The Promise of Collective Security.' International Security. 20(1). pp. 52-61; Thomas R. Cusack & Richard J. Stoll. (March 1994) 'Collective Security and State Survival in the Interstate System.' International Studies Quarterly. 38(1). pp. 33-59; Inis L. Claude (Ed.). (1984) Swords Into Plowshares: The Problems and Progress of International Organization. 4th Ed. New York, Random House. Harold K. Jacobson. (1984) Networks of Interdependence: International Organizations and the Global Political System. 2nd Ed. New York, Alfred E. Knopf; Saul H. Mendlovitz. (1975) On the Creation of a Just World Order: Preferred Worlds for the 1990's. New York, Free Press.

[23] Cicero. (1991) *On Duties*. M.T. Griffin & F.M. Atkins (Trans.). Cambridge, Cambridge University Press; Thomas Aquinas. (1988) *On Law, Morality, and Politics*. William P. Baumgarth & Richard J. Regan (Ed.). Indianapolis, Hackett Publishing Company; Hugo Grotius. (2008) 'Prolegomena.' <http://www.lonang.com/exlibris/grotius/gro-100.htm>; Jean Jacques Rousseau. (1762) 'The Social Contract or Principles of Political Right.' <http://www.constitution.org/jjr/socon.htm>.

[24] Kant. 'Perpetual Peace.' p. 135.

[25] Immanuel Kant. (1993) *Grounding for the Metaphysics of Morals*. James W. Ellington (Trans.). 3rd Ed. Indianapolis, Hackett Publishing Company. p. 36.

[26] Robert L. Rothstein. (1968) *Alliances and Small Powers*. New York, Columbia University Press. p. 29.

[27] Hedley Bull. (1977) *The Anarchical Society: A Study of Order in World Politics*. New York, Columbia University Press; Martin Wight. (1977) *Systems of States*. Leicester, Leicester University Press.

[28] Kaldor. *New and Old Wars*. p. 3.

[29] H. Brinton Milward & Keith G. Provan. (April 2000) 'Governing the Hollow State.' *Journal of Public Administration Research and Theory*. 10(2). pp. 359-379.

[30] William W. Boyer. (March 1990) 'Political Science and the 21st Century: From Government to Governance.' *PS: Political Science and Politics*. 23(1). pp. 50-54.

[31] cf. Michel Foucault. (1977). *Discipline and Punish: The Birth of the Prison*. London, Penguin Books. pp. 24-28; Michel Foucault. (1980) *Power/Knowledge*. New York, Pantheon. p. 39.

32 Bull. *The Anarchical Society.* p. 254; Mark Duffield. (Spring 1998) 'Post-modern Conflict: Warlords, Post-adjustment States and Private Protection.' *Civil Wars.* 1(1). pp. 65-102.
33 Mark Duffield. (2007) *Development, Security and Unending War: Governing the World of Peoples.* Cambridge, Polity Press.
34 Cherif Bassiouni. (1997) 'Organized crime and new wars.' *New Wars.* Mary Kaldor & Basker Vashee (Eds.). London, Pinter; Frank Cillufo & George Salmoiraghi. (Autumn 1999) 'And the Winner Is....the Albanian Mafia.' *Washington Quarterly.* 22(4); Michael Charles Pugh, Neil Cooper & Jonathan Goodhand M. and N. Cooper, (Eds.). (2004) *War Economies in a Regional Context: Challenges of Transformation.* London, Lynne Reiner Publishers, Inc.; Misha Glenny. (2008) *McMafia: A Journey Through the Global Criminal Underworld.* New York, Knopf.
35 Duffield. Global Governance and the New Wars. pp. 14, 44-74.
36 Herfried Münkler. (2005) *The New Wars.* Cambridge, Polity Press; Duffield. *Global Governance and the New Wars.* pp. 136-201; Kaldor. *New and Old Wars*; Martin van Creveld. (1991) *The Transformation of War.* New York, The Free Press.
37 Duffield. 'Governing the Borderlands.' p. 309.
38 Ibid. p. 309.
39 Duffield. *Development, Security and Unending War.*
40 Duffield. *Global Governance and the New Wars.* p. 2.
41 Milward & Provan. 'Governing the Hollow State.'
42 Duffield. *Global Governance and the New Wars.* p. 12.
43 Solomon Hughes. (2007) *War on Terror, Inc.: Corporate Profiteering from the Politics of Fear.* London, Verso. p. 7.
44 Naomi Klein. (2007) *The Shock Doctrine: The Rise of Disaster Capitalism.* London, Allen Lane. p. 12.
45 P.W. Singer. (2003) *Corporate Warriors: The Rise of the Privatized Military Industry.* Ithaca, Cornell University Press. pp. 69-70; Richard Sennet. (1976) *The Fall of Public Man.* New York, W.W. Norton.
46 James William Gibson. (1994) *Warrior Dreams: Paramilitary Culture in Post-Vietnam America.* New York, Hill and Wang.
47 Edward J. Blakely & Mary Gail Snyder. (1997) *Fortress America: Gated Communities in the United States.* Washington DC, The Brookings Institution. p. 2.
48 Ibid. p. 148.
49 Klein. *Shock Doctrine.* pp. 406-422; Jon Coaffee. (March 2004) 'Rings of Steel, Rings of Concrete and Rings of Confidence: Designing out Terrorism in Central London pre and post September 11th.' *International Journal of Urban and Regional Research.* 28(1).
50 James Ferguson. (2005) 'Seeing Like an Oil Company: Space, Security, and Global Capital in Neoliberal Africa.' *American Anthropologist.* 107(3). pp. 377-382.
51 Ulrich Jürgens & Martin Gnad. (2002) 'Gated communities in South Africa – experiences from Johannesburg.' *Environment and Planning B: Planning and Design.* 29(3). pp. 337-353; Charlotte Lemanski. (October 2004) 'A new apartheid? The spatial implications of fear of crime in Cape Town, South Africa.' *Environment and Urbanization.* 16(2). pp. 101-112; Derek Hook & Michele Vrdoljak. (2002) 'Gated communities, heterotopia and a 'rights' of privilege: a 'heterotopology' of the South African security-park.' *Geoforum.* 33. pp. 195-219.
52 Blakely & Snyder. *Fortress America.* p. 1.
53 Ibid.
54 Rajiv Chandrasekaran. (2007) *Imperial Life in the Emerald City: Inside Baghdad's Green Zone.* London, Bloomsbury Publishing. p. 12, 19.
55 Eyal Weizman. (2007) *Hollow Land: Israel's Architecture of Occupation.* London, Verso. p. 176.

56 cf. Duffield. *Development, Security and Unending War.*
57 Friedrich Hayek. (1994) *The Road to Serfdom.* 50th anniversary edition. Chicago, University of Chicago Press.
58 Milton Friedman. (2002) *Capitalism and Freedom: Fortieth Anniversary Edition.* Chicago, University of Chicago Press.
59 George J. Stigler. (1987) *The Theory of Price.* 2nd Rvsd Ed. London, Collier Macmillan.
60 E.S. Savas. (1996) 'Applications in Protective and Human Services.' *Privatization: Critical Perspectives on the World Economy.* Vol. II. George Yarrow & Piotr Jasinski (Eds.). London, Routledge. pp. 324-331; E.S. Savas. (1982) *Privatizing the Public Sector: How to Shrink Government.* Chatham, NJ, Chatham House.
61 Milward & Provan. 'Governing the Hollow State.'
62 Christopher Moraff. (3 October 2006) 'Along Came a Spider.' *The American Prospect.* <http://www.prospect.org/cs/articles?article=along_came_a_spider>.
63 Reubén Berríos. (2000) *Contracting for Development: The Role of For-Profit Contractors in U.S. Foreign Development Assistance.* Westport, CT, Praeger.
64 Ibid. p. 58.
65 Chris Horwood. (March 2000) 'Humanitarian Mine Action: The First Decade of a New Sector in Humanitarian Aid.' RRN Network Paper. 32. p. 31.
66 Ann Fitz-Gerald & Derrick J. Neal. (2000) 'Dispelling the Myth Between Humanitarian and Commercial Mine Action Activity.' *Journal of Mine Action.* 4(3). <http://maic.jmu.edu/Journal/4.3/features/myth/myth.htm>.
67 Eddie Banks. (August 2003) 'In the Name of Humanity.' *Journal of Mine Action.* 7(2). <http://maic.jmu.edu/journal/7.2/focus/banks/banks.htm>.
68 Hughes. *War on Terror, Inc.* p. 6.
69 Ibid. p. 71-92.
70 Singer. *Corporate Warriors.* pp. 49-70; Duffield. *Global Governance and the New Wars.* p. 65-68.
71 Singer. *Corporate Warriors.* pp. 22-29.
72 Ibid. pp. 42-48.
73 Hughes. *War on Terror, Inc.* p. 121, 201; Singer. *Corporate Warriors.* pp. 209-211.
74 Deborah Avant. (Spring 2006) 'The Privatization of Security: Lessons from Iraq.' *Orbis.* 50(2). p. 330.
75 For details on the private security industry, see: Singer. *Corporate Warriors*; Christopher Kinsey. (2007) *Corporate Soldiers and International Security: The Rise of Private Military Companies.* London, Routledge; Deborah Avant. (2005) *The Market for Force: The Consequences of Privatizing Security.* Cambridge, Cambridge University Press.
76 Hughes. *War on Terror, Inc.* pp. 56-70; Singer. *Corporate Warriors.* pp. 49-70.
77 Josh Manchester. (19 December 2006) 'Al Qaeda for the Good Guys: The Road to Anti-Qaeda.' *TCSDaily.* <http://www.tcsdaily.com/Article.aspx?id=121606A>.
78 Christopher Spearin. (November 2001) 'Ends and Means: Assessing the Humanitarian Impact of Commercialised Security on the Ottawa Convention Banning Anti-Personnel Mines.' YCISS Occasional Paper Number. 69.
<http://www.yorku.ca/yciss/publications/OP69-Spearin.pdf>.
79 Garrett Hardin. (September 1974) 'Lifeboat Ethics: The Case Against Helping the Poor.' *Psychology Today.*
<http://www.garretthardinsociety.org/articles/art_lifeboat_ethics_case_against_helping_poor.html>.
80 Blakely & Snyder. *Fortress America.* p. 129.
81 James H. Davis, F. David Schoorman & Lex Donaldson. (1997) 'Toward a Stewardship Theory of Management.' *The Academy of Management Review.* 22(1). p. 22.

82 John W. Pratt & Richard J. Zeckhauser, eds. (1984) *Principals and agents: The structure of business*. Boston, MA: Harvard Business School Press. pp. 1-35; Kathleen M. Eisenhardt. (1989) 'Agency theory: An assessment and review.' *Academy of Management Review*. 14. pp. 57–74; Dietmar Braun. (April-June 1993) 'Who Governs Intermediary Agencies? Principal-Agent Relations in Research Policy-Making.' *Journal of Public Policy*. 13(2). pp. 135-162.

83 David M. Van Slyke. (April 2007) 'Agents or Stewards: Using Theory to Understand the Government-Nonprofit Social Service Contracting Relationship.' *Journal of Public Administration Research and Theory*. 17(2). pp. 162-164.

84 Klein. *The Shock Doctrine*. p. 15.

85 Berríos. *Contracting for Development*. pp. 24-25, 30-33; Joseph E. Stiglitz. (2002) *Globalization and Its Discontents*. London, Penguin Books; Steven R. Smith & Michael Lipsky. (1993) *Nonprofits for Hire: The Welfare State in the Age of Contracting*. Cambridge, Mass., Harvard University Press.

86 Banks. 'In the Name of Humanity.'

87 Matthew Bolton & Hugh Griffiths. (September 2006) *Bosnia's Political Landmines: A Call for Socially Responsible and Conflict-Sensitive Mine Action*. London, Landmine Action. <http://www.landmineaction.org/resources/Bosnias_Political_Landmines.pdf>.

88 Transparency International. (2005) *Global Corruption Report 2005: Corruption In Construction And Post-Conflict Reconstruction*. <http://www.transparency.org/publications/gcr/download_gcr/download_gcr_2005>. pp. 73-92; Avant. 'The Privatization of Security.' p. 332-337.

89 Timothy Donais. (2005) *The Political Economy of Peacebuilding in Post-Dayton Bosnia*. London, Routledge. p. 108.

90 Jacques Buré & Pierre Pont. (November 2003) 'Landmine Clearance Projects: A Task Manager's Guide.' Social *Development Papers: Conflict Prevention & Reconstruction 10*. <http://go.worldbank.org/H9XPUBHKP0>. pp. 18-19.

91 Geneva International Centre for Humanitarian Demining (GICHD). (2004) *A Study of Local Organisations in Mine Action*. Geneva, GICHD. pp. 45-89.

92 Bolton & Griffiths. *Bosnia's Political Landmines*.

93 Hughes. *War on Terror, Inc.* London, Verso. pp. 158-169.

94 Singer. *Corporate Warriors*. pp. 220-221; Nils Roseman. (2005) 'The Privatization of Human Rights Violations – Business' Impunity or Corporate Responsibility? The Case of Human Rights Abuses and Torture in Iraq.' *Non-State Actors and International Law*. 5. pp. 77-100; Avant. 'The Privatization of Security.' pp. 338-340.

95 Dwight D. Eisenhower. (17 January 1961) 'Farewell Address.' <http://www.eisenhower.archives.gov/speeches/farewell_address.html>.

96 Klein. *Shock Doctrine*. pp. 306-307; Hughes. (2007) *War on Terror, Inc.* pp. 135-137; Avant. 'The Privatization of Security.' p. 341; Singer. *Corporate Warriors*. pp. 213-215.

97 Adam Roberts. (2006) *The Wonga Coup*. New York, PublicAffairs; Robert Young Pelton. (2006) *Licensed to Kill: Hired Guns in the War on Terror*. New York, Random House. pp. 302-333.

98 cf. Bull. *The Anarchical Society*.

99 Kenneth R. Rutherford, Stefan Brem & Richard A. Matthew (Eds.). (2003) *Reframing the Agenda: the Impact of NGO and Middle Power Cooperation in International Security Policy*. London, Praeger.

100 Øyvind Østerud. (September 2005) 'Introduction: The Peculiarities of Norway.' *West European Politics*. 28(4). p. 714.

101 Mary Kaldor. (2007) *Human Security: Reflections on Globalization and Intervention*. Cambridge, Polity Press. p. 8.

[102] Ibid. p. 182.
[103] Mary Kaldor. (2003) *Global Civil Society: An Answer to War*. Cambridge, Polity Press. p. 3.
[104] Middle powers are defined here as high and middle income countries that have neither a permanent seat on the UN Security Council nor nuclear weapons. A few examples include: Canada, Mexico, Brazil, Germany, Austria, Italy, the Scandinavian states, Egypt, South Africa, Japan, South Korea and Australia.
[105] Christine Ingebritsen. (2002) 'Norm Entrepreneurs: Scandinavia's Role in World Politics.' *Cooperation and Conflict: Journal of the Nordic International Studies Association*. 37(1). pp. 11-23; Olav Stokke (Ed.). (1989) *Western Middle Powers and Global Poverty: The Determinants of the Aid Policies of Canada, Denmark, the Netherlands, Norway and Sweden*. Uppsala, Scandinavia Institute of African Studies; Andrew Cooper, Richard Higgott & Kim Nossal. (1993) *Relocating Middle Powers: Australia and Canada in a Changing World Order*. Vancouver, University of British Columbia Press; Andrew Cooper (Ed.). (1997) *Niche Diplomacy: Middle Powers after the Cold War*. New York, St. Martin's Press.
[106] Rutherford, Brem & Matthew (Eds.). *Reframing the Agenda*. p. 8.
[107] Joseph S. Nye. (2004) *Soft Power: The Means to Success in World Politics*. New York, PublicAffairs.
[108] Richard A. Matthew. (2003) 'Middle Power and NGO Partnerships: The Expansion of World Politics.' In: Rutherford, Brem & Matthew (Eds.). *Reframing the Agenda*. p. 7.
[109] Kaldor. *Global Civil Society*. pp. 50-77.
[110] Kaldor. *Human Security*. p. 16-72; Jonathan Goodhand. (2006) *Aiding Peace? The Role of NGOs in Armed Conflict*. Bourton on Dunsmore, Intermediate Technology Publications. pp. 77-81.
[111] Boutros Boutros Ghali. (17 June 1992) 'An Agenda for Peace: Preventive diplomacy, peacemaking and peace-keeping.' <http://www.un.org/Docs/SG/agpeace.html>.
[112] UN General Assembly. (15 September 2005) '2005 World Summit Outcome.' A/RES/60/1. <http://www.un.org/summit2005/documents.html>. p. 31.
[113] cf. UNDP. (1994) *Human Development Report 1994: New dimensions of human security*. New York, Oxford University Press.
[114] Human Security Centre. (2005) 'What Is Human Security?' <http://www.humansecurityreport.org/index.php?option=content&task=view&id=24&Itemid=59>.
[115] Human security advocates differ on whether it should have a narrow definition, focusing on overt violence or a broad definition including poverty, environmental degradation, natural disasters and disease. The author leans toward the narrow definition, but believes the primary value-added of the concept comes from its movement of the locus of security-building from the state to the human.
[116] Kaldor. *Global Civil Society*. pp. 6, 12.
[117] Duffield. Global Governance and the New Wars. p. 53.
[118] e.g. Thomas Richard Davies. (December 2006) 'The Rise and Fall of Transnational Civil Society.' <http://www.bisa.ac.uk/2006/pps/davies.pdf>.
[119] Lester Salamon. (1994) 'The Rise of the Non-Profit Sector.' *Foreign Affairs*. 73(4). pp. 109–122; Kaldor. *Global Civil Society*. pp. 86-95.
[120] e.g. Bill Howell. (May 1997) 'NGOs perform vital role.' *Landmines*. 22. pp. 10, 11, 13.
[121] Ann Kelleher & James Larry Taulbee. (October 2006) 'Bridging the Gap: Building Peace Norwegian Style.' *Peace & Change*. 31(4); John Stephen Moolakkattu. (2005) 'Peace Facilitation by Small States: Norway in Sri Lanka.' *Cooperation and Conflict: Journal of the Nordic International Studies Association*. 40(4).
[122] Kaldor. *Human Security*.

123 Rutherford, Brem & Matthew. *Reframing the Agenda*; Kelleher & Taulbee. 'Bridging the Gap.'; Moolakkattu. 'Peace Facilitation.'
124 Davis, Schoorman & Donaldson. 'Toward a Stewardship Theory of Management.' p. 24.
125 Davis, Schoorman & Donaldson. 'Toward a Stewardship Theory of Management.'; Van Slyke. 'Agents or Stewards.' pp. 164-167; Lisa A Dicke. (2002) 'Ensuring accountability in human services contracting: Can stewardship theory fill the bill?' *American Review of Public Administration* 32; Lisa A. Dicke & Steven J. Ott. (2002) 'A test: Can stewardship theory serve as a second conceptual foundation for accountability methods in contracted human services?' *International Journal of Public Administration*. 25. pp. 463–87.
126126 Milward & Provan. 'Governing the Hollow State.' pp. 368-376.
127 Davis, Schoorman & Donaldson. 'Toward a Stewardship Theory of Management.' p. 26.
128 David Rieff. (2002) *A Bed for the Night: Humanitarianism in Crisis*. London, Vintage; Alex de Waal. (1998) *Famine Crimes: Politics and the Disaster Relief Industry in Africa*. Bloomington, Indiana University Press; Mary B. Anderson. (1999) *Do No Harm: How Aid Can Support Peace – or War*. Boulder, Lynne Rienner Publishers.
129 Wolf-Dieter Eberwein. (October 2001) 'Realism or Idealism, or Both? Security Policy and Humanitarianism.' Arbeitsgruppe Internationale Politik Discussion Paper P 01-307. <http://skylla.wzb.eu/pdf/2001/p01-307.pdf >. p. 14; Richard Perle. (21 March 2003) 'Thank God for the death of the UN.' *The Guardian* (Manchester). <http://www.guardian.co.uk/politics/2003/mar/21/foreignpolicy.iraq1>.
130 For more on the 'Asian Values' debate, see: Chan Heng Chee. (1993) 'Democracy: Evolution and Implementation: An Asian Perspective.' *Democracy and Capitalism: Asian and American Perspectives*. Robert Bartley, Chan Heng Chee, Samuel P. Huntington and Shijuro Ogata (Eds.). Singapore, Institute of Southeast Asian Studies; Amartya Sen. (July 14-July 21, 1997) 'Human Rights and Asian Values.' *The New Republic*.
131 A recent Russian NGO law declared that NGOs can 'create a threat to the sovereignty, political independence, territorial integrity, national unity, unique character, cultural heritage and national interests of the Russian Federation.' Quoted in: Carl Gershman & Michael Allen. (2006) 'The Assault on Democracy Assistance.' *Journal of Democracy*. 17(2). p. 39.
132 Angela Charlton. (10 August 2000) 'Russia accuses British mine-clearing charity of aiding Chechens.' *The Independent*. <http://www.independent.co.uk/news/world/europe/russia-accuses-british-mineclearing-charity-of-aiding-chechens-711627.html>.
133 Monica Kathina Juma & Astri Suhrke (Eds.). (2002) *Eroding Local Capacity: International Humanitarian Action in Africa*. Uppsala, Nordiska Afrikainstitutet.
134 Duffield. *Development, Security and Unending War*. pp. 184-222.
135 Milton Friedman. (1979) *Tax Limitation, Inflation, and the Role of Government*. Dallas, Texas, The Fisher Institute. p. 7.
136 Thomas W. Dichter. (2003) *Despite Good Intentions: Why Development Assistance to the Third World Has Failed*. Amherst, University of Massachusetts Press; Ian Smillie. (1995) *The Alms Bazaar: Altruism Under Fire—Non-Profit Organizations and International Development*. West Hartford, Kumarian; Goodhand. *Aiding Peace?* pp. 91-92.
137 Terje Tvedt. (1994) 'The Collapse of the State in Southern Sudan after the Addis Ababa Agreement: A Study of Internal Causes and the Role of the NGOs.' *Short-Cut to Decay: The Case of the Sudan*. Sharif Harir & Terje Tvedt (Eds.). Uppsala, Scandinavian Institute of African Studies; Volker Riehl. (2001) *Who Is Ruling in South Sudan? The role of NGOs in rebuilding socio-political order*. Uppsala, Nordiska Afrikainstitutet; Michel Leezenberg. (2000) 'Humanitarian Aid in Iraqi Kurdistan.' *Cahiers d'études sur la Méditerranée orientale et le monde*

turco-iranien. 29. <http://www.ceri-sciencespo.com/publica/cemoti/textes29/leezenbe.pdf>. pp. 1–18.
138 Maurice N. Amutabi. (2006) *The NGO Factor in Africa: The Case of Arrested Development in Kenya.* New York: Routledge. p. xxiii.
139 David Hulme & Michael Edwards. (1997) *NGOs, States and Donors Too Close for Comfort?* Macmillan, Basingstoke.
140 Amutabi. The NGO Factor in Africa. p. 34.
141 Duffield. *Development, Security and Unending War.*
142 Ibid.
143 William Easterly. (2006) *The White Man's Burden: Why the West's Efforts to Aid the Rest Have Done So Much Ill and So Little Good.* London, Penguin Books.
144 Maier. 'Consigning the Twentieth Century to History'; Milward & Provan. 'Governing the Hollow State.'

3. Donor Policy Making in the US and Norway

1 US Department of State. (n.d.) 'Humanitarian Demining Program (HDP).' <http://www.state.gov/t/pm/65535.htm>.
2 Jonas Gahr Støre. (17 September 2007) 'Norway's commitment to mine action and human security.' <http://www.regjeringen.no/en/dep/ud/About-the-Ministry/Minister-of-Foreign-Affairs-Jonas-Gahr-S/Speeches-and-articles/2007/mineaction.html?id=481024>.
3 Jan Egeland. (1988) *Impotent Superpower—Potent Small State: Potentials and Limitations of Human Rights Objectives in the Foreign Polices of the United States and Norway.* Oslo, Norwegian University Press. p. 185. Another interesting comparative study of Norway and the US is: Wayne S. Cole. (1989) *Norway and the United States, 1905-1955: Two Democracies in Peace and War.* Ames, Iowa State University Press.
4 International Campaign to Ban Landmines (ICBL). (2007) 'Mine Action Funding.' *Landmine Monitor 2007.* <http://www.icbl.org/lm/2007/es/mine_action_funding.html>.
5 Egeland. *Impotent Superpower.* p. 5.
6 Chalmers Johnson. (2004) *The Sorrows of Empire: Militarism, Secrecy, and the End of the Republic.* New York, Metropolitan Books. pp. 124, 132-134; Dana Priest. (2004) *The Mission: Waging War and Keeping Peace with America's Military.* New York, W.W. Norton and Company. p. 75.
7 Rachel Stohl. (January/February 2008) 'Questionable Reward: Arms Sales and the War on Terrorism.' *Arms Control Today.* <http://www.armscontrol.org/act/2008_01-02/stohl.asp>.
8 Jeffrey E. Garten. (May/June 1997) 'Business and Foreign Policy.' *Foreign Affairs.* <http://www.foreignaffairs.org/19970501faessay3772/jeffrey-e-garten/business-and-foreign-policy.html>.
9 Carol Lancaster. (2007) *Foreign Aid: Diplomacy, Development, Domestic Politics.* Chicago, Chicago University Press. p. 102.
10 Vernon W. Ruttan. (1996) *United States Development Assistance Policy: The Domestic Politics of Foreign Economic Aid.* Baltimore, Johns Hopkins University Press. p. xviii.
11 Curt Tarnoff & Larry Nowels. (15 April 2004) 'Foreign Aid: An Introductory Overview of U.S. Programs and Policy.' *CRS Report for Congress.* Washington DC, Congressional Research Service. <http://usinfo.state.gov/usa/infousa/trade/files/98-916.pdf>. p. 2.
12 Egeland. *Impotent Superpower.* pp. 13-19.

[13] Helge Pharo. (2003) 'Altruism, Security and the Impact of Oil: Norway's Foreign Economic Assistance Policy, 1958-1971.' *Contemporary European History*. 12(4). p. 546.

[14] Ann Kelleher & James Larry Taulbee. (October 2006) 'Bridging the Gap: Building Peace Norwegian Style.' *Peace & Change*. 31(4); John Stephen Moolakkattu. (2005) 'Peace Facilitation by Small States: Norway in Sri Lanka.' *Cooperation and Conflict: Journal of the Nordic International Studies Association*. 40(4).

[15] Bergen Museum. (n.d.) 'Norwegian Missionaries Practice and Representation in the Formation
of 'Self' and 'Other', 1870-2005.'
<http://bergenmuseum.uib.no/nettutstillinger/mission/index.htm>.

[16] Morten Bøås. (2002) 'Public attitudes to aid in Norway and Japan.' University of Oslo Centre for Development and the Environment Working Paper 2002/03.
<http://www.sum.uio.no/publications/pdf_fulltekst/wp2002_03_boas.pdf>. p. 5.

[17] Olav Riste. (2001) *Norway's Foreign Relations - A History*. Oslo, Universitetsforlaget, p. 257.

[18] Pharo. 'Altruism, Security and the Impact of Oil.' p. 542.

[19] Organization for Economic Cooperation and Development (OECD). (2008) *Development Co-Operation Report 2007*. Paris, OECD. p. 137.

[20] Egeland. *Impotent Superpower*. p. 14.

[21] Egeland. *Impotent Superpower*. pp. 87-104; Tore Linné Eriksen & Anita Kristensen Kroken. (2000) "Fuelling the Apartheid War Machine': A Case Study of Shipowners, Sanctions and Solidarity Movements.' *Norway and National Liberation in Southern Africa*. Tore Linné Eriksen (Ed.). Stockholm, Nordiska Afrikainstitutet. pp. 193-210.

[22] Riste. *Norway's Foreign Relations*. p. 10.

[23] Joseph S. Nye. (2004) *Soft Power: The Means to Success in World Politics*. New York, PublicAffairs. pp. 9-10, 89, 112. See also: Mark Leonard. (September-October 2002) 'Diplomacy by Other Means.' *Foreign Policy*. 132. p. 53.

[24] Jan Braatha. (8 March 2007) Personal interview with author in Sarajevo.

[25] Brand Management Group. (18 December 2003) 'Norway's Public Diplomacy: a Strategy.' <http://www.brandmanagement.no/merkevareutvikling/hoyre/dbaFile12106.html>.

[26] Rubén Berríos. (2000) *Contracting for Development: The Role of For-Profit Contractors in U.S. Foreign Development Assistance*. Westport, Connecticut, Praeger Publishers.

[27] Mark Duffield. (2001) *Global Governance and the New Wars: The Merging of Development and Security*. London, Zed Books. pp. 75-107.

[28] Colin Powell. (26 October 2001) 'Remarks to the National Foreign Policy Conference for Leaders of Nongovernmental Organizations.'
<http://www.yale.edu/lawweb/avalon/sept_11/powell_brief31.htm>.

[29] Traci Hukill. (30 June 2003) 'U.S.: AID Chief Outlines Change In Strategy Since 2001 Terrorist Attacks.' *UN Wire*. <http://comunica.org/pipermail/cr-afghan_comunica.org/2003-June/000043.html>.

[30] NORAD. (September 2005) 'Civil society as a channel for Norwegian development assistance.' <http://www.norad.no/default.asp?V_ITEM_ID=3371>.

[31] Øyvind Østerud. (September 2005) 'Introduction: The Peculiarities of Norway.' *West European Politics*. 28(4). p. 714; Riste. *Norway's Foreign Relations*. p. 264.

[32] Kelleher & Taulbee. 'Bridging the Gap.'

[33] Kelleher & Taulbee. 'Bridging the Gap.' Moolakkattu. 'Peace Facilitation by Small States.' pp. 387-388.

[34] The Arms Project of Human Rights Watch & Physicians for Human Rights. (1993) *Landmines: A Deadly Legacy*. New York, Human Rights Watch. p. 17.

[35] Ibid. p. 28-29.

[36] Ibid. p. 61. Similar statements continue to be made by the State Department.

37 Ibid. pp. 63-77.
38 International Campaign to Ban Landmines (ICBL). (1999) 'United States of America.' *Landmine Monitor 1999*. <http://www.icbl.org/lm/1999/usa>.
39 J. Antonio Ohe. (2004) 'Are Landmines Still Needed to Defend South Korea?: A Mine Use Case Study.' *Landmines and Human Security: International Politics and War's Hidden Legacy*. Richard A. Matthew, Bryan McDonald and Kenneth R. Rutherford (Eds.). Albany, State University of New York Press.
40 International Campaign to Ban Landmines (ICBL). (1999) 'Norway.' *Landmine Monitor 1999*. <http://www.icbl.org/lm/1999/Norway>; International Campaign to Ban Landmines (ICBL). (2004) 'Norway.' *Landmine Monitor 2004*. <http://www.icbl.org/lm/2004/norway>.
41 Norwegian Ministry of Foreign Affairs. (n.d.) 'The Norwegian Government's initiative for a ban on cluster munitions - Questions and answers.' <http://www.regjeringen.no/en/dep/ud/selected-topics/Humanitarian-efforts/The-Norwegian-Governments-initiative-for/The-Norwegian-Governments-initiative-for.html?id=449708>.
42 Leon V. Sigal. (2006) *Negotiating Minefields: The Landmines Ban in American Politics*. New York, Routledge. pp. 14-16; Mary Wareham. (1998) 'Rhetoric and Policy Realities in the United States.' *To Walk without Fear: The Global Movement to Ban Landmines*. Maxwell A. Cameron, Robert J. Lawson & Brian W. Tomlin (Eds.). Oxford, Oxford University Press. p. 213.
43 Sigal. *Negotiating Minefields*. pp. 16-19; United States of America. (2 January 2006) 'Title 22 - Foreign Relations and Intercourse: Chapter 39 - Arms Export Control, Subchapter III - Military Export Controls.' 22 US Code Section 2778. <http://uscode.house.gov/search/criteria.shtml>.
44 Sigal. *Negotiating Minefields*. pp. 47-51; UN General Assembly. (15 December 1994) 'Moratorium on the export of anti-personnel land-mines.' Document A/RES/49/75. <http://www.un.org/Depts/dhl/res/resa49.htm>. p. 6.
45 Patrick Leahy. (Fall 1999) 'Leahy Amendment Moratorium on Use of Anti-personnel Landmines.' *Journal of Mine Action*. 3(3). <http://maic.jmu.edu/Journal/3.3/profiles/leahy_war_victims_fund.htm>.
46 Steffen Kongstad. (3 September 2007) Personal interview with author in Oslo.
47 Per Breivik. (28 February 2007) Personal interview with author in Sarajevo.
48 ICBL. (1999) 'Norway.'
49 Ibid.
50 J. Ashley Roach. (1984) 'Certain Conventional Weapons Convention: Arms Control or Humanitarian Law?' *Military Law Review*. 105. p. 4.
51 Wareham. 'Rhetoric and Policy Realities.' p. 214.
52 e.g. Sarah B. Sewall & Carl Kaysen (Eds). (2000) *The United States and the International Criminal Court: National Security and International Law*. Oxford, Rowman & Littlefield.
53 Wareham. 'Rhetoric and Policy Realities.' p. 222.
54 For background on Canada's decision to initiate this process, see: Sigal. *Negotiating Minefields*. pp. 89-102.
55 Støre. 'Norway's commitment to mine action.'
56 Robert J. Lawson, Mark Gwozdecky, Jill Sinclair & Ralph Lysyshyn. (1998) 'The Ottawa Process and the International Movement to Ban Anti-Personnel Mines.' In: Cameron, Lawson & Tomlin (Eds.). *To Walk without Fear*. p. 171.
57 Støre. 'Norway's commitment.'
58 (2004) '1997 Ottawa Convention on the Prohibition of the Use, Stockpiling, Production and Transfer of Anti-Personnel Mines and on their Destruction.' *Documents on the Laws of*

War. Adam Roberts & Richard Guelff (Eds.). 3rd Ed. Oxford, Oxford University Press. pp. 645-666.
59 ICBL. (1999) 'Norway.'
60 Wareham. 'Rhetoric and Policy Realities.' pp. 226, 230-233.
61 For a detailed discussion of the debate within the US government on Ottawa, see: Sigal. *Negotiating Minefields.* pp. 103-154.
62 Sigal. *Negotiating Minefields.* pp. 194-199; Jody Williams & Stephen Goose. (1998) 'The International Campaign to Ban Landmines.' In: Cameron, Lawson & Tomlin (Eds.). *To Walk without Fear.* p. 43.
63 Williams & Goose. 'International Campaign to Ban Landmines.' p. 44.
64 Ibid. p. 44.
65 ICBL. (1999) 'United States of America.'
66 Richard Kidd. (21 November 2007) 'U.S. Landmine Policy and the Ottawa Convention Ban on Anti-Personnel Landmines: Similar Path.' <http://www.state.gov/t/pm/rls/rm/95596.htm>.
67 US Department of State. (27 February 2004) 'New United States Policy on Landmines: Reducing Humanitarian Risk and Saving Lives of United States Soldiers.' <http://ccwtreaty.state.gov/022704landmines.htm>.
68 Ibid.
69 Ibid.
70 Lincoln P. Bloomfield. (27 February 2004) 'New Developments in the U.S. Approach to Landmines: Lincoln P. Bloomfield, Jr., Assistant Secretary for Political-Military Affairs: On-The-Record Briefing.' <http://www.state.gov/t/pm/rls/rm/29976.htm>.
71 US Department of State. 'New United States Policy on Landmines.'
72 Richard Kidd. (4 April 2007) 'State Official Discusses U.S. Policy on Land Mines.' <http://usinfo.state.gov/usinfo/Archive/2007/Apr/04-991121.html>.
73 US Department of State. 'New United States Policy on Landmines.'
74 Bloomfield. 'New Developments in the U.S. Approach.'
75 Christopher Moraff. (3 October 2006) 'Along Came a Spider.' *The American Prospect.* <http://www.prospect.org/cs/articles?article=along_came_a_spider>; ICBL. (2007) 'United States of America.' *Landmine Monitor 2007.* <http://www.icbl.org/lm/2007/usa.html>; US Campaign to Ban Landmines. (2006) 'Evolution of U.S. Policy on Antipersonnel Landmines.' <http://www.banminesusa.org/policy/evolution.html>.
76 ICBL. (2007) 'United States of America.'
77 (28 November 2003) 'Protocol on Explosive Remnants of War (Protocol V to the 1980 Convention).' <http://www.mineaction.org/docs/1850_.asp>. Article 4. For further details see: Rosy Cave. (2006) 'Disarmament as Humanitarian Action? Comparing Negotiations on Anti-Personnel Mines and Explosive Remnants of War.' *Disarmament as Humanitarian Action: From Perspective to Practice.* Geneva, UNIDIR.
78 Ronald Bettaur. (6 November 2006) 'Third Review Conference of the Convention on Conventional Weapons: Press Statement by Ronald Bettauer.' <http://www.state.gov/t/pm/rls/rm/75707.htm>.
79 Roald Næss. (7 November 2006) 'General statement by Norway at the Third Review Conference of the CCW.' <http://www.regjeringen.no/nb/dep/ud/dep/org/avdelinger/Avdeling_for_sikkerhets politikk_og_bilat/Seksjon_for_nedrustning_og_ikke_sprednin/Seniorradgiver-Roald-Nass/taler_artikler/2006/General-statement-by-Norway-at-the-Third-Review-Conference-of-the-CCW.html?id=436988>.

80 Ronald J. Bettauer. (17 November 2006) 'Statement to the Closing Plenary Session of the Third Review Conference of the Convention on Certain Conventional Weapons (CCW).' <http://www.state.gov/t/pm/rls/rm/76261.htm>.

81 Norwegian Ministry of Foreign Affairs. 'The Norwegian Government's initiative.'

82 Oslo Conference on Cluster Munitions. (22 – 23 February 2007) 'Declaration.' <http://www.stopclustermunitions.org/files/Oslo%20declaration.pdf>.

83 (18 March 2009) 'Convention on Cluster Munitions.' <http://www.clusterconvention.org>.

84 Bettauer. 'Statement to the Closing Plenary Session.'

85 Richard Kidd. (20 June 2007) 'U.S. Intervention on Humanitarian Impacts of Cluster Munitions.' <http://www.state.gov/t/pm/rls/rm/87303.htm>. For a refutation of this speech see: Richard Moyes. (Autumn 2007) 'Spiked!' *Landmine Action Campaign*. 13. <http://www.landmineaction.org/resources/Campaign%2013.pdf>. p. 10.

86 Bettauer. 'Statement to the Closing Plenary Session.'

87 Task Force 65. (May 1975) 'Suez Canal Clearance Operation, Task Force 65.' DTIC Accession No. ADA010261. <http://stinet.dtic.mil/oai/oai?verb=getRecord&metadataPrefix=html&identifier=ADA010261>.

88 US Department of State. (29 July 2003) 'Milestones in Humanitarian Mine Action: Development of the Landmine Threat and the Discipline of Humanitarian Mine Action.' <http://www.state.gov/t/pm/rls/fs/22948.htm>; US Department of State. (2003) 'Operation Lempira.' <http://www.state.gov/r/pa/ei/pix/b/22970.htm>.

89 RONCO. (June 1998) 'Humanitarian Demining: Ten Years of Lessons.' *The Journal of Humanitarian Demining*. 2(2). <http://maic.jmu.edu/journal/2.2/field/ronco.htm>; Margaret Buse. (June 2000) 'RONCO Executives Talk About Demining, Integration and the IMAS Contract: (An Interview with Lawrence Crandall, Stephen Edelmann and A. David Lundberg).' *Journal of Mine Action*. 4(2). <http://maic.jmu.edu/journal/4.2/Features/ronco/ronco.htm>; Dave McCracken. (2001) 'Thailand: The Land of Smiles (But Be Careful Where You Step!)' *Journal of Mine Action*. 5(1); Dan Hayter. (April 2003) 'The Evolution of Mine Detection Dog Training.' *Journal of Mine Action*. Issue 7(1). <http://maic.jmu.edu/journal/7.1/features/hayter/hayter.htm>.

90 John Prados. (1996) *Presidents' Secret Wars: CIA and Pentagon Covert Operations from World War II through the Persian Gulf*. Rvsd Ed. Chicago, Elephant Paperbacks. p. 480.

91 James M. Scott. (1996) *Deciding to Intervene: The Reagan Doctrine and American Foreign Policy*. Durham, North Carolina, Duke University Press. pp. 34-35.

92 US Agency for International Development (USAID). (24 October 1988) 'A.I.D. Strategy: Afghan Resettlement and Rehabilitation.' *Development Experience Clearinghouse*. Document PN-ABR-629. <http://www.dec.org/pdf_docs/PNABR629.pdf>. p. 7; US Agency for International Development (USAID). (1 June 1989) 'Afghanistan: Briefing for the Deputy Administrator-designate.' *Development Experience Clearinghouse*. Document PN-ABR-629. <http://www.dec.org/pdf_docs/PNABR629.pdf>. pp. 3-4.

93 Kurt Lohbeck. (1993) *Holy War, Unholy Victory: Eyewitness to the CIA's Secret War in Afghanistan*. Washington DC, Regnery Gateway. pp. 92-93; RONCO. (1994) 'USAID/Afghanistan Commodity Export Program (CEP) Contract No. 306-0205-C-00-9384-00, March 1, 1989 through February 28, 1994: Final Report.' Available from USAID Development Experience Clearinghouse. Document PD-ABI-329. pp. 3-4; George Crile. (2004) *Charlie Wilson's War*. New York, Grove Press. p. 369; Steve Coll. (2004) *Ghost Wars: The Secret History of the CIA, Afghanistan and bin Laden, From the Soviet Invasion to September 10, 2001*. London, Penguin Books. p. 218.

NOTES 213

94 Scott. *Deciding to Intervene.* p. 58.
95 USAID. 'A.I.D. Strategy: Afghan Resettlement.' p. 11.
96 Martin Barber. (28 November 2006) Personal interview with the author in Kabul.
97 US Department of State. (1994) 'Chapter VII: Case Studies.' *Hidden Killers 1994: The Global Landmine Crisis.*
<http://www.state.gov/www/global/arms/rpt_9401_demine_ch7.html>.
98 US Department of State. (1994) 'Chapter V: Demining – The U.S. Response.' *Hidden Killers 1994: The Global Landmine Crisis.*
<http://www.state.gov/www/global/arms/rpt_9401_demine_ch5.html>.
99 Norway Mission to the UN. (2007) 'Norwegian Contributions to the UN.'
<http://www.norway-un.org/NorwayandtheUN/NorwegianContributions/Norwegian+contributions+to+the+UN.htm>.
100 Norway Mission to the UN. (2007) 'Norway and peacekeeping.' <http://www.norway-un.org/NorwayandtheUN/PeacekeepingOperations/Norway+and+Peace-Keeping.htm>.
101 Boutros Boutros Ghali. (17 June 1992) 'An Agenda for Peace: Preventive diplomacy, peacemaking and peace-keeping.' <http://www.un.org/Docs/SG/agpeace.html>.
102 UNOCA. (30 April 1990) '1990 Annual Reports Demining Headquarters Peshawar and Quetta.' Document PC-AAA-512. USAID Development Experience Clearinghouse. p. 22.
103 Hårvard Bach. (20 September 2006) Personal interview with author in Geneva.
104 NPA. (2008) 'NPA Humanitarian Mine Action 2007-2008.'
<http://www.npaid.org/filestore/5_3portfolio_A4_11-07.pdf>. p. 7.
105 Hildegard Scheu. (August 2002) 'Humanitarian Mine Action in Mozambique.' *Journal of Mine Action.* 6(2).
<http://maic.jmu.edu/journal/6.2/focus/hildegardscheu/hildegardscheu.htm>.
106 International Campaign to Ban Landmines (ICBL). (1999) 'Mozambique.' *Landmine Monitor.* <http://www.icbl.org/lm/1999/mozambique.html>.
107 Nina Monsen. (2001) 'The peace which never came.' *A Mine-Free World is Possible.* Norwegian People's Aid (Ed.). Oslo, NPA. p. 24.
108 US Department of State. (1 October 2007) '10th Anniversary of Public-Private Partnerships to Reinforce Humanitarian Mine Action.'
<http://www.state.gov/r/pa/prs/ps/2007/oct/93023.htm>.
109 ICBL. 'Mine Action Funding.'
110 US Department of State. 'Humanitarian Demining Program (HDP).' See also: Stacy Bernard Davis & Donald F. Patierno. (2004) 'Tackling the Global Landmine Problem: The United States Perspective.' *Landmines and Human Security: International Politics and War's Hidden Legacy.* Richard A. Matthew, Bryan McDonald and Kenneth R. Rutherford (Eds.). Albany, State University of New York Press. pp. 126-127.
111 Interagency Working Group on Humanitarian Demining n.d.
112 US Department of State. (n.d.) 'PCC Subgroup on Humanitarian Mine Action.'
<http://www.state.gov/t/pm/wra/c4306.htm>.
113 Ibid.
114 Ibid.
115 US Department of State. (June 2006) 'To Walk the Earth in Safety: The U.S. Commitment to Humanitarian Mine Action.'
<http://www.state.gov/t/pm/rls/rpt/walkearth/2006/68014.htm>.
116 Sigal. *Negotiating Minefields.* p. 220; Wareham. 'Rhetoric and Policy Realities.' p. 240.
117 Kidd. 'U.S. Landmine Policy and the Ottawa Convention.'

118 Richard Kidd. (8 August 2006) Personal interview with author in Washington DC.
119 (2004) '1997 Ottawa Convention.' Article 5.1. p. 650.
120 Kidd. 'U.S. Landmine Policy and the Ottawa Convention.'
121 Jacquelyn S. Porth. (14 December 2007) 'U.S. Still Top Financial Contributor to Humanitarian Mine Action.' USINFO. <http://www.america.gov/st/washfile-english/2007/December/20071214110213sjhtrop0.828869.html>.
122 Kidd. Personal interview.
123 Tarnoff & Nowels. 'Foreign Aid.' p. 22.
124 US Department of State. (5 April 2002) 'The U.S. Humanitarian Demining Program and NADR Funding.' <http://www.state.gov/t/pm/rls/fs/2002/9183.htm>.
125 Kidd. 'State Official Discusses.'
126 John Stevens. (5 December 2007) 'Re: Reply to your query about PM/WRA's contract and grant funding.' Personal email to author.
127 Kidd. Personal interview.
128 John Stevens. (13 December 2007) 'Re: Reply to your query about PM/WRA's contract and grant funding.' Personal email to author.
129 Kidd. Personal interview.
130 US Department of State. (9 May 2005) 'U.S. Department of State Awards Multiple Contracts to Clean Up Battlefields and Control Conventional Weapons.' <http://www.state.gov/r/pa/prs/ps/2005/45859.htm>.
131 Stevens. (13 December 2007) 'Re: Reply to your query.'
132 John Stevens. (4 December 2007) 'Reply to your query about PM/WRA's contract and grant funding.' Personal email to author.
133 Ibid.
134 Lois Carter Fay. (23 August 2006) Telephone conversation with author.
135 Stevens. (5 December 2007) 'Re: Reply to your query.'
136 Ibid.
137 Stevens. (4 December 2007) 'Reply to your query.'
138 US Department of State. (n.d.) 'International Trust Fund.' <http://www.state.gov/t/pm/65536.htm>.
139 e.g. US Department of Defense. (28 March 2007) 'Contracts: Defense Logistics Agency.' <http://www.defenselink.mil/contracts/contract.aspx?contractid=3483>.
140 Tarnoff & Nowels. 'Foreign Aid.' p. 22.
141 US Department of Defense. (March 2006) 'Defense Security Cooperation Agency.' <http://www.dsca.osd.mil/Default.htm>.
142 US Department of Defense. (17 April 2006) 'Humanitarian Demining Training Center.' <http://www.wood.army.mil/hdtc/default.htm>.
143 US Department of State. 'The U.S. Humanitarian Demining Program.'
144 Tom Smith. (August 2006) Personal interview with author in Crystal City, Virginia.
145 Ibid.
146 Defense Security Cooperation Agency. (February 2006) 'Humanitarian and Civic Assistance (HCA) and Humanitarian Mine Action (HMA) Programs of the Department of Defense.' Document obtained by author through Freedom of Information Act Request 06-F-2708 to the Department of Defense. p. 1.
147 Tom Smith. Personal interview.
148 Stuart Maslen. (2004) *Mine Action After Diana: Progress in the Struggle Against Landmines.* London, Pluto Press. pp. 138-140.
149 US Department of State. 'The U.S. Humanitarian Demining Program.'
150 Maslen. *Mine Action After Diana.* pp. 49-53, 151-152.
151 US Department of State. 'To Walk the Earth in Safety.'

NOTES

152 US Department of State. 'The U.S. Humanitarian Demining Program.'; US Agency for International Development (USAID). (5 May 2005) 'Leahy War Victims Fund.' <http://www.usaid.gov/our_work/humanitarian_assistance/the_funds/lwvf/index.html>.

153 Patrick Leahy. (1993) 'Congressional Record – Senate. S9290.' *Landmines: A Deadly Legacy*. The Arms Project of Human Rights Watch & Physicians for Human Rights. New York, Human Rights Watch. Appendix 8.

154 Centers for Disease Control and Prevention. (November 2004) 'War-related Injury Prevention.' <http://www.cdc.gov/nceh/publications/factsheets/War-relatedInjuryPrevention.pdf>.

155 Amy Joyce. (4 January 2002) 'A Mission to Remove Land Mines; D.C. Firm Is Sending Team to Afghanistan to Help Defuse Devices.' *The Washington Post*. p. E5.

156 Michael Dobbs. (11 December 2000) 'U.S. Advice Guided Milosevic Opposition.' *The Washington Post*. <http://www.washingtonpost.com/ac2/wp-dyn/A18395-2000Dec3?language=printer>. p. A01.

157 RONCO. (April 2003) 'Mine Detection Dogs: An Integral Tool in RONCO Mine Clearance Operations' *Journal of Mine Action*. 7(1). <http://maic.jmu.edu/journal/7.1/features/ronco/ronco.htm>. John Lundberg. (August 2005) 'Reflecting on 10 Years of RONCO Operations in Mine Action.' *Journal of Mine Action*. 9(1).

158 Lundberg. 'Reflecting on 10 Years of RONCO.'

159 Michael R. Gordon. (21 October 2004) 'Abolishing the Iraq Army: The Fallout.' *International Herald Tribune*.

160 Kris Hundley. (5 July 2004) 'Money, and Worries, in Iraq.' *St Petersburg Times*. <http://www.stpetetimes.com/2004/07/05/news_pf/Business/Money_and_worries_i.shtml>.

161 US Department of State. (20 August 1999) 'State Department Statement on Mine Action Support Contract.' *USIS Washington File*. <http://canberra.usembassy.gov/hyper/WF990823/epf105.htm>.

162 US Department of State. 'U.S. Department of State Awards.'

163 Group 4 Securicor. (4 April 2008) 'G4S plc Acquisition of RONCO Consulting Corporation.' <http://www.g4s.com/home/home-news_and_media/home-news_and_media-pr.htm?id=46702>.

164 Wackenhut Services Incorporated. (n.d.) 'U.S. Government Services.' <http://www.wsihq.com/>.

165 Solomon Hughes. (2007) *War on Terror, Inc.: Corporate Profiteering from the Politics of Fear*. London, Verso. pp. 13-38; Service Employees International Union. (n.d.) *Focus on G4S*. <http://www.focusong4s.org/>; Service Employees International Union. (n.d.) *Eye on Wackenhut*. <http://www.eyeonwackenhut.com/>; Gregory Palast. (26 September 1999) 'US: Wackenhut's Free Market in Human Misery.' *The Observer*. <http://www.corpwatch.org/article.php?id=868>.

166 ICBL. (1999) 'Norway.'; Annette Abelsen. (26 April 2007) 'Cooperation and assistance in clearing mined areas - Priorities for Norway in resource allocation.' <http://www.regjeringen.no/se/dep/ud/Departemeantta-birra/Organisauvdna/Ossodagat/ONa-rafi-ja-humanitara-gaaldagaid-ossodat/Seksjon-for-humanitare-sporsmal/aab/taler/minedareas.html?id=467173>.

167 ICBL. 'Mine Action Funding.'

168 International Campaign to Ban Landmines (ICBL). (2006) 'Norway.' *Landmine Monitor 2006*. <http://www.icbl.org/lm/2006/norway>.

169 ICBL. (2004) 'Norway.'

170 Abelsen. 'Cooperation and assistance.'
171 Ibid.
172 ICBL. (2004) 'Norway.' A similar breakdown is unfortunately not available for later years.
173 Yngvild Berggrav. (3 September 2007) Personal interview with author in Oslo.
174 Stig Traavik. (5 September 2007) Personal interview with the author in Oslo.
175 Berggrav. Personal interview.
176 Susan Eckey. (3 September 2007) Personal interview with author in Oslo.
177 Kongstad. Personal interview.
178 Abelsen. 'Cooperation and assistance.'
179 Kongstad. Personal interview.
180 Braatha. Personal interview.
181 Traavik. Personal interview.
182 Compiled by author from: ICBL. (2004) 'Norway.' A similar breakdown is unfortunately not available for later years.
183 Peace Research Institute, Oslo (PRIO). (n.d.) 'Assistance to Mine-Affected Communities (AMAC).' <http://www.prio.no/amac>.
184 Abelsen. 'Cooperation and assistance.'
185 Kongstad. Personal interview.
186 Abelsen. 'Cooperation and assistance.'
187 Braatha. Personal interview.
188 Norwegian People's Aid (NPA). (2007) 'Annual Report 2006.' <http://www.npaid.org/filestore/FolkehjelpEngelsk.pdf>. pp. 14, 17.
189 Aid Watch. (6 March 2006) 'Norwegian People's Aid [Norsk Folkehjelp].' <http://www.observatoire-humanitaire.org/fusion.php?l=GB&id=75>.
190 Norwegian People's Aid (NPA). (22 June 2003) 'Solidarity: principles and value basis for Norwegian Peoples's Aid.' Oslo, NPA.
191 NPA internal document quoted in: Aid Watch. 'Norwegian People's Aid.'
192 Eva Helene Østbye. (2000) 'The South African Liberation Struggle: Official Norwegian Support.' In: Eriksen (Ed.). *Norway and National Liberation*. pp. 148-149; Vesla Vetlesen. (2000) 'Trade Union Support to the Struggle Against Apartheid: the Role of the Norwegian Confederation of Trade Unions.' In: Eriksen (Ed.). *Norway and National Liberation*. pp. 340-342.
193 Aid Watch. (6 March 2006) 'Norwegian People's Aid [Norsk Folkehjelp].' <http://www.observatoire-humanitaire.org/fusion.php?l=GB&id=75>.
194 Ketil Volden. (2001) 'Working among a forgotten people.' *A Mine-Free World is Possible.* NPA (Ed.). Oslo, NPA. pp. 8-9.
195 Norwegian People's Aid (NPA). (8 March 1999) 'Strategy for Mine Action 1999-2002 for Norwegian People's Aid.' In NPA archives. p. 9.
196 NPA. 'NPA Humanitarian Mine Action 2007-2008.' p. 4.
197 Norwegian People's Aid (NPA). (2004) 'NPA Humanitarian Mine Action 2004.' Oslo, NPA.
198 NPA. 'Annual Report 2006.' p. 11.
199 Breivik. (28 February 2007) Personal interview; Bach. Personal interview; Laila Nikolaisen. (4 September 2007) Personal interview with author in Oslo.
200 Aid Watch. 'Norwegian People's Aid.'
201 Max Weber. (1958) 'Religions Rejections of the World and Their Directions.' *From Max Weber: Essays in Sociology*. Hans Heinrich Gerth and C. Wright Mills (Trans.). New York, Oxford University Press. p. 280.
202 Wareham. 'Rhetoric and Policy Realities.' p. 212.
203 Sigal. *Negotiating Minefields*. p. 62.

4. Implementation in Afghanistan, Bosnia and Sudan

[1] This section is based on: Matthew Bolton. (2008) 'Goldmine? A Critical Look at the Commercialization of Afghan Demining.' Centre for the Study of Global Governance (LSE) Research Paper 01/2008. <http://www.lse.ac.uk/Depts/global/Publications/ResearchPapers/RP_0108.pdf>.

[2] For more details about the landmine situation in Afghanistan see: Survey Action Center. (2006) *Landmine Impact Survey – Islamic Republic of Afghanistan*. Takoma Park, MD, Survey Action Center.

[3] US Agency for International Development (USAID). (24 October 1988) 'A.I.D. Strategy: Afghan Resettlement and Rehabilitation.' *Development Experience Clearinghouse*. Document PN-ABR-629. <http://www.dec.org/pdf_docs/PNABR629.pdf>. p. 7; US Agency for International Development (USAID). (1 June 1989) 'Afghanistan: Briefing for the Deputy Administrator-designate.' *Development Experience Clearinghouse*. Document PN-ABR-629. <http://www.dec.org/pdf_docs/PNABR629.pdf>. p. 3-4.

[4] Larry Crandall. (1 July 2006) Personal interview with author in McLean, Virginia. cf. Kurt Lohbeck. (1993) *Holy War, Unholy Victory: Eyewitness to the CIA's Secret War in Afghanistan*. Washington DC, Regnery Gateway. p. 92.

[5] Crandall. Personal interview.

[6] RONCO. (18 April 1991) 'Proposal to Transfer Management of MDD Program from RONCO Consulting Corporation to Afghan Technical Consultants, UNOCA.' Available from USAID Development Experience Clearinghouse. Document PD-ABJ-299. p. iii.

[7] Dan Hayter. (April 2003) 'The Evolution of Mine Detection Dog Training.' *Journal of Mine Action*. 7(1). <http://maic.jmu.edu/journal/7.1/features/hayter/hayter.htm>; Margaret Buse. (June 2000) 'RONCO Executives Talk About Demining, Integration and the IMAS Contract: (An Interview with Lawrence Crandall, Stephen Edelmann and A. David Lundberg).' *Journal of Mine Action*. 4(2). <http://maic.jmu.edu/journal/4.2/Features/ronco/ronco.htm>.

[8] Adam Kelliher. (9 March 1989) 'Anti-personnel mines maim Afghans.' *United Press International*.

[9] Aid Watch. (10 June 2003) 'Handicap International.' <http://www.observatoire-humanitaire.org/fusion.php?l=GB&id=21>.

[10] Rae McGrath. (2000) *Landmines and Unexploded Ordnance: A Resource Book*. London, Pluto Press. pp. 116, 238; Sayed Aqa. (August 2005) 'Mine Action: Success and Challenges.' *Journal of Mine Action: A Retrospective on Mine Action*. 9.1. <http://maic.jmu.edu/JOURNAL/9.1/Focus/aqa/aqa.htm>.

[11] Guy Willoughby. (21 September 2007) Personal interview with author in Thornhill, Scotland.

[12] Farid Homayoun. (19 November 2006) Personal interview with author in Kabul.

[13] HALO Trust. (2 November 2006) 'The HALO Trust Afghanistan.' PowerPoint Presentation given to author by Dr. Farid Homayoun. Slide 7; Homayoun. Personal interview.

[14] UNOCA. (December 1988) *Operation Salam News*. 1.

[15] UNOCA. (1992) 'UNOCA Demining Programme.' Document PC-AAA-509. USAID Development Experience Clearinghouse. pp. 2, 7.

[16] Demining Headquarters Peshawar and Quetta. (1990) '1990 Annual Reports: Demining Headquarters Peshawar and Quetta.' Document PC-AAA-512. USAID Development Experience Clearinghouse. p. 17.

17 Martin Barber. (28 November 2006) Personal interview with author in Kabul.
18 Ibid.
19 Robert Eaton, Chris Horwood & Norah Niland. (1997) *Afghanistan: The Development of Indigenous Mine Action Capacities*. New York, UN Department of Humanitarian Affairs Lessons Learned Unit. p. 14, footnote 13.
20 UNOCHA (1993) 'Afghanistan: Mine Clearance Programme for 1993: Annual Report 1992.' Document PC-AAA-511. USAID Development Experience Clearinghouse. p. 6.
21 Barber. Personal interview.
22 UN Mine Action Center for Afghanistan. (UNMACA). (2006) 'Mine Action Programme for Afghanistan (MAPA) funding 1994 / 2004 - Voluntary Trust Fund and Bilateral funds.' Excel spreadsheet given to author.
23 Mohammed Shohab Hakimi. (6 November 2006) Personal interview with author in Kabul.
24 Eaton, Horwood & Niland. *Afghanistan*.
25 For an analysis of Taliban and North Alliance attitudes on landmines see: Geneva Call. (2005) *Armed Non-State Actors and Landmines*. Geneva, Program for the Study of International Organization(s). pp. 65-67. See also: Shalini Chawla. (June 2000) 'Diffusion of Landmines in Afghanistan.' *Strategic Analysis: A Monthly Journal of the IDSA*. 24(3). <http://www.ciaonet.org/olj/sa/sa_jun00chs01.html#note8>.
26 US Agency for International Development (USAID). (2006) 'U.S. Overseas Loans and Grants [Greenbook].' <http://qesdb.usaid.gov/gbk/>. Donation data from: UNMACA. 'Mine Action Programme for Afghanistan (MAPA) funding.'
27 Statistics Norway. (2007) 'Public expenditure on development aid etc. Bilateral and multi-bi assistance, by recipient countries. NOK million.' <http://www.ssb.no/english/yearbook/tab/tab-481.html>. Donation data from: UNMACA. 'Mine Action Programme for Afghanistan (MAPA) funding.'
28 Donation data from: UNMACA. 'Mine Action Programme for Afghanistan (MAPA) funding.'
29 Article 8 prohibits NGOs from 'Participation in construction projects and contracts' except in 'exceptional cases' granted by the Minister of Economy 'Government of Afghanistan. (June 2005) 'Law on Non-Governmental Organizations.' *Official Gazette*. 857/2005. <http://www.usig.org/countryinfo/laws/Afghanistan/Afghan%20NGO%20Law%20Final%20ENG%20(10.July.05).pdf>.
30 Dean Hutson. (30 November 2006) Personal interview with author in Kabul.
31 US Agency for International Development (USAID). (May 2005) 'USAID/Afghanistan Strategic Plan: 2005-2010.' <http://www.usaid.gov/locations/asia_near_east/afghanistan/Afghanistan_2005-2010_Strategy.pdf>. pp. 1, 5.
32 John Lundberg. (August 2005) 'Reflecting on 10 Years of RONCO Operations in Mine Action.' *Journal of Mine Action*. 9(1).
33 John L. Wilkinson. (December 2002) 'Demining During Operation Enduring Freedom in Afghanistan.' *Journal of Mine Action*. 6(3). <http://maic.jmu.edu/JOURNAL/6.3/notes/wilkinson/wilkinson.htm>.
34 Arlington National Cemetery Website. 'Justin Joseph Galewski.' <http://www.arlingtoncemetery.net/jjgalewski.htm>.
35 Lundberg. 'Reflecting on 10 Years of RONCO Operations.'; RONCO. (April 2003) 'Mine Detection Dogs: An Integral Tool in RONCO Mine Clearance Operations' *Journal of Mine Action*. 7(1). <http://maic.jmu.edu/journal/7.1/features/ronco/ronco.htm>.

36 US Department of Defense. (28 March 2007) 'Contracts.'
<http://www.defenselink.mil/contracts/contract.aspx?contractid=3483>.
37 Colin Wanley. (30 October 2006) Personal interview with author in Kabul.
38 Lundberg. 'Reflecting on 10 Years of RONCO Operations.'; RONCO. 'Mine Detection Dogs.'; Robert Gannon. (20 November 2006) Personal interview with author in Kabul.
39 Lundberg. 'Reflecting on 10 Years of RONCO Operations.'
40 Anonymous. (8 September 2005) 'DynCorp International to Remove Land Mines In Afghanistan.' *Business Wire*.
<http://www.findarticles.com/p/articles/mi_m0EIN/is_2005_Sept_8/ai_n15375022>.
The US has also provided funding to the HALO Trust for weapons destruction.
41 Kenneth Katzman. (10 September 2007) 'Afghanistan: Post-War Governance, Security, and U.S. Policy.' *CRS Report for Congress*.
<http://www.fas.org/sgp/crs/row/RL30588.pdf>. p. 20.
42 George W. Bush. (15 February 2007) 'President Bush Discusses Progress in Afghanistan, Global War on Terror.' <http://www.state.gov/p/sca/rls/rm/2007/80548.htm>.
43 US Agency for International Development (USAID). (22 September 2006) 'USAID Awards $1.4 Billion Contract for Infrastructure in Afghanistan.'
<http://www.usaid.gov/press/releases/2006/pr060922.html>.
44 US Agency for International Development (USAID). (1 August 2007) 'Strategic Provincial Roads-South & East Afghanistan (SPR-SEA).'
<http://www07.grants.gov/search/search.do;jsessionid=Gk3WKnGGcWj24tbJ19hMgy51KwXFPgJ1695s7cgQ5XtLpJhHczyY!696297004?oppId=14852&flag2006=true&mode=VIEW >.
45 USAID. 'USAID/Afghanistan Strategic Plan: 2005-2010.' p. 7.
46 Chris Johnson & Jolyon Leslie. (2004) *Afghanistan: The Mirage of Peace*. London, Zed Books. p. 106.
47 Hutson. Personal interview.
48 Lloyd Carpenter. (21 November 2006) Personal interview with author in Kabul.
49 Cranfield University. (December 2006) 'Minutes of the Strategic Organizational development Workshop.' p. 7-8. Document given to author.
50 Anne-Grethe Strøm-Erichsen. (January 2007) 'Norway's Interests in the North.' *Norwegian Defence Review: Status of Norwegian Defence 2007*. p. 7.
51 Tom Robertsen. (2007) 'Making New Ambitions Work: The Transformation of Norwegian Special Operations Forces.' *Defence and Security Studies* (Norwegian Institute for Defence Studies). 1/2007. pp. 42, 59, 61.
52 Jan Reksten. (January 2007) 'International Operations – Where Are We Now?' *Norwegian Defence Review: Status of Norwegian Defence 2007*. pp. 26-28.
53 Stig Traavik. (5 September 2007) Personal interview with the author in Oslo.
54 News of Norway. (16 January 2002) 'Additional Norwegian support for Afghanistan.'
<http://www.norway.org/News/archive/2001/200105afghanistan.htm>; US Department of Defense. (February 2002) 'Multinational Effort Aims to Rid War-torn Afghanistan of Mines.' *Defend America*.
<http://www.defendamerica.mil/articles/feb2002/a021902b.html>.
55 Royal Norwegian Embassy in Kabul. (2007) 'Norwegian led PRT in Faryab.'
<http://www.norway.org.af/prt/faryab/>. Nonetheless, the following report questioned whether Norway was doing enough to draw a clear line between military and civilian activities: Petter Bauck, Arne Strand, Mohammad Hakim and Arghawan Akbari. (2007) 'Afghanistan: An Assessment of Conflict and Actors in Faryab Province to Establish a Basis for Increased Norwegian Civilian Involvement.' *CMIReport*. R2007:1.

<http://www.cmi.no/publications/publication/?2594=afghanistan-an-assessment-of-conflict-and-actors>.
56 Homayoun. Personal interview.
57 This section, though substantially rewritten and paraphrasing Hugh Griffiths' contributions, draws largely on: Matthew Bolton & Hugh Griffiths. (September 2006) *Bosnia's Political Landmines: A Call for Socially Responsible and Conflict-Sensitive Mine Action.* London, Landmine Action.
58 For an extremely detailed look at Bosnia's mine problem, see: Darvin Lisica. (2006) *Risk Management in Mine Action Planning.* Sarajevo, Bosnia and Herzegovina, Ministry of Civil Affairs & Norwegian People's Aid. pp. 42-45, 62-94.
59 General Framework Agreement for Peace in Bosnia and Herzegovina. (14 December 1995) 'Annex 1A: Agreement on the Military Aspects of the Peace Settlement.' <http://www.ohr.int/dpa/default.asp?content_id=368>. Article IV, 2e.
60 Lisica. *Risk Management.* pp. 115-150; Financial Police Head Inspectors Office Sarajevo. (12 December 2000) 'Prvi Izvejestaj: O utvrdjenim cinjenicama I prikupljenim dokazima u postupku istrage, nacina raspolaganja sredstvima namijenjenim programima deminiranja u Bosni i Hercegovini odnosno Federaciji Bosne i Hercegovine, zloupotrebama utvrdjenim u dosadasnjem toku istrage te njihovim nosiocima.' ('First report: On established facts and gathered evidence during the investigation on ways of handling funds meant for demining programs in BiH, that is FBiH, abuses proved during the investigation and on their perpetrators.') Sarajevo, Ministry of Finance. Obtained by author. pp. 2-10.
61 For details on the US intervention, see: Richard Holbrooke. (1999) *To End a War.* New York, Random House; Brendan Simms. (2003) *Unfinest Hour: Britain and the Destruction of Bosnia.* New Ed. New York, Penguin Global.
62 Tracy Wilkinson. (11 November 1997) 'In Bosnia, U.S. Creeps Deeper.' *Los Angeles Times.* p. A1.
63 Philip P. Pan. (31 December 1995) 'Rockville Man Is First American Injured in Bosnian Peace Mission.' *The Washington Post.* p. A24.
64 Jacquetta A. Dunmyer & Adele Nicholson. (1996) 'A Note on the Operational Environment in Bosnia.' *Drawing a Line in the Mud.* Fort Leavenworth, Kansas, Center for Army Lessons Learned. <http://www.globalsecurity.org/military/library/report/call/call_96-5_sec1zos.htm>.
65 HQ SFOR. (August 2003) 'A Joint Demining Battalion for BiH: A Proposal for the Reorganisation of Armed Forces Demining in Bosnia-Herzegovina.' Unpublished document obtained by author. See also: Presidency of Bosnia and Herzegovina. (June 2005) *Defense White Paper of Bosnia and Herzegovina.* Sarajevo, Bosnia and Herzegovina. pp. 31-33.
66 Geneva International Center for Humanitarian Demining (GICHD). (June 2003) *The Role of the Military in Mine Action.* Geneva, GICHD. <http://www.gichd.org/fileadmin/pdf/publications/Role_Military_MA.pdf>. p. 37.
67 Ibid.
68 Chalmers Johnson. (2004) *The Sorrows of Empire: Militarism, Secrecy, and the End of the Republic.* New York, Metropolitan Books. pp. 124, 132-134.
69 Dana Priest. (2004) *The Mission: Waging War and Keeping Peace with America's Military.* New York, W.W. Norton and Company. p. 75.
70 Johnson. *Sorrows of Empire.* pp. 124.
71 GICHD. *Role of the Military.* p. 39.
72 Tom Smith. (August 2006) Personal interview with author in Crystal City, Virginia.

73 Mine.ba. (21 September 2007) 'Thirty-Six Members of BiH Armed Forces Travel to Iraq.' <http://mine.ba/?PID=7&RID=64>; Reuters. (15 April 2005) 'Bosnia to send explosives unit to Iraq in June.' <http://www.reliefweb.int/rw/RWB.NSF/db900SID/ACIO-6BGJTC?OpenDocument>.

74 UXB Balkans. (2006) 'UXB Balkans d.o.o.' Unpublished document given to author by UXB Sarajevo office.

75 Parsons Global Services, Inc. (October 2004) 'Community Reintegration and Stabilization Program (CRSP) in Bosnia-Herzegovina: Final Progress Report.' USAID Development Experience Clearinghouse, Document No. PD-ACF-581. Available from <http://www.dec.org>. p. 4.

76 Geneva International Center for Humanitarian Demining (GICHD). (November 2004) *A Study of Local Organisations in Mine Action*. Geneva, GICHD. <http://www.isn.ethz.ch/pubs/ph/details.cfm?lng=en&id=26816>. pp. 164-166.

77 Nick van Praag. (2004) Email correspondence with Hugh Griffiths. Shown to author in June 2005. Van Praag was responsible for external affairs for the Europe and Central Asia Region at the World Bank.

78 Buse. 'RONCO Executives Talk.'

79 International Crisis Group (ICG). (18 July 1997) 'Ridding Bosnia of Landmines: The Urgent Need for a Sustainable Policy.' *ICG Bosnia Report No. 25*. pp. 14-22.

80 Details of the nature of corruption in the demining sector are outlined in: State Court of Bosnia and Herzegovina. (22 November 2006) Custody decision against Radomir Kojic and Radoslav Ilic. Number: X-KRN-06/250; Cantonal Prosecutor's Office Mostar. (5 July 2001) Criminal charges against Berislav Pusic and Davor Kolenda. Number: 700-213/01; Financial Police Department Sarajevo. (27 March 2001) Criminal Complaint against Mirzad Gradascevic. Number: 101-511/01; Financial Police Department Sarajevo. (23 March 2001) Criminal Complaint against Damir Kunsten. Number 101-492-/01; Financial Police Department Sarajevo. (22 March 2001) Criminal Complaint against Stjepan Strbad. Number 700-76/01; Financial Police Head Inspectors Office Sarajevo. 'Prvi Izvještaj.'; GICHD. *Role of the Military*; Melissa Eddy. (22 May 2000) 'Mine removal snarled in conflicting interests, mismanagement.' *The Associated Press*.

81 World Bank. (19 March 2004) 'Project Performance Assessment Report: Bosnia and Herzegovina.' Report No. 28288. p. v.

82 Office of the High Representative (OHR). (12 October 2000) 'Decision removing Milos Krstic from his position as a member of the Demining Commission and banning him from holding any official or appointive public office.' <http://www.ohr.int/decisions/removalssdec/default.asp?content_id=317>; Office of the High Representative (OHR). (12 October 2000) 'Decision removing Enes Cengic from his position as a member of the Demining Commission and banning him from holding any official or appointive public office.' <http://www.ohr.int/decisions/removalssdec/default.asp?content_id=318>; Office of the High Representative (OHR). (12 October 2000) 'Decision removing Berislav Pusic from his position as a member of the Demining Commission and banning him from holding any official or appointive public office.' <http://www.ohr.int/decisions/removalssdec/default.asp?content_id=319>.

83 Danka Savic & Mirsad Fazlic. (19 October 2000) 'Demining Program Abuses in BiH.' *Slobodna Bosna*. English translation available from *Adopt-A-Minefield*. <http://www.landmines.org.uk/NewsWire_Article/231>.

84 RONCO. (27 October 2000) 'Official Press Release: Humanitarian Demining in Bosnia 'Smear Campaign.'' Document obtained by author.

⁸⁵ Bolton & Griffiths. *Bosnia's Political Landmines*. pp. 12-13.
⁸⁶ US Department of State. (11 April 2007) 'United States Funding Support for Humanitarian Demining in Bosnia and Herzegovina.' <http://www.state.gov/t/pm/rls/fs/82833.htm>.
⁸⁷ ICG. 'Ridding Bosnia of Landmines.'
⁸⁸ Bolton & Griffiths. *Bosnia's Political Landmines*. pp. 12-13.
⁸⁹ X (Name withheld). (12 June 1996) "Situation Report #13." Unclassified report for State Dapratment/PM & Mr. Richard G. Stickles, Jr. A115 B6. Released to author by US State Department in 2008. p. 2.
⁹⁰ David Keen. (1994) *The Benefits of Famine*. Princeton, Princeton University Press; David Keen. (1998) *The Economic Functions of Violence in Civil Wars*. Adelphi Paper 320. London, Oxford University Press.
⁹¹ Edward J. Clay & Bernard B. Schaffer. (1984) *Room for Manoeuvre: An Exploration of Public Policy in Agriculture and Rural Development*. London, Heinemann Educational Books. pp. 142-190.
⁹² Chris Hedges. (17 August 1999) 'Leaders in Bosnia Are Said to Steal Up to $1 Billion.' *New York Times*.
⁹³ Amer Kapetanovic. (2000) 'Kako Je Nestalo 6,7 Miliona Dolara.' ('How 6.7 Million Dollars Disappeared.') *Dani*. <http://www.bhdani.com/arhiva/177/t17705.htm>; Savic & Fazlic. 'Demining Program Abuses in BiH.'
⁹⁴ cf. Mushtaq Kahn. (1995) 'State Failure in Weak States: A critique of new institutionalist explanations.' *The New Institutional Economics and Third World Development*. John Harriss, Janet Hunter & Colin Lewis (Eds.). London, Routledge.
⁹⁵ Wolfgang Petrisch. (1 December 2006) Personal interview with author in Kabul.
⁹⁶ Samantha Power. (July 2005) Keynote address at 'The International Scientific Conference on the Genocide over Bosniaks of the UN Safe Area in Srebrenica, July 1995' at the Holiday Inn, Sarajevo.
⁹⁷ World Bank. (2002) *World Development Report 2002: Building Institutions for Markets*. New York, Oxford University Press.
⁹⁸ International Crisis Group (ICG). (2 November 2000) 'War Criminals in Bosnia's Republika Srpska: Who are the People in Your Neighborhood.' *ICG Balkans Report No. 103*. pp. iii, 70, 76-77.
⁹⁹ Vera Devine. (2005) 'Corruption in Post-War Reconstruction: The Experience of Bosnia and Herzegovina.' *Corruption in Post-War Reconstruction: Confronting the Vicious Circle*. Beirut, Lebanese Transparency Association. <http://www.transparency-lebanon.org/2006/Publications/PWR/4.pdf>.
¹⁰⁰ Ibid.
¹⁰¹ GICHD. *Role of the Military*. p. 44.
¹⁰² Ibid. p. 40.
¹⁰³ Norwegian People's Aid (NPA). (2002) 'MFA Report 2002: Project Presentation.' In NPA archives. p. 5.
¹⁰⁴ Lisa Kirkengen. (2006) 'Norwegian Housing and Return Projects in Bosnia and Herzegovina.' *NORDEM Report*. 18/2006. <http://www.humanrights.uio.no/forskning/publ/nr/2006/1806.pdf >. p. 18.
¹⁰⁵ Norwegian People's Aid (NPA). (25 March 1996) 'Project Presentation: NPA De-Mining Project, Former Yugoslavia.' Unpublished document in NPA archives.
¹⁰⁶ Norwegian People's Aid (NPA). (February 1998) 'Final Report: Mine disposal in Bosnia Herzegovina.' Report to Royal Netherlands Embassy Sarajevo in NPA archives.
¹⁰⁷ UN Mine Action Center (UNMAC). (4 October 1997) 'Monthly Report for September 1997 Sekondering av Instruktorer UNMAC.' Fax to Per Nergaard in NPA archives.

108 Jan Braatha. (8 March 2007) Personal interview with author in Sarajevo.
109 Ibid.
110 Per Breivik. (28 February 2007) Personal interview with author in Sarajevo.
111 UN Mine Action Center (UNMAC). (17 May 1998) 'Mine Action Plan for Sarajevo Canton.' In NPA archives. p. 11.
112 Ibid. p. 17.
113 John Rodsted. (2005) *Mine Action in Bosnia-Herzegovina 1996-2005*. Oslo, Norwegian Peoples Aid.
114 Matthew Bolton. (August 2003) 'Mine Action in Bosnia's Special District: A Case Study.' *Journal of Mine Action*. 7(2). <http://maic.jmu.edu/JOURNAL/7.2/focus/bolton/bolton.htm>.
115 Office of the High Representative (OHR). (8 March 2002) 'Report from the Supervisor of Brcko to the PIC Steering Board on the Progress of Implementation of the Final Award of the International Arbitral Tribunal for Brcko (8 March 2001 - 8 March 2002).' Brcko District, Bosnia and Herzegovina, OHR. p. 12. See also: Alex Jeffrey. (February 2006) 'Building state capacity in post-conflict Bosnia and Herzegovina: The case of Brcko District.' *Political Geography*. 25(2).
116 Norwegian People's Aid (NPA). (n.d.) 'Mine Action Programme.' <http://www.npa-bosnia.org/MAIN/page%202/mine%20action%20team%202.htm>; Norwegian People's Aid (NPA). (2004) 'What is TAP.' <http://www.npa-bosnia.org/operations/TAP/mine%20cont%20map.htm>.
117 Lisica. *Risk Management*. pp. 42-45, 62-94.
118 Parts of this section are from: Matthew Bolton. (2008) 'Sudan's Expensive Landmines: An Evaluation of Political and Economic Problems in Sudanese Mine Clearance.' *Political Minefields*. <http://politicalminefields.wordpress.com/2008/07/sudansexpensivelandmines1.pdf>.
119 UN Mine Action Office (UNMAO). (July 2007) 'IMSMA Monthly Report.' Obtained by author from UNMAO. p. 8.
120 Survey Action Center (SAC), Mines Advisory Group (MAG) and Sudanese Association for Combating Landmines (JASMAR). (September 2007) 'Landmine Impact Survey Sudan: Kassala, Red Sea, Gadaref and Sennar States.' <http://www.sac-na.org/pdf_text/sudan/ES_Report_Sep07.pdf>;
Survey Action Center (SAC), Mines Advisory Group (MAG) and Sudanese Association for Combating Landmines (JASMAR). (March 2007) 'Landmine Impact Survey Sudan: Blue Nile State.' <http://www.sac-na.org/pdf_text/sudan/BN_Report_Mar07.pdf>;
Survey Action Center (SAC) & Mines Advisory Group (MAG). (March 2006) 'Landmine Impact Survey Sudan: Eastern Equatoria.' < http://www.sac-na.org/pdf_text/sudan/EE_Report_Sep06.pdf>.
121 International Campaign to Ban Landmines (ICBL). 'Sudan.' *Landmine Monitor Report 1999: Toward a Mine Free World*. <http://www.icbl.org/lm/1999/sudan.html>.
122 Peter Moszynski. (27 November 2005) 'Mine Action in the Midst of Internal Conflict: The Case of Sudan.' *Mine Action in the Midst of Internal Conflict*. Geneva, Geneva Call. p. 31.
123 World Food Program (WFP). (September 2005) 'W.F.P. Sudan Progress Report Road Repair and Demining Activities As At End September 2005.' <http://www.unjlc.org/sudan/infrastructure/roads/wfp_roads_progress_report/2006-05-17.9085524273/view>. p. 4.
124 Steve Crosskey. (11 August 2007) Personal interview with author in Juba.
125 Rebecca Roberts & Mads Frilander. (2004) 'Preparing for Peace Mine Action's Investment in the Future of Sudan.' *Preparing the Ground for Peace: Mine Action in Support of*

Peacebuilding. PRIO Report 2/2004. Kristian Berg Harpviken & Rebecca Roberts (Eds.). Oslo, PRIO. p. 14.

126 Matt Murphy. (11 January 2007) 'The U.S. Humanitarian Mine Action Program in Sudan.' Unpublished paper written in answer to author's email questions to the State Department. p. 1.

127 Tempas Camilo Moses. (7 August 2007) Personal interview with author in Juba.

128 Murphy. 'U.S. Humanitarian Mine Action Program in Sudan.' p. 2.

129 US Agency for International Development (USAID). (2 June 2006) 'USAID/Sudan Operational Plan FY2006.' <http://ww.dec.org>. p. 5; US Agency for International Development (USAID). (16 June 2005) 'USAID/Sudan Annual Report FY2005.' <http://ww.dec.org>.

130 Berger Group. (16 October 2006) 'Berger Awarded Five-Year USAID Contract for Southern Sudan Reconstruction.'
<http://www.louisberger.com/phpscr/press_list_one.php3?idnum=180>.

131 Andy Bailey. (28 August 2007) Personal email to author.

132 Murphy. 'U.S. Humanitarian Mine Action Program in Sudan.' p. 1.

133 Fred Maio. (20 August 2007) Personal interview with author in Khartoum. Evy Van Weezendonk and Christine Murphy. (3 August 2008) Personal interview with author in Yei.

134 DanChurchAid. (5 November 2005) 'Humanitarian Mine Action in Sudan.'
<http://www.danchurchaid.org/sider_paa_hjemmesiden/what_we_do/issues_we_work_on/hma/read_more/humanitarian_mine_action_in_sudan>.

135 Braatha. Personal interview.

136 GICHD. *Role of the Military*. p. 98.

137 Ibid. p. 82.

138 Barnett Rubin. (2002) *The Fragmentation of Afghanistan: State Formation and Collapse in the International System*. 2nd Ed. New Haven, Yale University Press. p. x.

139 Ahmed Rashid. (2001) *Taliban: The Story of the Afghan Warlords*. London, Pan Books.

140 Alex de Waal. (April 2007) 'Sudan: International Dimensions to the State and Its Crisis.' *LSE Crisis States Research Centre Occasional Paper*. 3.
<http://www.crisisstates.com/download/op/op3.DeWaal.pdf>. p. 18; Séverine Autesserre. (January 2002) 'United States 'Humanitarian Diplomacy' in South Sudan.' *Journal of Humanitarian Assistance*. <http://www.jha.ac/articles/a085.htm>.

141 Ted Dagne. (12 April 2006) 'Sudan: Humanitarian Crisis, Peace Talks, Terrorism, and U.S. Policy.' *CRS Issue Brief for Congress*. Washington DC, Congressional Research Service. <http://www.fas.org/sgp/crs/row/IB98043.pdf>. p. 14; Kenneth Silverstein. (29 April 2005) 'Official Pariah Sudan Valuable to America's War on Terrorism.' *Los Angeles Times*. <http://www.globalpolicy.org/empire/terrorwar/analysis/2005/0429sudan.htm>.

142 John Young. (2005) 'Sudan: A Flawed Peace Process Leading to a Flawed Peace.' *Review of African Political Economy*. 103. p. 103.

143 de Waal. 'Sudan: International Dimensions.' p. 7.

144 Neil Middleton & Phil O'Keefe. (2006) 'Politics, History & Problems of Humanitarian Assistance in Sudan.' *Review of African Political Economy*. 109. p. 555.

145 Lauren Landis. (10 April 2007) Personal interview in Washington DC.

146 Autesserre. 'United States 'Humanitarian Diplomacy.''

147 Ibid.

148 US Agency for International Development (USAID). (1 July 2002) 'USAID/REDSO/ESA FY2002 Annual Report Sudan Program.' <http://ww.dec.org>. p. 6.

[149] John C. Danforth. (26 April 2002) 'Report to the President of the United States on the Outlook for Peace in Sudan.' <http://www.whitehouse.gov/news/releases/2002/05/outlook_for_peace_in_sudan.pdf>. p. 31.

[150] International Republican Institute. (2005) 'Southern Sudan Political Party Development & Legislative Strengthening Program USAID CEPPS II Agreement 623-A-00-04-00072-00: USAID Semi-Annual Narrative Report September 15, 2004 – March 15, 2005.' <http://ww.dec.org>; USAID. 'USAID/REDSO/ESA FY2002.' p. 4.

[151] Vesla Vetlesen. (2000) 'Trade Union Support to the Struggle Against Apartheid: The Role of the Norwegian Confederation of Trade Unions.' *Norway and National Liberation in Southern Africa*. Tore Linné Eriksen (Ed.). Stockholm, Nordiska Afrikainstitutet. pp. 148-149; Eva Helene Østbye. 'The South African Liberation Struggle.' In: Eriksen. *Norway and National Liberation*. pp. 340-342.

[152] Larry Minnear. (1991) *Humanitarianism under Siege: A Critical Review of Operation Lifeline Sudan*. Trenton, New Jersey. p. 85. Emphasis in original.

[153] Michael Griffin. (Third Quarter 2004) 'Mirror Launches in New Sudan.' *Global Journalist*. <http://www.globaljournalist.org/magazine/2004-3/sudan.html>.

[154] Bertha Kang'ong'oi. (6 August 2005) 'Dying in the Crash Was a Great Irony.' *The Nation* (Nairobi). <http://www.gurtong.com/forums/index.php?showtopic=2040&st=25>.

[155] Quoted in: Aid Watch. (6 March 2006) 'Norwegian People's Aid [Norsk Folkehjelp].' <http://www.observatoire-humanitaire.org/fusion.php?l=GB&id=75>.

[156] NRK Television. (17 November 1999) 'Vapensmuglerne I Sudan.' ('Weapons Smuggling in Sudan.') Brennpunkt. NRK Television. Though probably a form of grey propaganda produced by a pro-Sudan Government lobby group, the following report compiles some of the allegations and reports about NPA's closeness to the SPLA from other sources: The European-Sudanese Public Affairs Council. (December 1999) 'Perpetuating Conflict and Sustaining Repression: Norwegian People's Aid and the Militarisation of Aid in Sudan.' <http://www.espac.org/pdf/Perpetuating%20Conflict%20and%20Sustaining.pdf>.

[157] Norwegian People's Aid (NPA). (30 July – 12 August 2007) 'Norwegian People's Aid.' *Sudan Mirror*. p. 12.

[158] Young. 'Sudan: A Flawed Peace.' pp. 103-104.

[159] USAID. 'USAID/Sudan Operational Plan FY2006.' p. 3.

[160] US Africa Command. (October 2007) 'Questions and Answers about AFRICOM.' <http://www.africom.mil/africomFAQs.asp>.

[161] cf. Mark Duffield. (December 2001) 'Governing the Borderlands: Decoding the Power of Aid.' *Disasters*. 25(4); Alex de Waal. (1997) *Famine Crimes: Politics & the Disaster Relief Industry in Africa*. Oxford, James Currey; Autesserre. 'United States 'Humanitarian Diplomacy.''

5. Comparing the Performance of Tenders and Grants

[1] Timothy Donais. (2005) *The Political Economy of Peacebuilding in Post-Dayton Bosnia*. London, Routledge. p. 117.

[2] Eddie Banks. (August 2003) 'In the Name of Humanity.' *Journal of Mine Action*. 7(2). <http://maic.jmu.edu/journal/7.2/focus/banks/banks.htm>.

[3] Graeme A. Hodge. (2000) *Privatization: An International Review of Performance*. Boulder, Colorado, Westview Press. p. 107; World Bank. (1997) *World Development Report 1997: The State in a Changing World*. New York, Oxford University Press. p. 6; Robert Carnaghan &

Barry Bracewell-Milnes. (1996) 'Conclusions and Recommendations.' *Privatization: Critical Perspectives on the World Economy.* Vol. II. George Yarrow & Piotr Jasinski (Eds.). London, Routledge. p. 318-319.

[4] Data on NGO clearance from: UN Mine Action Center for Afghanistan (UNMACA). (November 2006) IMSMA Database. Obtained by author. Data queries on clearance area. Approximation of NGO cost of clearance from list of KAIA NGO contracts shown to author by UNMACA source. Data on commercial clearance and cost was self-reported by email to the author by the companies.

[5] Banks. 'In the Name of Humanity.'

[6] Darvin Lisica & David Rowe. (2004) 'Strategic Analysis of Mine Action in Bosnia and Herzegovina.' Available from the Bosnia and Herzegovina Mine Action Center, Sarajevo. p. 53. See also: Ann Fitz-Gerald & Derrick J. Neal. (2000) 'Dispelling the Myth Between Humanitarian and Commercial Mine Action Activity.' *Journal of Mine Action.* 4(3). <http://maic.jmu.edu/Journal/4.3/features/myth/myth.htm>.

[7] Adapted from: Lisica & Rowe. 'Strategic Analysis.' p. 53.

[8] Geneva International Center for Humanitarian Demining (GICHD). (June 2003) *The Role of the Military in Mine Action.* Geneva, GICHD. <http://www.gichd.org/fileadmin/pdf/publications/Role_Military_MA.pdf>. p. 45.

[9] Steve Crosskey. (11 August 2007) Personal interview with author in Juba.

[10] Russell Gasser. (November 2006) 'Landmine Action Nuba Mountains Evaluation.' Unpublished document obtained by author. p. 31.

[11] Jamie Franklin. (31 July 2007) Personal interview with author in Juba.

[12] Dataset compiled by author from NPA and ITF reports.

[13] International Campaign to Ban Landmines (ICBL). (1999) 'Mozambique.' *Landmine Monitor 1999.* <http://www.icbl.org/lm/1999/mozambique.html#Heading1108>. The unusually high costs for Handicap and CIDEV are explained in part by high start-up costs.

[14] Robert Gannon. (20 November 2006) Personal interview with author in Kabul.

[15] Data on NGO clearance from: UNMACA. IMSMA Database. Data queries on clearance area and dates. Data on commercial clearance and cost was self-reported by email to the author by the companies.

[16] Darvin Lisica. (22 February 2007) Personal interview with author in Sarajevo.

[17] Data from: Bosnia and Herzegovina Mine Action Center (BHMAC). (2002) Report on Mine Action in Bosnia and Herzegovina. Sarajevo, BHMAC. p. 16. Error bars represent the interquartile range of variation between different organizations within the category in 2002. The points are plotted at the median.

[18] Data from UN Mine Action Office (UNMAO). (July 2007) IMSMA database. Obtained by author. Error bars indicate the variation in average speed per organization per year. The points represent the median.

[19] Eva Veble. (20 September 2008) Personal interview with author in Geneva.

[20] Per Breivik in: Matthew Bolton & Hugh Griffiths. (September 2006) *Bosnia's Political Landmines: A Call for Socially Responsible and Conflict-Sensitive Mine Action.* London, Landmine Action. <http://www.landmineaction.org/resources/Bosnias_Political_Landmines.pdf>. p. 14.

[21] Larry Crandall in: Margaret Buse. (June 2000) 'RONCO Executives Talk About Demining, Integration and the IMAS Contract: (An Interview with Lawrence Crandall, Stephen Edelmann and A. David Lundberg).' *Journal of Mine Action.* 4(2). <http://maic.jmu.edu/journal/4.2/Features/ronco/ronco.htm>.

[22] Marc Bendick, Jr. (1989) 'Privatizing the Delivery of Social Welfare Services: An Idea to Be Taken Seriously.' *Privatization and the Welfare State*. Sheila B. Kamerman & Alfred J. Kahn (Eds.). Princeton, Princeton University Press. p. 107.

[23] Paul Molam. (7 December 2006) Comment at presentation of preliminary research findings by author to the UN Mine Action Center for Afghanistan (UNMACA), Kabul.

[24] Guy Willoughby. (21 September 2007) Personal interview with author in Thornhill, Scotland.

[25] Per Breivik. (22 June 2005) Interview with author in Sarajevo; Melissa Eddy. (22 May 2000) 'Mine removal snarled in conflicting interests, mismanagement.' *The Associated Press*; Stuart Maslen. (2004) *Mine Action After Diana: Progress in the Struggle Against Landmines*. London, Landmine Action/Pluto Press. p. 148.

[26] Crosskey. Personal interview.

[27] Data on NGO clearance from: UNMACA. IMSMA Database. Data queries on clearance area. Data on commercial clearance was self-reported by email to the author by the companies. Error bars indicate the interquartile range in variation of average density per organization per year. The points represent the median.

[28] Data from BHMAC Annual Reports and Bosnia and Herzegovina Mine Action Center (BHMAC). (2007) IMSMA Database. Obtained by author. Error bars represented the interquartile range of average values per year. The dots are plotted at the median value per year.

[29] Data from UNMAO. IMSMA database. Error bars indicate the interquartile range in variation of average density per organization per year. The points represent the median.

[30] Lisica. Personal interview.

[31] Survey Action Center. (2006) *Landmine Impact Survey – Islamic Republic of Afghanistan*. Takoma Park, Maryland, Survey Action Center. p. 21.

[32] UNMACA. IMSMA Database. Data queries on clearance area and devices.

[33] One should note that there was anecdotal evidence of quality management inspectors having a bias against commercial companies, because they were unfamiliar with new commercial technologies used. The capacity of the quality management teams has also been questioned.

[34] Quality assurance report data provided to the author by UNMACA's Kabul Area Mine Action Center (AMAC).

[35] Banks. 'In the Name of Humanity.'

[36] Percentage of Quality Assurance Reports that report errors. Note the method of determining what was an error changed in June 2003, hence the need for two graphs. Note that from June 2003, a 'Critical Error' was one that could endanger a person's life, whether a deminer or a future user of the land. From BHMAC. IMSMA Database.

[37] Percentage of Quality Assurance Reports that report errors. Note the method of determining what was an error changed in June 2003, hence the need for two graphs. Note that from June 2003, a 'Critical Error' was one that could endanger a person's life, whether a deminer or a future user of the land. From BHMAC. IMSMA Database.

[38] UN Mine Action Office (UNMAO). (July 2007) 'IMSMA Monthly Report.' Obtained by author from UNMAO. p. 13. This data is generated by UNMAO quality assurance personnel who conduct random spotchecks on demining tasks.

[39] Andy Smith. (June 2000) 'The Facts on Protection Needs in Humanitarian Demining.' *Journal of Mine Action*. 4(2). <http://maic.jmu.edu/journal/4.2/focus/PN/protectneeds.htm>. cf. James Trevelyan. (June 2000) 'Reducing Accidents in Demining: Achievements in Afghanistan.' *Journal of Mine Action*. 4(2).

<http://maic.jmu.edu/journal/4.2/Focus/Accidents/accidents.htm>; Maslen. *Mine Action After Diana.* p. 50.
40 Data from: BHMAC. 2002 Annual Report; BHMAC. IMSMA Database; Author's compiled demining accident dataset.
41 Mine and UXO data is for ordnance found in clearance only, not stockpile destruction or EOD operations. Data on accidents from: UN Mine Action Center for Afghanistan (UNMACA). (November 2006) 'Demining Accidents, Incidents, IED attacks and Major Non-Demining Accidents.' Spreadsheet given to author. Data on NGO clearance from: UNMACA. IMSMA Database. Data queries on clearance area. Data on commercial clearance was self-reported by email to the author by the companies.
42 Clearance data from BHMAC. IMSMA Database. Author's compiled accident database using records from BHMAC, ICRC, GICHD and the press. This graph displays accidents as unweighted because correcting for the lower probability of accidents in later years does not change the overall picture of the graph significantly.
43 Marco Buono. (15 August 2007) Personal interview with author in Khartoum.
44 Gannon. Personal interview.
45 UN Mine Action Center (UNMAC). (11 December 1997) 'Demining Accidents by Mine-Tech in Bosnia and Herzegovina.' p. 3.
46 James Trevelyan. (n.d.) 'Alternatives to the 99.6% demining standard.' <http://www.mech.uwa.edu.au/jpt/demining/quality/standards4.pdf>.p. 2; James Trevelyan. (August 1999) 'Landmines in Bosnia-Herzegovina and Croatia.' <http://www.mech.uwa.edu.au/jpt/demining/countries/balkans/cro-bos.html>; UNMAC. 'Demining Accidents by Mine-Tech.'; International Crisis Group (ICG). (18 July 1997) 'Ridding Bosnia of Landmines: The Urgent Need for a Sustainable Policy.' *ICG Bosnia Report No. 25.* p. 14; Eddy. 'Mine removal snarled.'
47 Daniel Bellamy. (22 November 2006) Personal interview with author in Kabul.
48 UNMAC. 'Demining Accidents by Mine-Tech.'; Eddy. 'Mine removal snarled.'; Neil McKenzie. (7 July 2005) Interview with author in Sarajevo.
49 UN Mine Action Center (UNMAC). (23 July 1997) 'Report on Accident at Sevarlije, Near Doboj.' p. 7.
50 UNMAC. 'Demining Accidents by Mine-Tech.' p. 3.
51 UN Mine Action Center (UNMAC). (14 April 1999) 'Report of Board of Inquiry into Accident 08 April 1999.' p. 9.
52 UN Mine Action Center (UNMAC). (27 July 1999) 'Report of Board of Inquiry into Accident of 27 July 1999.' p. 7.
53 UN Mine Action Center (UNMAC). (11 August 1999) 'Report of Board of Inquiry into an Accident on 5 August 1999.' p. 7.
54 Source X. (24 November 2006) Personal interview with the author in Kabul.
55 Kabul Central Area Mine Action Center (AMAC) Quality Management Investigation Team (QMIT) 15. (3 August 2006) 'Observation Form.' Available from Kabul AMAC.
56 Ibid.
57 Mohammed Sediq. (23 November 2006) Personal interview with author in Kabul.
58 Chris Stephens. (16 April 2007) Personal interview with author in New York.
59 Data on NGO clearance from: UNMACA. IMSMA Database. Data queries on clearance area. Data on commercial clearance was self-reported by email to the author by the companies. Quality assurance report data provided to the author by UNMACA's Kabul Area Mine Action Center (AMAC).
60 Donais. *Political Economy of Peacebuilding.* p. 111.

6. Impact on Peacebuilding

[1] Timothy Donais. (2005) *The Political Economy of Peacebuilding in Post-Dayton Bosnia*. London, Routledge p. 81.
[2] David J. Keen. (2008) *Complex Emergencies*. Cambridge, Polity; Fiona Terry. (2002) *Condemned to Repeat?: The Paradox of Humanitarian Action*. Ithaca, Cornell University Press; Alex de Waal. (1998) *Famine Crimes: Politics and the Disaster Relief Industry in Africa*. Bloomington, Indiana University Press; Mark Duffield. (2001) *Global Governance and the New Wars: The Merging of Development and Security*. London, Zed Books; Mark Duffield. (2007) *Development, Security and Unending War: Governing the World of Peoples*. Cambridge, Polity Press.
[3] Mary B. Anderson. (1999) *Do No Harm: How Aid Can Support Peace – or War*. Boulder, Lynne Rienner Publishers.
[4] Jonathan Goodhand. (2006) *Aiding Peace? The Role of NGOs in Armed Conflict*. Bourton on Dunsmore, Intermediate Technology Publications.
[5] Robert Gannon. (20 November 2006) Personal interview with author in Kabul.
[6] Terje Tvedt. (1994) 'The Collapse of the State in Southern Sudan after the Addis Ababa Agreement: A Study of Internal Causes and the Role of the NGOs.' *Short-Cut to Decay: The Case of the Sudan*. Sharif Harir & Terje Tvedt (Eds.). Uppsala, Scandinavian Institute of African Studies; Volker Riehl. (2001) *Who Is Ruling in South Sudan? The role of NGOs in rebuilding socio-political order*. Uppsala, Nordiska Afrikainstitutet.
[7] Christopher Spearin. (November 2001) 'Ends and Means: Assessing the Humanitarian Impact of Commercialized Security on the Ottawa Convention Banning Anti-Personnel Mines.' YCISS Occasional Paper Number 69.
<http://www.ucalgary.ca/~zamans/SMSS/pdf/spearin_smss2001.pdf>.
[8] P.W. Singer. (2003) *Corporate Warriors: The Rise of the Privatized Military Industry*. Ithaca, New York, Cornell University Press.
[9] Sayed Yaqub Ibrahimi. (6 December 2007) 'Security Firms in Afghanistan: Part of the Problem?' *Institute for War and Peace Reporting*.
<http://www.iwpr.net/index.php?m=p&o=341232&s=f&apc_state=henfarr341232>.
[10] Mohammed Sediq. (23 November 2006) Personal interview with author in Kabul.
[11] Lloyd Carpenter. (21 November 2006) Personal interview with author in Kabul.
[12] e.g. Fariba Nawa. (6 October 2006) *Afghanistan, Inc.: A Corpwatch Investigative Report*.
<http://s3.amazonaws.com/corpwatch.org/downloads/AfghanistanINCfinalsmall.pdf>. pp. 17-20.
[13] David Robertson. (3 April 2007) 'US Embassy calls in ArmorGroup.' *The Times* (London). <http://business.timesonline.co.uk/tol/business/industry_sectors/support_services/article1605075.ece>.
[14] Gannon. Personal interview.
[15] Dean Hutson. (30 November 2006) Personal interview with author in Kabul.
[16] Lisa Rimli & Susanne Schmeidl. (November 2007) *Private Security Companies and Local Populations. An exploratory study of Afghanistan and Angola*. Bern, Switzerland, SwissPeace.
<http://www.swisspeace.ch/typo3/fileadmin/user_upload/pdf/PSC.pdf>. p. 28.
[17] Rimli & Schmeidl. *Private Security Companies*. p. 36.
[18] Ibrahimi. 'Security Firms.'
[19] Swissinfo. (13 November 2007) 'Study criticises security firms in Afghanistan.' *Swissinfo*. <http://www.swissinfo.ch/eng/top_news/detail/Study_criticises_security_firms_in_Afghanistan.html?siteSect=106&sid=8413697&cKey=1194937728000&ty=st>.
[20] Dusan Gavran. (16 February 2007) Personal interview with author in Sarajevo.

21 Frontier Medical. (n.d.) 'Projects in Africa.'
<http://www.frontiermedical.co.uk/projects_africa.html>.
22 e.g. see ArmorGroup's Sudan strategy: Andrew MacKinnon. (12-14 February 2006) 'Sudan Development Program: ArmorGroup.'
<http://www.sudandevelopmentprogram.org/sp/events/sdpd06/transport/armorgroup.pdf>.
23 For more on the UNDP demobilization program, see: Arne Strand. (2004) 'Transforming Local Relationships: Reintegration of Combatants through Mine Action in Afghanistan.' *Preparing the Ground for Peace: Mine Action in Support of Peacebuilding*. Kristian Berg Harpviken & Rebecca Roberts (Eds.). Oslo, PRIO.
24 Mohammed Shohab Hakimi. (6 November 2006) Personal interview with author in Kabul. Also: Fazel Karim Fazel. (29 November 2006) Personal interview with author in Kabul.
25 Kerei Ruru. (29 November 2006) Personal interview with author in Kabul.
26 See, for instance, this damning report on the privatization of the reconstruction process in Afghanistan: Nawa. *Afghanistan, Inc.* Also: William Maley. (2006) *Rescuing Afghanistan*. London, C. Hurst & Co. p. 98-99.
27 Geneva International Center for Humanitarian Demining (GICHD). (2004) *A Study of Local Organisations in Mine Action*. Geneva, GICHD. pp. 104-105. Mine Clearance and Planning Agency (MCPA). (December 1999) *Socio-Economic Impact Study of Landmines and Mine Action Operations in Afghanistan*. Islamabad, Pakistan, UNMAPA.
28 Christian Aid. (2004) *The Politics of Poverty: Aid in the New Cold War*. London, Christian Aid. <http://christianaid.org.uk/indepth/404caweek/index.htm>.pp. 40-51.
29 The Development Initiative. (n.d.) 'TDI Projects.'
<http://thedevelopmentinitiative.com/projects.asp>.
30 Fazel. Personal interview.
31 Jim Pansegrouw. (14 August 2007) Personal interview with author in Khartoum.
32 Tempas Camilo Moses. (7 August 2007) Personal interview with author in Juba.
33 D. Michael Shafer. (January 1990) 'Sectors, States, and Social Forces: Korea and Zambia Confront Economic Restructuring.' *Comparative Politics*. 22(2). p. 127.
34 Ibid. p. 128.
35 David Keen. (1998) *The Economic Functions of Violence in Civil Wars*. Adelphi Paper 320. London, Oxford University Press; de Waal. *Famine Crimes*; Michael Charles Pugh, Neil Cooper & Jonathan Goodhand (Eds.). (2004) *War Economies in a Regional Context: Challenges of Transformation*. London, Lynne Reiner Publishers, Inc.; Karen Ballentine & Jake Sherman. (2003) *The Political Economy of Armed Conflict: Beyond Greed and Grievance*. Boulder, Co., Lynne Reiner Publishers.
36 cf. Mary Kaldor & Vesna Bojicic. (1999) 'The 'Abnormal' Economy of Bosnia-Herzegovina.' *Scramble for the Balkans: Nationalism, Globalism and the Political Economy of Reconstruction*. Carl-Ulrick Schierup (Ed.). New York, Macmillan; Michael Pugh. (2002) 'Postwar Political Economy in Bosnia and Herzegovina: The Spoils of Peace.' *Global Governance*. 8; Peter Andreas. (2004) 'The Clandestine Political Economy of War and Peace in Bosnia.' *International Studies Quarterly*. 48.
37 Anderson. *Do No Harm*.
38 Larry Crandall. (1 July 2006) Personal interview with author in McLean, Virginia.
39 Colin Wanley. (30 October 2006) Personal interview with author in Kabul.
40 Donais. *Political Economy of Peacebuilding*.
41 Ibid.
42 Hugh Griffiths. (Summer 1999) 'A Political Economy of Ethnic Conflict, Ethno-nationalism and Organised Crime.' *Civil Wars*. 2(2). pp. 56-73; Vesna Bojicic-Dzelilovic.

(2000) 'From Humanitarianism to Reconstruction: Towards an Alternative Approach to Economic and Social Recovery from War.' *Global Insecurity: Restructuring the Global Military Sector.* Vol. III. Mary Kaldor (Ed.). London, Pinter. p. 115; General Accounting Office. (June 2000) 'Crime and Corruption Threaten Successful Implementation of the Dayton Peace Agreement.' GAO/NSIAD-00-156; Pugh. 'Postwar Political Economy.' pp. 467-482; Andreas. 'Clandestine Political Economy.' pp. 29-51; Vera Devine. (2005) 'Corruption in Post-War Reconstruction: The Experience of Bosnia and Herzegovina.' *Corruption in Post-War Reconstruction: Confronting the Vicious Circle.* Beirut, Lebanese Transparency Association. <http://www.transparency-lebanon.org/2006/Publications/PWR/4.pdf>.

43 Donais. *Political Economy of Peacebuilding.* p. 119.

44 cf. International Crisis Group (ICG). (18 July 1997) 'Ridding Bosnia of Landmines: The Urgent Need for a Sustainable Policy.' *ICG Bosnia Report No. 25.* pp. 14; Don Hubert. (1998) 'The Challenge of Humanitarian Mine Clearance.' *To Walk without Fear: The Global Movement to Ban Landmines.* Maxwell A. Cameron, Robert J. Lawson and Brian W. Tomlin (Eds.). Oxford, Oxford University Press. pp. 323-324; Melissa Eddy. (22 May 2000) 'Mine removal snarled in conflicting interests, mismanagement.' *The Associated Press*; Danka Savic & Mirsad Fazlic. (19 October 2000) 'Demining Program Abuses in BiH.' *Slobodna Bosna.* English translation available from *Adopt-A-Minefield.* <http://www.landmines.org.uk/NewsWire_Article/231>.

45 Sulejman Crncalo. (2 September 2004) Transcript of witness testimony in 'Prosecutor vs. Momcilo Krajisnik.' *International Criminal Tribunal for the Former Yugoslavia.* <http://www.un.org/icty/transe39/040902IT.htm>. pp. 5292, 5321, 5348)

46 UN Human Rights Committee. (27 April 1993) 'Document submitted in compliance with a special decision of the Committee: Bosnia and Herzegovina. 27/04/93. CCPR/C/89.' <http://www.unhchr.ch/tbs/doc.nsf/0/333378630589b6d680256674005bc280?Opendocument>. para. 54.

47 Ian Traynor, Yigal Chazan & Ian Black. (19 February 1994) 'Serbs' Retreat Ruffles NATO: UN says pullback is going ahead.' *The Guardian*; American Broadcasting Companies, Inc. (28 September 1994, 11:30 pm ET) Transcript 3483. *Nightline*; Julijana Mojsilovic. (12 February 1994) 'If NATO Bombs, 'The People Will Kill Those Frenchmen' With Yugoslavia.' *Associated Press Worldstream.*

48 Les Courrir des Balkans. (12 March 2002) 'Bosnie: Comment l'arrestation de Karadzic a échoué.' ('Bosnia: How the Arrest of Karadzic Failed.') *Les Courrir des Balkans.* <http://www.balkans.eu.org/article148.html>; European Council. (27 June 2003) 'COUNCIL DECISION 2003/484/CFSP of 27 June 2003 implementing Common Position 2003/280/CFSP in support of the effective implementation of the mandate of the International Criminal Tribunal of the former Yugoslavia (ICTY).' <http://europa.eu.int/eur-lex/pri/en/oj/dat/2003/l_162/l_16220030701en00770079.pdf>; US Department of the Treasury. (9 February 2004) 'Operation Balkan Vice III: Treasury Designation of Thirteen Individuals Obstructing the Dayton Peace Accords in Bosnia.' <http://www.treas.gov/press/releases/js1162.htm>; Hugh Griffiths & Nerma Jelacic. (5 February 2004) 'Investigative Report: Karadzic Protective Shield Cracking.' *Institute for War and Peace Reporting: Balkan Crisis Report.* <http://www.iwpr.net/index.pl?archive/bcr3/bcr3_200402_479_4_eng.txt>; Agence France Presse (AFP). (5 November 2004) 'New NATO operation in stronghold of wanted Bosnian Serb.' *Agence France Presse*; Hidajet Delic. (5 November 2004) 'NATO troops search homes of alleged supporters of war-crimes suspects in Bosnia.' *Associated Press.*

49 Court of Bosnia and Herzegovina. (22 November 2006) Custody decision against Radomir Kojic and Radislav Ilic. No. X-KRN-06/250. Obtained by author. p. 2.
50 Court of Bosnia and Herzegovina. (28 August 2006) 'Custody ordered for Radomir Kojic.' <http://www.sudbih.gov.ba/?id=218&jezik=e>; Court of Bosnia and Herzegovina. Custody decision against Radomir Kojic.
51 Court of Bosnia and Herzegovina. Custody decision against Radomir Kojic. p. 2-3.
52 Anonymous. (11 February 2007) 'Vojnici šest sati pretresali kucu Radomira Kojica.' ('Soldiers raid Radomir Kojic's house at 6 a.m..') *Dnevni Avaz* (Sarajevo). p. 3.
53 International Criminal Tribunal for the Former Yugoslavia (ICTY). (2 March 2004) 'The Office of the Prosecutor Against Prlic et al.' <http://www.un.org/icty/indictment/english/prl-ii040304e.htm>.
54 Cantonal Prosecutor's Office Mostar. (5 July 2001) Criminal Charges against Berislav Pusic and Davor Kolenda. No. 700 – 213/01. p.4 .
55 Financial Police Department Sarajevo. (22 March 2001) Criminal Complaint against Stjepan Strbad. Number 700-76/01; Financial Police Head Inspectors Office Sarajevo. (12 December 2000) 'Prvi Izvejestaj: O utvrdjenim cinjenicama I prikupljenim dokazima u postupku istrage, nacina raspolaganja sredstvima namijenjenim programima deminiranja u Bosni i Hercegovini odnosno Federaciji Bosne i Hercegovine, zloupotrebama utvrdjenim u dosadasnjem toku istrage te njihovim nosiocima.' ('First report: On established facts and gathered evidence during the investigation on ways of handling funds meant for demining programs in BiH, that is FBiH, abuses proved during the investigation and on their perpetrators.') Sarajevo, Ministry of Finance. Obtained by author. pp. 172-173.
56 Financial Police Department Sarajevo. Criminal Complaint against Stjepan Strbad; Financial Police Head Inspectors Office Sarajevo. 'Prvi Izvještaj.' pp. 180-182.
57 Cantonal Prosecutor's Office Mostar. Criminal Charges against Berislav Pusic. pp. 3-4, 13-20.
58 Ibid. p. 16.
59 Ibid. p. 14.
60 bpo. (19 April 1994) 'Sarajevo wird nie eine geteilte Stadt sein.' ('Sarajevo will not be a divided city.') *taz, die tageszeitung*. p. 22; Office of the High Representative (OHR). (27 March 1997) 'Side Agreement on the Implementation of Sarajevo Protocol.' <http://www.ohr.int/other-doc/fed-mtng/default.asp?content_id=3610>.
61 Financial Police Head Inspectors Office Sarajevo. 'Prvi Izvještaj.' pp. 14-16.
62 Financial Police Department Sarajevo. (27 March 2001) Criminal Complaint against Mirzad Gradascevic. Number: 101-511/01; Financial Police Head Inspectors Office Sarajevo. 'Prvi Izvještaj.' pp. 136-171.
63 For more details, see: Matthew Bolton & Hugh Griffiths. (September 2006) *Bosnia's Political Landmines: A Call for Socially Responsible and Conflict-Sensitive Mine Action*. London, Landmine Action. <http://www.landmineaction.org/resources/Bosnias_Political_Landmines.pdf>.
64 Duffield. *Global Governance and the New Wars*. pp. 14-15, 161-201.
65 Donais. *Political Economy of Peacebuilding*. p. 107.
66 Truth and Reconciliation Commission. (29 October 1998) *Truth and Reconciliation Commission of South Africa Report*. Vol. 3. pp. 11, 182-183. <http://www.doj.gov.za/trc/report/finalreport/TRC%20VOLUME%203.pdf >. Truth and Reconciliation Commission. (18 June 1999) 'The Truth and Reconciliation Commission Amnesty Decision Regarding the Pebco Killings.' <http://www.info.gov.za/speeches/1999/9906211042a1003.htm>.

67 David Masunda. (1 November 2001) 'Dyck heads for war-torn Afghan territory.' *Financial Gazette.* International Campaign to Ban Landmines (ICBL). (1999) 'Mozambique.' *Landmine Monitor 1999.* <http://www.icbl.org/lm/1999/mozambique>. International Campaign to Ban Landmines (ICBL). (1999) 'Zimbabwe.' *Landmine Monitor 1999.* <http://www.icbl.org/lm/1999/zimbabwe>.

68 David Rowe. (2 February 2007) Personal interview with author in Sarajevo.

69 Col. Dyck said 'Kojic was certainly involved in the protective screen [of Karadzic] and this intelligence we passed on to the UNMACC [sic].' Lionel Dyck. (n.d.) Email to Hugh Griffiths shown to author on 20 June 2005.

70 Peter Moszynski. (27 November 2005) 'Mine Action in the Midst of Internal Conflict: The Case of Sudan.' Geneva Call (Ed.). *Mine Action in the Midst of Internal Conflict.* Geneva, Geneva Call. pp. 29-39.

71 John Young. (2005) 'Sudan: A Flawed Peace Process Leading to a Flawed Peace.' *Review of African Political Economy.* 103. pp. 103-104.

72 Meeting of the States Parties, Zagreb. (2 December 2005) 'List of Participants.' <http://www.apminebanconvention.org/fileadmin/pdf/mbc/MSP/6MSP/6MSP_List_of_participants.pdf>;
Meeting of the States Parties, Geneva. (22 September 2006) 'List of Participants.' <http://www.apminebanconvention.org/fileadmin/pdf/mbc/MSP/7MSP/7MSP_List_of_Participants.pdf>.

73 BBC. (21 September 2007) 'Prosecutor demands Sudan arrests.' *BBC News.* <http://news.bbc.co.uk/2/hi/africa/7006180.stm>.

74 UN Mine Action Office (UNMAO). (2006) 'Capacity Building.' <http://www.sudanmap.org/CapacityBuilding.html>.

75 Human Rights Watch (HRW). (August 1998) 'Arms Transfers to the Government Of Sudan.' *Sudan: Global Trade, Local Impact: Arms Transfers to all Sides in the Civil War in Sudan.* <http://www.hrw.org/reports98/sudan/Sudarm988-05.htm>.

76 Piet Van Niekerk. (8 February 2008) 'SA to train pilots and technicians from Sudan.' *Daily Dispatch.* <http://www.dispatch.co.za/article.aspx?id=173789>.

77 Andrew Marshall. (1 November 1998) 'Terror 'Blowback' Burns CIA.' *The Independent.* p. 17; Chalmers Johnson. (2000) *Blowback: The Costs and Consequences of American Empire.* New York, Henry Holt and Company. pp. 8, 11 & 2002, p 23.

78 Christine Johnson. (2004) 'Aid and Recovery.' *Afghanistan.* Edward Girardet & Jonathan Walter (Eds.). 2nd Ed. Geneva, CROSSLINES Publications. p 22.

79 Edward Girardet & Jonathan Walter (Eds.). (2004) 'Aid and Recovery.' *Afghanistan.* 2nd Ed. Geneva, CROSSLINES Publications.

80 Agence France Presse (AFP). (8 April 2007) '7 Die in Taliban Raid on Mine-Clearing Team.' *The New York Times.* <http://www.nytimes.com/2007/04/08/world/asia/08afghan.html>.

81 David Fox. (24 June 2007) 'Taliban seize 18 Afghan mine clearing experts.' <http://www.reuters.com/article/featuredCrisis/idUSL24706480>.

82 IRIN. (30 August 2007) 'AFGHANISTAN: Deminers demand security guarantees before resuming work in Kandahar.' <http://www.alertnet.org/thenews/newsdesk/IRIN/bc7808a3be51e841bb79c46582717391.htm>.

83 Sediq. Personal interview.

84 Survey Action Center. (2006) *Landmine Impact Survey – Islamic Republic of Afghanistan.* Takoma Park, MD, Survey Action Center.

85 David Isaksson. (4 June 1998) 'Doghandlers from Mozambique transfer knowledge to Bosnia.' *Global Reporting.* In NPA archives.

[86] John Rodsted. (2005) *Mine Action in Bosnia-Herzegovina 1996-2005*. Oslo, Norwegian Peoples Aid. See also: Samuel Gruber. 'U.S. Commission Urges Sarajevo Cemetery Restoration.' <http://www.isjm.org/Links/Sarajevo.htm>.
[87] Jan Braatha. (8 March 2007) Personal interview with author in Sarajevo.
[88] Charles Frisby. (2 August 2008) Personal interview with the author in Yei.
[89] Marco Buono. (15 August 2007) Personal interview with author in Khartoum.
[90] HALO Trust. (2 November 2006) 'The HALO Trust Afghanistan.' PowerPoint Presentation given to author by Dr. Farid Homayoun. Slide 59-64.
[91] e.g. Mines Advisory Group (MAG). (n.d.) 'SUDAN: Clearing Small Arms & Light Weapons.' <http://www.mag.org.uk/news.php?s=2&p=2775>.
[92] Frisby. Personal interview.
[93] Rune Kristian Andersen. (11 June 2007) 'Capacity Building Plan.' Obtained by author from NPA.
[94] Frisby. Personal interview.
[95] Margaret Mathiang. (21 August 2007) Personal interview with author in Juba.
[96] Kefayatullah Eblagh. (26 November 2006) Personal interview with author in Kabul.
[97] O. Flem. (1998) 'Proposal for Manual Demining in Republika Srpska 1998.' In NPA archives.
[98] Frisby. Personal interview.
[99] Mathiang. Personal interview.
[100] Guy Willoughby. (21 September 2007) Personal interview with author in Thornhill, Scotland.
[101] Al-Qu'ran 5:32.
[102] Sediq. Personal interview.
[103] Ibid.
[104] Rimli & Schmeidl. *Private Security Companies*. p. 42.
[105] Per Breivik. (16 February 2007) Personal interview with author in Sarajevo.
[106] Goodhand. *Aiding Peace?*
[107] Rebecca Roberts & Mads Frilander. (2004) 'Preparing for Peace: Mine Action's Investment in the Future of Sudan.' *Preparing the Ground for Peace: Mine Action in Support of Peacebuilding*. PRIO Report 2/2004. Kristian Berg Harpviken & Rebecca Roberts (Eds.). Oslo, PRIO. p. 18.
[108] e.g. John L. Wilkinson. (December 2002) 'Demining During Operation Enduring Freedom in Afghanistan.' *Journal of Mine Action*. 6.3. <http://maic.jmu.edu/JOURNAL/6.3/notes/wilkinson/wilkinson.htm>.
[109] Kristian Berg Harpviken. (2002) 'Breaking new ground: Afghanistan's response to landmines and unexploded ordnance.' *Third World Quarterly*. 23(5). p. 940.
[110] cf. Chris Johnson & Jolyon Leslie. (2005) *Afghanistan: The Mirage of Peace*. London, Zed Books. p. 106.
[111] Stephen Edelmann. (3 August 2006) Telephone conversation with author.
[112] Carpenter. Personal interview.
[113] Norwegian People's Aid (NPA). (15 February 2007) 'Update on NPA's work against cluster munitions.' <http://www.npaid.org/./?module=Articles;action=Article.publicShow;ID=4533>.
[114] Save the Children Sweden. (December 2004) 'Organizational Capacity and Impact Assessment to Selected Partners in Sudan.' Unpublished document obtained by author.

Conclusions and Reflections

[1] Jody Williams. (2008) 'New Approaches in a Changing World: The Human Security Agenda.' *Banning Landmines: Disarmament, Citizen Diplomacy, and Human Security.* Jody Williams, Stephen D. Goose & Mary Wareham (Eds.). Lanham, Rowman & Littlefield. pp. 293-294.

[2] Mark Duffield. (2001) *Global Governance and the New Wars: The Merging of Development and Security.* London, Zed Books. p. 2.

[3] Giovanni Andrea Cornia, Richard Jolly and Frances Stewart (Eds.) (1987) *Adjustment with a Human Face: Protecting the Vulnerable and Promoting Growth.* Oxford, Clarendon Press.

[4] Ibid. p. 2.

[5] UNMAS. (2005) 'Mine Action and Effective Coordination: The United Nations Policy.' p. 3.
<http://www.undp.org/cpr/documents/mine_action/role_undp/UN_Mine_Action_Policy.pdf>.

[6] ICBL. (June 1999) 'Mine Action Programmes from a development-oriented point of view ('The Bad Honnef Framework').'
<http://www.apminebanconvention.org/index.php?id=424>.

[7] James Madison University Mine Action Information Center. (n.d.) 'Search the JMA.'
<http://maic.jmu.edu/journal/index/search2.htm>.

[8] Mary B. Anderson. (1999) *Do No Harm: How Aid Can Support Peace – or War.* Boulder, Colorado: Lynne Rienner Publishers.

[9] Timothy Donais. (2005) *The Political Economy of Peacebuilding in Post-Dayton Bosnia.* London, Routledge. p. 41.

[10] ICBL. 'Mine Action Programmes.'

[11] Ibid.

[12] Eddie Banks. (August 2003) 'In the Name of Humanity.' *Journal of Mine Action.* 7(2).
<http://maic.jmu.edu/journal/7.2/focus/banks/banks.htm>.

[13] Ananda S. Millard, Kristian Berg Harpviken & Kjell E. Kjellman. (2002) 'Risk Removed? Steps Towards Building Trust in Humanitarian Mine Action.' *Disasters.* 26(2).

[14] ICBL. 'Mine Action Programmes.'

[15] Stephen D. Goose, Mary Wareham & Jody Williams. (2008) 'Banning Landmines and Beyond.' In: Williams, Goose & Wareham. *Banning Landmines.* p. 12.

SELECT BIBLIOGRAPHY

This bibliography only lists the works considered most important and influential to this book, under key topical headings. The full list of references is in the endnotes.

Mine Action

Banks, Eddie. (August 2003) "In the Name of Humanity." *Journal of Mine Action.* 7(2). <http://maic.jmu.edu/journal/7.2/focus/banks/banks.htm>.

Bolton, Matthew & Hugh Griffiths. (September 2006) *Bosnia's Political Landmines: A Call for Socially Responsible and Conflict-Sensitive Mine Action.* London, Landmine Action. <http://www.landmineaction.org/resources/Bosnias_Political_Landmines.pdf>.

Buré, Jacques & Pierre Pont. (November 2003) "Landmine Clearance Projects: A Task Manager's Guide." Social Development Papers: Conflict Prevention & Reconstruction 10. <http://go.worldbank.org/H9XPUBHKP0>.

Cahill, Kevin M. (Ed.). (1995) *Clearing the Fields: Solutions to the Global Land Mines Crisis.* New York, Basic Books.

Cameron, Maxwell A., Robert J. Lawson & Brian W. Tomlin (Eds.). (1998) *To Walk without Fear: The Global Movement to Ban Landmines.* Toronto, Oxford University Press.

Croll, Mike. (1998) *The History of Landmines.* Barnesly, Leo Cooper.

Geneva Call. (4 October 2001) "Deed of Commitment under Geneva Call for Adherence to a Total Ban on Anti-Personnel Mines and for Cooperation in Mine Action." <http://www.genevacall.org/about/testi-mission/gc-04oct01-deed.htm>.

--------. (2005) *Armed Non-State Actors and Landmines.* Geneva, Program for the Study of International Organization(s).

Geneva International Centre for Humanitarian Demining (GICHD). (June 2003) *The Role of the Military in Mine Action.* Geneva, GICHD. <http://www.gichd.org/fileadmin/pdf/publications/Role_Military_MA.pdf>.

--------. (July 2003) *A Guide to Mine Action.* Geneva, GICHD.

--------. (2004) *A Study of Local Organisations in Mine Action.* Geneva, GICHD.

--------. (August 2005) *A Study of Manual Mine Clearance: 1. History, Summary and Conclusions of a Study of Manual Mine Clearance.* Geneva, GICHD.

Harpviken. Kristian Berg (Ed.). (2004) *The Future of Humanitarian Mine Action.* New York, Palgrave Macmillan.

Harpviken, Kristian Berg & Rebecca Roberts. (2004) *Preparing the Ground for Peace: Mine Action in Support of Peacebuilding.* Oslo, PRIO.

Horwood, Chris. (March 2000) "Humanitarian Mine Action: The First Decade of a New Sector in Humanitarian Aid." RRN Network Paper. 32.

Human Rights Watch. (1992) *Hidden Death: Landmines and Civilian Casualties in Iraqi Kurdistan.* New York, Human Rights Watch.

--------. (1994) *Landmines in Mozambique.* New York, Human Rights Watch.

Human Rights Watch & Physicians for Human Rights. (1991) *Land Mines in Cambodia: The Coward's War.* New York, Human Rights Watch.

--------. (1993) *Landmines: A Deadly Legacy.* New York, Human Rights Watch.

International Committee of the Red Cross (ICRC). (March 1996) *Anti-Personnel Land Mines: Friend or Foe?* Geneva, ICRC.

International Committee to Ban Landmines (ICBL). (June 1999) "Mine Action Programmes from a development-oriented point of view ('The Bad Honnef Framework')." <http://www.apminebanconvention.org/index.php?id=424>.

Matthew, Richard A. Bryan McDonald and Kenneth R. Rutherford (Eds.). (2004) *Landmines and Human Security: International Politics and War's Hidden Legacy.* Albany, State University of New York Press.

Maslen, Stuart. (2004) *Mine Action After Diana: Progress in the Struggle Against Landmines.* London, Landmine Action/Pluto Press.

McGrath, Rae. (2000) *Landmines and Unexploded Ordnance: A Resource Book.* London, Pluto Press.

McGrath, Rae & Human Rights Watch. (1993) *Land Mines in Angola: An Africa Watch Report.* New York, Human Rights Watch.

Millard, Ananda S., Kristian Berg Harpviken & Kjell E. Kjellman. (2002) "Risk Removed? Steps Towards Building Trust in Humanitarian Mine Action." *Disasters.* 26(2).

Monin, Lydia & Andrew Gallimore. (2002) *The Devil's Gardens: A History of Landmines.* London, Pimlico.

Roberts, Shawn & Jody Williams (1995) *After the Guns Fall Silent: The Enduring Legacy of Landmines.* Washington, Vietnam Veterans of America Foundation.

Spearin, Christopher. (November 2001) "Ends and Means: Assessing the Humanitarian Impact of Commercialised Security on the Ottawa Convention Banning Anti-Personnel Mines." YCISS Occasional Paper Number. 69. <http://www.yorku.ca/yciss/publications/OP69-Spearin.pdf>.

Stiff, Peter. (1986) *Taming the Landmine.* Alberton, South Africa, Galago.

The Arms Project of Human Rights Watch & Physicians for Human Rights. (1993) *Landmines: A Deadly Legacy.* New York, Human Rights Watch.

United Nations Development Programme (UNDP) & Geneva International Centre for Humanitarian Demining (GICHD). (2001) *A Study of Socio-Economic Approaches to Mine Action.* Geneva, GICHD.

United States Department of State. (29 July 2003) "Milestones in Humanitarian Mine Action: Development of the Landmine Threat and the Discipline of Humanitarian Mine Action." <http://www.state.gov/t/pm/rls/fs/22948.htm>.

--------. (June 2006) "To Walk the Earth in Safety: The U.S. Commitment to Humanitarian Mine Action." <http://www.state.gov/t/pm/rls/rpt/walkearth/2006/68014.htm>.

Webster, Donovan. (1998) *Aftermath: The Remnants of War.* New York, Vintage Books.

Westing, Arthur H. (Ed.). (1985) *Explosive Remnants of War: Mitigating the Environmental Effects.* London, Taylor & Francis.

Williams, Jody, Stephen D. Goose & Mary Wareham (Eds.). (2008) *"Banning Landmines: Disarmament, Citizen Diplomacy, and Human Security.* Lanham, Rowman & Littlefield.

International Humanitarian Law and Disarmament

(18 September 1997) "Convention on the Prohibition of the Use, Stockpiling, Production and Transfer of Anti-Personnel Mines and on Their Destruction." <http://www.un.org/Depts/mine/UNDocs/ban_trty.htm>.

(28 November 2003) "Protocol on Explosive Remnants of War (Protocol V to the 1980 Convention)." <http://www.mineaction.org/docs/1850_.asp>.

(18 March 2009) 'Convention on Cluster Munitions.' <http://www.clusterconvention.org>.

Anderson, Kenneth. (2000) "The Ottawa Convention Banning Landmines, the Role of International Non-governmental Organizations and the Idea of International Civil Society." *European Journal of International Law.* 11(1).

Best, Geoffrey. (1980) *Humanity in Warfare: The Modern History of the International Law of Armed Conflicts.* New York, Columbia University Press.

Cave, Rosy. (2006) "Disarmament as Humanitarian Action? Comparing Negotiations on Anti-Personnel Mines and Explosive Remnants of War." *Disarmament as Humanitarian Action: From Perspective to Practice.* Geneva, UNIDIR.

Hubert, Don. (2000) "The Landmine Ban: A Case Study in Humanitarian Advocacy." Thomas J Watson Jr. Institute for International Studies Occasional Paper. 42.

Karlshoven, Frits & Liesbeth Zegveld. (March 2001) *Constraints on the Waging of War: An Introduction to International Humanitarian Law.* Geneva, ICRC.

Landmine Action. (2006) *Foreseeable Harm: The use and impact of cluster munitions in Lebanon.* London, Landmine Action. <http://www.stopclustermunitions.org/files/Foreseeable%20harm.pdf>.

Latham, Andrew. (October 2000) "Global Cultural Change and the Transnational Campaign to Ban Antipersonnel Landmines: A Research Agenda." YCISS Occasional Paper Number 62. <http://www.yorku.ca/yciss/publications/OP62-Latham.pdf>.

Lavoyer, Jean-Philippe & Louis Maresca. (1999) "The Role of the ICRC in the Development of International Humanitarian Law." *International Negotiation.* 4(3).

Maresca, Louis & Stuart Maslen (Eds.). *The Banning of Anti-Personnel Landmines: The Legal Contribution of the International Committee of the Red Cross 1955-1999.* Cambridge, Cambridge University Press.

Maslen, Stuart. (2001) *Anti-Personnel Mines under Humanitarian Law: A View from the Vanishing Point.* Antwerp, Intersentia.

McGrath, Rae. (2000) *Cluster Bombs: The Military Effectiveness and Impact on Civilians of Cluster Munitions.* London, Landmine Action. <http://www.stopclustermunitions.org//files/Cluster_Bombs%20LMA.pdf>.

Prokosch, Eric. (1995) *The Technology of Killing: A Military and Political History of Antipersonnel Weapons.* London, Zed Books.

Roach, J. Ashley. (1984) "Certain Conventional Weapons Convention: Arms Control or Humanitarian Law?" *Military Law Review.* 105.

Roberts, Adam & Richard Guelff (Eds.). (2004) *Documents on the Laws of War*. 3rd Ed. Oxford, Oxford University Press.

Rutherford, Kenneth Robin. (2000) "The Evolving Arms Control Agenda: Implications of the Role of NGOs in Banning Antipersonnel Landmines." *World Politics*. 53(1).

--------. (March 2000) "Internet Activism: NGOs and the Mine Ban Treaty." *International Journal on Grey Literature*. 1(3).

Shoemaker, Bruce. (March 1994) "Legacy of the Secret War." Mennonite Central Committee. <http://www.mcc.org/clusterbomb/secret_war>.

Short, Nicola. (March 1999) "The Role of NGOs in the Ottawa Process to Ban Landmines." *International Negotiation*. 4(3).

Wiebe, Virgil. (Fall 2000) "Footprints of Death: Cluster Bombs as Indiscriminate Weapons under International Humanitarian Law." *Michigan Journal of International Law*. 22(1).

International Relations and Political Theory

Boutros Ghali, Boutros. (17 June 1992) "An Agenda for Peace: Preventive diplomacy, peacemaking and peace-keeping." <http://www.un.org/Docs/SG/agpeace.html>.

Boyer, William W. (March 1990) "Political Science and the 21st Century: From Government to Governance." *PS: Political Science and Politics*. 23(1).

Bull, Hedley. (1977) *The Anarchical Society: A Study of Order in World Politics*. New York, Columbia University Press.

Claude, Inis L. (Ed.). (1984) *Swords Into Plowshares: The Problems and Progress of International Organization*. 4th Ed. New York, Random House.

Clay, Edward J. & Bernard B. Schaffer. (1984) *Room for Manoeuvre: An Exploration of Public Policy in Agriculture and Rural Development*. London, Heinemann Educational Books.

Cooper, Andrew, Richard Higgott & Kim Nossal. (1993) *Relocating Middle Powers: Australia and Canada in a Changing World Order*. Vancouver, University of British Columbia Press.

Cooper, Andrew (Ed.). (1997) *Niche Diplomacy: Middle Powers after the Cold War*. New York, St. Martin's Press.

Cusack, Thomas R. & Richard J. Stoll. (March 1994) "Collective Security and State Survival in the Interstate System." *International Studies Quarterly*. 38(1).

Davies, Thomas Richard. (December 2006) "The Rise and Fall of Transnational Civil Society." <http://www.bisa.ac.uk/2006/pps/davies.pdf>.

Hardin, Garrett. (September 1974) "Lifeboat Ethics: The Case Against Helping the Poor." *Psychology Today*. <http://www.garretthardinsociety.org/articles/art_lifeboat_ethics_case_against_helping_poor.html>.

Herz, John H. (1951) *Political Realism and Political Idealism*. Chicago, University of Chicago Press.

Ingebritsen, Christine. (2002) "Norm Entrepreneurs: Scandinavia's Role in World Politics." *Cooperation and Conflict: Journal of the Nordic International Studies Association*. 37(1).

Jacobson, Harold K. (1984) *Networks of Interdependence: International Organizations and the Global Political System*. 2nd Ed. New York, Alfred E. Knopf.

Kant, Immanuel. (1983) "To Perpetual Peace: A Philosophical Sketch." *Perpetual Peace and Other Essays*. Ted Humphrey (Trans.). Indianapolis, Hackett Publishing Company.

Kupchan, Charles A. & Clifford A. Kupchan. (Summer 1995) "The Promise of Collective Security." *International Security.* 20(1).

Leonard, Mark. (September-October 2002) "Diplomacy by Other Means." *Foreign Policy.* 132.

Levy, Jack S. (1983) *War in the Modern Great Power System, 1495-1975.* Lexington, Kentucky, University Press of Kentucky.

Kaldor, Mary. (2003) *Global Civil Society: An Answer to War.* Cambridge, Polity Press.

--------. (2007) *Human Security: Reflections on Globalization and Intervention.* Cambridge, Polity Press.

Machiavelli, Niccolo. (1997) *The Prince.* Ware, Hertfordshire, Wordsworth Editions Ltd.

Maier, Charles S. (2000) "Consigning the Twentieth Century to History: Alternative Narratives for the Modern Era." *The American Historical Review.* 105(3). <http://www.historycooperative.org/journals/ahr/105.3/ah000807.html>.

Mearsheimer, John J. (2001). *The Tragedy of Great Power Politics.* New York, Norton.

Milward, H. Brinton & Keith G. Provan. (April 2000) "Governing the Hollow State." *Journal of Public Administration Research and Theory.* 10(2).

Niou, Emerson M.S., Peter C. Ordeshook and Gregory F. Rose. (1989) *The Balance of Power: Stability and Instability in International Systems.* New York, Cambridge.

Nye, Joseph S. (2004) *Soft Power: The Means to Success in World Politics.* New York, PublicAffairs.

Oneal, John R. & Bruce Russett. (1999) "The Kantian Peace: The Pacific Benefits of Democracy, Interdependence, and International Organizations, 1885-1992." *World Politics.* 52(1).

Rice, Condoleezza. (January/February 2000) "Campaign 2000: Promoting the National Interest." *Foreign Affairs.* <http://www.foreignaffairs.org/20000101faessay5/condoleezza-rice/campaign-2000-promoting-the-national-interest.html >.

--------. (July/August 2008) "Rethinking the National Interest: American Realism for a New World." *Foreign Affairs.* <http://www.foreignaffairs.org/20080701faessay87401/condoleezza-rice/rethinking-the-national-interest.html>.

Rothstein, Robert L.. (1968) *Alliances and Small Powers.* New York, Columbia University Press.

Rutherford, Kenneth R. Stefan Brem & Richard A. Matthew (Eds.). (2003) *Reframing the Agenda: the Impact of NGO and Middle Power Cooperation in International Security Policy.* London, Praeger.

Shafer, D. Michael. (January 1990) "Sectors, States, and Social Forces: Korea and Zambia Confront Economic Restructuring." *Comparative Politics.* 22(2).

Sennet, Richard. (1976) *The Fall of Public Man.* New York, W.W. Norton.

Sheehan, Michael. (2000) *The Balance of Power: History and Theory.* London, Routledge.

Tzu, Sun. (1998) *The Art of War.* Ware, Hertfordshire, Wordsworth Editions Ltd.

von Clausewitz, Carl. (1997) *On War.* J.J. Graham (Trans.). Ware, Hertfordshire, Wordsworth Editions Ltd.

Wight, Martin. (1977) *Systems of States.* Leicester, Leicester University Press.

Zolo, Danilo. (1998) "Hans Kelsen: International Peace through International Law." *European Journal of International Law.* 9(2).

Privatization and Private Security

Avant, Deborah. (2005) *The Market for Force: The Consequences of Privatizing Security.* Cambridge, Cambridge University Press.

Bendick, Marc, Jr. (1989) "Privatizing the Delivery of Social Welfare Services: An Idea to Be Taken Seriously." *Privatization and the Welfare State.* Sheila B. Kamerman & Alfred J. Kahn (Eds.). Princeton, Princeton University Press.

Berríos, Reubén. (2000) *Contracting for Development: The Role of For-Profit Contractors in U.S. Foreign Development Assistance.* Westport, CT, Praeger.

Blakely, Edward J. & Mary Gail Snyder. (1997) *Fortress America: Gated Communities in the United States.* Washington DC, The Brookings Institution.

Braun, Dietmar. (April-June 1993) "Who Governs Intermediary Agencies? Principal-Agent Relations in Research Policy-Making." *Journal of Public Policy.* 13(2).

George Yarrow & Piotr Jasinski (Eds.). (1996) *Privatization: Critical Perspectives on the World Economy.* Vol. II. London, Routledge.

Davis, James H., F. David Schoorman & Lex Donaldson. (1997) "Toward a Stewardship Theory of Management." *The Academy of Management* Review. 22(1).

Dicke, Lisa A. (2002) "Ensuring accountability in human services contracting: Can stewardship theory fill the bill?" *American Review of Public Administration* 32.

Dicke, Lisa A. & Steven J. Ott. (2002) "A test: Can stewardship theory serve as a second conceptual foundation for accountability methods in contracted human services?" *International Journal of Public Administration.* 25.

Eisenhardt, Kathleen M. (1989) "Agency theory: An assessment and review." *Academy of Management Review.* 14.

Hodge, Graeme A. (2000) *Privatization: An International Review of Performance.* Boulder, Colorado, Westview Press.

Hughes, Solomon. (2007) *War on Terror, Inc.: Corporate Profiteering from the Politics of Fear.* London, Verso.

Kinsey, Christopher. (2007) *Corporate Soldiers and International Security: The Rise of Private Military Companies.* London, Routledge.

Klein, Naomi. (2007) *The Shock Doctrine: The Rise of Disaster Capitalism.* London, Allen Lane.

Pelton, Robert Young. (2006) *Licensed to Kill: Hired Guns in the War on Terror.* New York, Random House.

Pratt, John W. & Richard J. Zeckhauser (Eds.) (1984) *Principals and agents: The structure of business.* Boston, MA: Harvard Business School Press.

Savas, E.S. (1982) *Privatizing the Public Sector: How to Shrink Government.* Chatham, NJ, Chatham House.

Singer, P.W. (2003) *Corporate Warriors: The Rise of the Privatized Military Industry.* Ithaca, Cornell University Press.

Smith, Steven R. & Michael Lipsky. (1993) *Nonprofits for Hire: The Welfare State in the Age of Contracting.* Cambridge, Mass., Harvard University Press.

Stiglitz, Joseph E. (2002) *Globalization and Its Discontents.* London, Penguin Books.

Van Slyke, David M. (April 2007) "Agents or Stewards: Using Theory to Understand the Government-Nonprofit Social Service Contracting Relationship." *Journal of Public Administration Research and Theory.* 17(2).

World Bank. (1997) *World Development Report 1997: The State in a Changing World.* New York, Oxford University Press.

--------. (2002) *World Development Report 2002: Building Institutions for Markets.* New York, Oxford University Press.

Political Economy of Conflict

Ballentine, Karen & Jake Sherman. (2003) *The Political Economy of Armed Conflict: Beyond Greed and Grievance.* Boulder, Co., Lynne Reiner Publishers.

Bojicic-Dzelilovic, Vesna. (2000) "From Humanitarianism to Reconstruction: Towards an Alternative Approach to Economic and Social Recovery from War." *Global Insecurity: Restructuring the Global Military Sector.* Vol. III. Mary Kaldor (Ed.). London, Pinter.

Duffield, Mark. (Spring 1998) "Post-modern Conflict: Warlords, Post-adjustment States and Private Protection." *Civil Wars.* 1(1).

--------. (1999) "Globalization and War Economies: Promoting Order or the Return of History?" *Fletcher Forum on World Affairs.* 23(2).

Griffiths, Hugh. (Summer 1999) "A Political Economy of Ethnic Conflict, Ethnonationalism and Organised Crime." *Civil Wars.* 2(2).

Kaldor, Mary. (1999). *New and Old Wars: Organized Violence in a Global Era.* Cambridge, Polity Press.

Keen, David. (1998) *The Economic Functions of Violence in Civil Wars.* Adelphi Paper 320. London, Oxford University Press.

--------. (2008) *Complex Emergencies.* Cambridge, Polity.

Münkler, Herfried. (2005) *The New Wars.* Cambridge, Polity Press.

Pugh, Michael Charles, Neil Cooper & Jonathan Goodhand M. and N. Cooper, (Eds.). (2004) *War Economies in a Regional Context: Challenges of Transformation.* London, Lynne Reiner Publishers, Inc.

van Creveld, Martin. (1991) *The Transformation of War.* New York, The Free Press.

Weizman, Eyal. (2007) *Hollow Land: Israel's Architecture of Occupation.* London, Verso.

Foreign Aid

Anderson, Mary B. (1999) *Do No Harm: How Aid Can Support Peace – or War.* Boulder, Lynne Rienner Publishers.

Christian Aid. (2004) *The Politics of Poverty: Aid in the New Cold War.* London, Christian Aid. <http://christianaid.org.uk/indepth/404caweek/index.htm>.

Cornia, Giovanni Andrea, Richard Jolly and Frances Stewart (Eds.). (1987) *Adjustment With A Human Face: Protecting the Vulnerable and Promoting Growth.* Oxford, Clarendon Press.

de Waal, Alex. (1998) *Famine Crimes: Politics and the Disaster Relief Industry in Africa.* Bloomington, Indiana University Press.

Duffield, Mark. (2001) "Governing the Borderlands: Decoding the Power of Aid." *Disasters.* 25(4).

--------. (2001) *Global Governance and the New Wars: The Merging of Development and Security.* London, Zed Books.

--------. (2007) *Development, Security and Unending War: Governing the World of Peoples.* Cambridge, Polity Press.

Easterly, William. (2006) *The White Man's Burden: Why the West's Efforts to Aid the Rest Have Done So Much Ill and So Little Good.* London, Penguin Books.

Goodhand, Jonathan. (2006) *Aiding Peace? The Role of NGOs in Armed Conflict.* Bourton on Dunsmore, Intermediate Technology Publications.

Hulme, David & Michael Edwards. (1997) *NGOs, States and Donors Too Close for Comfort?* Macmillan, Basingstoke.

Juma, Monica Kathina & Astri Suhrke (Eds.). (2002) *Eroding Local Capacity: International Humanitarian Action in Africa.* Uppsala, Nordiska Afrikainstitutet.

Rieff, David. (2002) *A Bed for the Night: Humanitarianism in Crisis.* London, Vintage.

Terry, Fiona. (2002) *Condemned to Repeat?: The Paradox of Humanitarian Action.* Ithaca, Cornell University Press.

Transparency International. (2005) *Global Corruption Report 2005: Corruption In Construction And Post-Conflict Reconstruction.* <http://www.transparency.org/publications/gcr/download_gcr/download_gcr_2005>.

United Nations Development Programme (UNDP). (1994) *Human Development Report 1994: New dimensions of human security.* New York, Oxford University Press.

US Foreign Policy

Chandrasekaran, Rajiv. (2007) *Imperial Life in the Emerald City: Inside Baghdad's Green Zone.* London, Bloomsbury Publishing.

Johnson, Chalmers. (2000) *Blowback: The Costs and Consequences of American Empire.* New York, Henry Holt and Company.

--------. (2004) *The Sorrows of Empire: Militarism, Secrecy, and the End of the Republic.* New York, Metropolitan Books.

Prados, John. (1996) *Presidents' Secret Wars: CIA and Pentagon Covert Operations from World War II through the Persian Gulf.* Rvsd Ed. Chicago, Elephant Paperbacks.

Priest, Dana. (2004) *The Mission: Waging War and Keeping Peace with America's Military.* New York, W.W. Norton and Company.

Ruttan, Vernon W. (1996) *United States Development Assistance Policy: The Domestic Politics of Foreign Economic Aid.* Baltimore, Johns Hopkins University Press.

Scott, James M. (1996) *Deciding to Intervene: The Reagan Doctrine and American Foreign Policy.* Durham, North Carolina, Duke University Press.

Sigal, Leon V. (2006) *Negotiating Minefields: The Landmines Ban in American Politics.* New York, Routledge.

Tarnoff, Curt & Larry Nowels. (15 April 2004) "Foreign Aid: An Introductory Overview of U.S. Programs and Policy." *CRS Report for Congress.* Washington DC, Congressional Research Service. <http://usinfo.state.gov/usa/infousa/trade/files/98-916.pdf>.

Norwegian Foreign Policy

Bøås, Morten. (2002) "Public attitudes to aid in Norway and Japan." University of Oslo Centre for Development and the Environment Working Paper 2002/03. <http://www.sum.uio.no/publications/pdf_fulltekst/wp2002_03_boas.pdf>.

Cole, Wayne S. (1989) *Norway and the United States, 1905-1955: Two Democracies in Peace and War*. Ames, Iowa State University Press.

Eriksen, Tore Linné (Ed.). (2000) *Norway and National Liberation in Southern Africa*. Stockholm, Nordiska Afrikainstitutet.

Egeland, Jan. (1988) *Impotent Superpower—Potent Small State: Potentials and Limitations of Human Rights Objectives in the Foreign Polices of the United States and Norway*. Oslo, Norwegian University Press.

Kelleher, Ann & James Larry Taulbee. (October 2006) "Bridging the Gap: Building Peace Norwegian Style." *Peace & Change*. 31(4).

Moolakkattu, John Stephen. (2005) "Peace Facilitation by Small States: Norway in Sri Lanka." *Cooperation and Conflict: Journal of the Nordic International Studies Association*. 40(4).

Østerud, Øyvind. (September 2005) "Introduction: The Peculiarities of Norway." *West European Politics*. 28(4).

Pharo, Helge. (2003) "Altruism, Security and the Impact of Oil: Norway's Foreign Economic Assistance Policy, 1958-1971." *Contemporary European History*. 12(4).

Riste, Olav. (2001) *Norway's Foreign Relations - A History*. Oslo, Universitetsforlaget.

Stokke, Olav (Ed.). (1989) *Western Middle Powers and Global Poverty: The Determinants of the Aid Policies of Canada, Denmark, the Netherlands, Norway and Sweden*. Uppsala, Scandinavia Institute of African Studies.

Afghanistan

Coll, Steve. (2004) *Ghost Wars: The Secret History of the CIA, Afghanistan and bin Laden, From the Soviet Invasion to September 10, 2001*. London, Penguin Books.

Crile, George. (2004) *Charlie Wilson's War*. New York, Grove Press.

Eaton, Robert, Chris Horwood & Norah Niland. (1997) *Afghanistan: The Development of Indigenous Mine Action Capacities*. New York, UN Department of Humanitarian Affairs Lessons Learned Unit.

Girardet, Edward & Jonathan Walter (Eds.). (2004) *Afghanistan*. 2nd Ed. Geneva, CROSSLINES Publications.

Harpviken. Kristian Berg. (2002) "Breaking new ground: Afghanistan's response to landmines and unexploded ordnance." *Third World Quarterly*. 23(5).

Johnson, Chris & Jolyon Leslie. (2004) *Afghanistan: The Mirage of Peace*. London, Zed Books.

Katzman, Kenneth. (10 September 2007) "Afghanistan: Post-War Governance, Security, and U.S. Policy." *CRS Report for Congress*.
<http://www.fas.org/sgp/crs/row/RL30588.pdf>.

Maley, William. (2006) *Rescuing Afghanistan*. London, C. Hurst & Co.

Nawa, Fariba. (6 October 2006) *Afghanistan, Inc.: A Corpwatch Investigative Report*.
<http://s3.amazonaws.com/corpwatch.org/downloads/AfghanistanINCfinalsmall.pdf>

Rashid, Ahmed. (2001) *Taliban: The Story of the Afghan Warlords*. London, Pan Books.

Rimli, Lisa & Susanne Schmeidl. (November 2007) *Private Security Companies and Local Populations. An exploratory study of Afghanistan and Angola*. Bern, Switzerland, SwissPeace. <http://www.swisspeace.ch/typo3/fileadmin/user_upload/pdf/PSC.pdf>.

Rubin, Barnett. (2002) *The Fragmentation of Afghanistan: State Formation and Collapse in the International System*. 2nd Ed. New Haven, Yale University Press.

Bosnia

Andreas, Peter. (2004) "The Clandestine Political Economy of War and Peace in Bosnia." *International Studies Quarterly*. 48.

Devine, Vera. (2005) 'Corruption in Post-War Reconstruction: The Experience of Bosnia and Herzegovina.' *Corruption in Post-War Reconstruction: Confronting the Vicious Circle*. Beirut, Lebanese Transparency Association. <http://www.transparency-lebanon.org/2006/Publications/PWR/4.pdf>.

Donais, Timothy. (2005) *The Political Economy of Peacebuilding in Post-Dayton Bosnia*. London, Routledge.

Holbrooke, Richard. (1999) *To End a War*. New York, Random House.

International Crisis Group (ICG). (18 July 1997) "Ridding Bosnia of Landmines: The Urgent Need for a Sustainable Policy." ICG Bosnia Report No. 25.

Jeffrey, Alex. (February 2006) "Building state capacity in post-conflict Bosnia and Herzegovina: The case of Brcko District." *Political Geography*. 25(2).

Kaldor, Mary & Vesna Bojicic. (1999) "The 'Abnormal' Economy of Bosnia-Herzegovina." *Scramble for the Balkans: Nationalism, Globalism and the Political Economy of Reconstruction*. Carl-Ulrick Schierup (Ed.). New York, Macmillan.

Lisica, Darvin. (2006) *Risk Management in Mine Action Planning*. Sarajevo, Bosnia and Herzegovina, Ministry of Civil Affairs & Norwegian People's Aid.

Mitchell, Shannon K. (2004) "Death, Disability, Displaced Persons and Development: The case of Landmines in Bosnia and Herzegovina." *World Development*. 32(12).

Pugh, Michael. (2002) "Postwar Political Economy in Bosnia and Herzegovina: The Spoils of Peace." *Global Governance*. 8.

Simms, Brendan. (2003) *Unfinest Hour: Britain and the Destruction of Bosnia*. New Ed. New York, Penguin Global.

Sudan

Autesserre, Séverine. (January 2002) "United States 'Humanitarian Diplomacy' in South Sudan." *Journal of Humanitarian Assistance*. <http://www.jha.ac/articles/a085.htm>.

Dagne, Ted. (12 April 2006) "Sudan: Humanitarian Crisis, Peace Talks, Terrorism, and U.S. Policy." *CRS Issue Brief for Congress*. Washington DC, Congressional Research Service. <http://www.fas.org/sgp/crs/row/IB98043.pdf>.

Keen, David. (1994) *The Benefits of Famine*. Princeton, Princeton University Press.

Middleton, Neil & Phil O'Keefe. (2006) "Politics, History & Problems of Humanitarian Assistance in Sudan." *Review of African Political Economy*. 109.

Minnear, Larry. (1991) *Humanitarianism under Siege: A Critical Review of Operation Lifeline Sudan*. Trenton, New Jersey.

Riehl, Volker. (2001) *Who Is Ruling in South Sudan? The role of NGOs in rebuilding socio-political order.* Uppsala, Nordiska Afrikainstitutet.

Tvedt, Terje. (1994) "The Collapse of the State in Southern Sudan after the Addis Ababa Agreement: A Study of Internal Causes and the Role of the NGOs." *Short-Cut to Decay: The Case of the Sudan.* Sharif Harir & Terje Tvedt (Eds.). Uppsala, Scandinavian Institute of African Studies.

Young, John. (2005) "Sudan: A Flawed Peace Process Leading to a Flawed Peace." *Review of African Political Economy.* 103.

INTERVIEWS

Job titles were correct at the time the person was interviewed.

Afghanistan

Asalati, A. Ghani. Director, AIMEIC. (22 November 2006) Personal interview with author in Kabul.

Attiqullah, Haji. Director, MCPA. (29 November 2006) Personal interview with author in Kabul.

Barber, Martin. Former Director, UNMAS. (28 November 2006) Personal interview with author in Kabul.

Bellamy, Daniel. Programme Manager, UNMACA. (22 November 2006) Personal interview with author in Kabul.

Carpenter, Lloyd. Manager, DynCorp Afghanistan Mine Action. (21 November 2006) Personal interview with author in Kabul.

De Benedetti, Elizabeth. Senior External Relations Officer, UNMACA. (29 November 2006) Personal interview with author in Kabul.

Eblagh, Kefayatullah. Director, both Afghan Technical Consultants and Hemayatbrothers International Demining. (26 November 2006) Personal interview with author in Kabul.

Elliot, David. Technical Advisor, HALO Trust Afghanistan. (19 November 2006) Personal interview with author in Kabul.

Fazel, Fazel Karim. President, OMAR. (29 November 2006) Personal interview with author in Kabul.

Gannon, Robert. Country Manager, RONCO Afghanistan. (20 November 2006) Personal interview with author in Kabul.

Hakimi, Mohammed Shohab. Director, MDC. (6 November 2006) Personal interview with author in Kabul.

Hergault, Thierry. Country Director, Handicap International Afghanistan. (11 December 2006) Personal interview with author in Kabul.

Holroyd, Mark. Project Manager, RONCO Afghanistan. (26 October 2006) Personal interview with author in Kabul.

Homayoun, Farid. Country Director, HALO Trust Afghanistan (19 November 2006) Personal interview with author in Kabul, Afghanistan.

Hutson, Dean. Demining Coordinator, USAID Afghanistan. (30 November 2006) Personal interview with author in Kabul.

Jouvenal, Peter. Former journalist during Afghan-Soviet war. (24 October 2006) Personal interview with author in Kabul.

Karim, Mohammed. General manager, Afghan Construction and Logistics Unit. (30 November 2006) Personal interview with author in Kabul.

Løvold, Andreas. Second Secretary, Royal Norwegian Embassy in Kabul. (12 December 2006) Personal interview with author in Kabul.

Molam, Paul. Country Director, MineTech Afghanistan. (29 November 2006) Personal interview with author in Kabul.

--------. (7 December 2006) Comment at presentation of preliminary research findings by author to the UNMACA, Kabul.

Naseri, Ghulam Rahman. Deputy Quality Assurance Specialist, UNMACA. (27 November) Personal interview with author in Kabul.

Omeragic, Amir. Programme Officer, UNMACA. (22 November 2006) Personal interview with author in Kabul.

Payab, Zekria. Deputy Director, OMAR. (2 November 2006) Personal interview with author in Kabul.

Powell, Sandy. Chief of Operations, UNMACA Afghanistan. (1 November 2006) Personal interview with author in Kabul.

Robinson, Stephen. Program Manager, Danish Demining Group Afghanistan. (29 November 2006) Personal interview with author in Kabul.

Ruru, Kerei. Chief of Staff, UNMACA. (29 November 2006) Personal interview with author in Kabul.

Safi, Mumtaz. Training Manager, MDC. (31 October 2006) Personal interview with author in Kabul.

Sattar, Abdul. Director, DAFA. (4 December 2006) Personal interview with author in Kabul.

Sediq, Mohammed. Chief of Operations, UNMACA. (23 November 2006) Personal interview with author in Kabul.

Stevens, Barry. Demining Operations Manager, EODT Afghanistan. (24 November 2006) Personal interview with the author in Kabul.

Sudell, William. Project Manager, S3 AG Afghanistan. (30 October 2006) Personal interview with author in Kabul.

Sutcliffe, Peter. Mine Action Manager, ArmorGroup Afghanistan. (27 November 2006) Personal interview with author in Kabul.

Wanley, Colin. Country Director, UXB Afghanistan. (30 October 2006) Personal interview with author in Kabul.

Petrisch, Wolfgang. Former High Representative to Bosnia and Herzegovina. (1 December 2006) Personal interview with author in Kabul.

Yamamoto, Yoshiyuki. Head of Programme Section, UNMACA. (22 November 2006) Personal interview with author in Kabul.

Bosnia and Herzegovina

Atikovic, Damir. Community Liason, NPA Bosnia. (22 February 2007) Personal interview with author in Sarajevo.

Balic, Amela. Operations Manager, NPA Bosnia. (23 February 2007) Personal interview with author in Sarajevo.

Banks, Eddie. Former Consultant to Bosnian Demining Commission. (10 March 2007) Personal interview with author in Br☐ko.

Braatha, Jan. Ambassador. Royal Norwegian Embassy in Sarajevo. (8 March 2007) Personal interview with author in Sarajevo.

Breivik, Per. Program Manager, NPA Bosnia. (22 June 2005) Personal interview with author in Sarajevo.

--------. Program Manager, NPA Bosnia. (28 February 2007) Personal interview with author in Sarajevo.

Chappel, Derek. Spokesman, NATO HQ Sarajevo. (11 February 2007) Personal interview with author in Sarajevo.

Gavran, Dusan. Director, BHMAC. (16 February 2007) Personal interview with author in Sarajevo.

Griffiths, Hugh. Investigative Journalist, Institute for War and Peace Reporting. (20 June 2005) Personal interview with author in Sarajevo.

Grujic, Zoran. IT Chief, BHMAC. (13 February 2007) Personal interview with author in Sarajevo.

Hadzimujagic, Nermin. Director, Mine Detection Dog Center for South East Europe. (16 March 2007) Personal interview with author in Borci.

Isaacs, Peter. Former UNMAC employee. (August 2005) Telephone interview with author.

Ivic, Darko. Bureau of International Security and Nonproliferation, US Embassy Sarajevo. (6 February 2007) Personal interview with author in Sarajevo.

Jalovicic, Kristijan. Project Manager, MECHEM Bosnia (27 February 2007) Personal interview with author in Sarajevo.

Klisara, Petar. Manager, TerraProm. (9 March 2007) Personal interview with author in Pale.

Lardner, Tim. Mine Action Specialist, GICHD. (28 February 2007) Personal interview with author in Sarajevo.

Lawrence, William. Former UNMAC employee. (1 August 2005). Telephone interview with author.

Lisica, Darvin. Deputy Director, BHMAC. (8 August 2005) Personal interview with author in Sarajevo.

--------. Deputy Program Manager, NPA Bosnia. (22 February 2007) Personal interview with author in Sarajevo.

McKenzie, Neil. Regional Manager for ArmorGroup Balkans (7 July 2005) Personal interview with author in Sarajevo.

Moore, Jennifer. Political Military Affairs, US Embassy Sarajevo. (8 July 2005) Personal interview with author in Sarajevo.

Morete, Hemi. Consultant, US Department of State Office of Weapons Removal and Abatement. (24 February 2007) Personal interview with author in Sarajevo.

Music, Zlatan. Former Project Assistant, Counterpart International Br☐ko. (14 April 2005) Personal interview with author in Br☐ko.

Neven, Cica. Project Manager, UXB Balkans. (15 March 2007) Personal interview with author in Sarajevo.

Prevljak, Hidajet. Director, BH Demining. (22 February 2007) Personal interview with author in Sarajevo.

Prevost, Nathalie. Program Manager, UNICEF Bosnia Mine Action. (15 April 2005) Personal interview with author in Sarajevo.

Ratel, Jonathan. International Prosecutor, State Court of Bosnia and Herzegovina. (8 February 2007) Personal interview with author in Sarajevo.

Redfern, Jenet. Project Officer, OSCE Defence Reform Commission. (23 March 2007) Personal interview with author in Sarajevo.

Rowe, David. Program Manager, UNDP Bosnia Mine Action. (18 April 2005) Personal interview with author in Sarajevo.

--------. (2 February 2007) Personal interview with author in Sarajevo.

--------. (12 February 2007) Personal interview with author in Sarajevo.

Two Former Deminers. (12 April 2005) Personal interview with author in Br☐ko.

Zivkovic, Radosav. President, STOP Mines. (20 February 2007) Personal interview with author in Pale.

Kenya

Daubney, Mark. Commandant, International Mine Action Training Center. (24 August 2007) Personal interview with author in Nairobi.

Whitworth, Jeff. Coordinator, Cranfield University Sudan Project. (24 August 2007) Personal interview with author in Nairobi.

Norway

Berggrav, Yngvild. Advisor, Section for Humanitarian Affairs, Norwegian Ministry of Foreign Affairs. (3 September 2007) Personal interview with author in Oslo.

Eckey, Susan. Deputy Director General, Department of UN, Peace and Humanitarian Affairs, Norwegian Ministry of Foreign Affairs. (3 September 2007) Personal interview with author in Oslo.

Kongstad, Steffen. Director General, Department of UN, Peace and Humanitarian Affairs, Norwegian Ministry of Foreign Affairs. (3 September 2007) Personal interview with author in Oslo.

Nikolaisen, Laila. Special Advisor, NPA Headquarters. (4 September 2007) Personal interview with author in Oslo.

INTERVIEWS 253

Ruge, Christian. Consultant, FAFO. (5 September 2007) Personal interview with author in Oslo.

Traavik, Stig. Afghanistan Coordinator, Norwegian Ministry of Foreign Affairs. (5 September 2007) Personal interview with the author in Oslo.

Slovenia

Gacnik, Goran. Director, International Trust Fund for Demining and Mine Victim Assistance. (13 April 2007) Telephone interview with author.

Sudan

Abdelatie, Abdelkheir Eid. Director, Sudan Campaign to Ban Landmines. (15 August 2007) Personal interview with author in Khartoum.

Akol, Madut. Manager, SIMAS. (9 August 2007) Personal interview with author in Juba.

Andersen, Rune Kristian. Programme Manager, NPA Sudan Mine Action. (9 August 2007) Personal interview with author in Juba.

Augustino, Pacifico L. Programme Manager, UNDP South Sudan Mine Action. (10 August 2007) Personal interview with author in Juba.

Birge, Cameron. Liaison Officer, WFP South Sudan Road Rehabilitation Project. (6 August 2007) Personal interview with author in Juba.

Boshoff, Rian. Operations Officer, ArmorGroup Mine Action Sudan. (1 August 2007) Personal interview with author in Juba.

Bosman, Nicholas. Business Development Manager, MineTech Sudan. (14 August 2007) Personal interview with author in Khartoum.

Buono, Marco. Program Manager, DanChurchAid Sudan Mine Action. (15 August 2007) Personal interview with author in Khartoum.

Crosskey, Steve. Project Manager, WFP South Sudan Road Rehabilitation Project. (11 August 2007) Personal interview with author in Juba.

Eldred, Paul. Regional Operations Coordinator, UNMAO Juba. (27 July 2007) Personal interview with author in Juba.

Elobeid, Hussein. General Manager, JASMAR. (13 August 2007) Personal interview with author in Khartoum.

Franklin, Jamie. Manager, MAG South Sudan. (31 July 2007) Personal interview with author in Juba.

Frilander, Mads. Programme Officer, DanChurchAid Kenya/Sudan. (7 August 2007) Personal interview with author in Juba.

Frisby, Charles. Deputy Programme Manager, NPA Sudan Mine Action. (2 August 2008) Personal interview with the author in Yei.

Giha, Ahmed. Technical Advisor, Sudan National Mine Action Authority. (14 August 2007) Personal interview with author in Khartoum.

Jacobsen, Bodil. Representative, Danish Demining Group. (31 July 2007) Personal interview with author in Juba.

Ledang, Jan. Consul-General, Royal Norwegian Consulate Juba. (21 August 2007) Personal interview with author in Juba.

Ludick, Norman. Manager, MECHEM North Sudan. (13 August 2007) Personal interview with author in Khartoum.

Ludwig, Achim. First Secretary, European Commission Sudan. Personal interview with author in Khartoum.

Manager. The Development Initiative South Sudan. (22 August 2007) Personal interview with author in Juba.

Maio, Fred. Project Manager, MAG Sudan-North. (20 August 2007) Personal interview with author in Khartoum.

Maree, Johan. Operations Officer, UNMAO Juba. (26 July 2007) Personal interview with author in Juba.

Mathiang, Margaret. Deputy Commissioner, South Sudan Demining Commission. (21 August 2007) Personal interview with author in Juba.

Moses, Tempas Camilo. Supervisor, RONCO Sudan. (7 August 2007) Personal interview with author in Juba.

Pansegrouw, Jim. Director, UNMAO. (14 August 2007) Personal interview with author in Khartoum.

Razak, Abdul. Technical Adisor, FSD/SIMAS. (10 August 2007) Personal interview with author in Juba.

Shakhawat, Major. Bangladesh Demining Company South Sudan. (9 August 2007) Personal interview with author in Juba.

Tariq, Qadeem Khan. Senior Technical Advisor, UNDP Sudan Mine Action. (14 August 2007) Personal interview with author in Khartoum.

Tesfaye, Makonnen. Programme Officer, UNHCR Juba. (1 August 2007) Personal interview in Juba.

Two employees. SLR Yei. (4 August 2007) Personal interview with author in Yei.

Van Weezendonk, Evy & Christine Murphy. (3 August 2008) Personal interview with author in Yei.

Vukovic, Bojan. Mine Risk Education Coordinator. UNMAO Juba. (25 July 2007) Personal interview with author in Juba.

Switzerland

Veble, Eva. Programme Officer, DanChurchAid Humanitarian Mine Action Unit. (20 September 2008) Personal interview with author in Geneva.

Bach, Hårvard. Operational Analysis Expert, GICHD. (20 September 2006) Personal interview with author in Geneva.

Lardner, Tim. Mine Action Specialist, GICHD. (25 October 2006) Interview with author in Geneva.

Paterson, Ted. Head of Evaluation, GICHD. (25 October 2006) Personal interview with author in Geneva.

United Kingdom

Willoughby, Guy. Director, HALO Trust. (21 September 2007) Personal interview with author in Thornhill, Scotland.

United States of America

Barron, William E. Director of Information Management and Mine Action Programs, Vietnam Veterans of America Foundation. (August 2006)Personal interview with author in Washington DC.

Brooks, Doug. President, International Peace Operations Association. (18 August 2006) Personal interview with author in Washington DC.

Conley, Chuck. Senior Information Management Officer, Vietnam Veterans of America Foundation. (August 2006) Personal interview with author in Washington DC.

Crandall, Larry. Former Country Director, USAID Cross Border Humanitarian Assistance Program for Afghanistan. (1 July 2006) Personal interview with author in McLean, Virginia.

Dorcus, Harry. Former Finance Officer, USAID Nicaragua. (25 October 2006) Interview with author in Washington DC.

Eaton, Bob. Executive Director, Survey Action Center. (2 August 2006) Personal interview with the author in Takoma Park, Maryland.

Edelmann, Stephen. President, RONCO. (3 August 2006) Telephone conversation with author.

Fay, Lois Carter. Editor, Journal of Mine Action. (23 August 2006) Telephone conversation with author.

Flanagan, John. Chief of Programme Support Section, UNMAS. (7 August 2006) Personal interview with author in New York.

Gouttierre, Thomas. Dean of International Studies and Programs, University of Nebraska at Omaha. (18 August 2006) Telephone interview with author.

Harris, Stu. Deputy Director, Programs, US Department of State Office of Weapons Removal and Abatement. (8 August 2006) Personal interview with author in Washington DC.

Kendellen, Mike. Director for Survey, Survey Action Center. (29 July 2006) Interview with author in Takoma Park, Maryland.

Kidd, Richard. Acting Deputy Assistant Secretary for Security Operations, US Department of State Office of Weapons Removal and Abatement. (8 August 2006) Personal interview with author in Washington DC.

Landis, Lauren. Senior Representative to Sudan, US Department of State African Affairs Bureau. (10 April 2007) Personal interview in Washington DC.

Long, Philip. Former EOD Technician, US Air Force. (15 August 2006) Telephone interview with the author.

Mahan, Val. Former Program Manager, USAID Cross Border Humanitarian Assistance Program for Afghanistan. (16 August 2006). Personal interview with author in Virginia, USA.

McCarthy, Rueben. Project Officer, Landmines and Small Arms Team, UNICEF. (7 August 2006) Personal interview with author in New York.

--------. (16 April 2007) Personal interview with author in New York.

McCloy, H. Murphey. (26 August 2005) Personal interview with author in Washington, DC.

Menzies, John. Former Ambassador, US Embassy Sarajevo. (9 August 2006) Telephone interview with author.

Morete, Hemi. Coordinator, UNOPS Mine Action. (7 August 2006) Personal interview with author in New York.

Peachey, Titus. Director of Peace Education, Mennonite Central Committee. (15 August 2006) Telephone interview with the author.

Smith, Tom. Program Manager, US Department of Defense Humanitarian Assistance and Mine Action. (August 2006) Personal interview with author in Crystal City, Virginia.

Stephens, Chris. Programme Officer, UNMAS. (16 April 2007) Personal interview with author in New York.

White, Jerry. Co-Founder and Executive Director, Landmine Survivors Network. (August 2006) Personal interview with author in Washington DC.

Wilson, Steve. Director for International Relations, MAG America. (August 2006) Personal interview with author in Washington DC.

INDEX

9/11, 45, 93, 96, 99, 100, 113, 115, 184
Accidents, Demining, 18, 29, 83, 91, 105, 117, 132-136, 140, 142, 144, 162, 168
Advocacy, 1, 6, 19, 24, 25, 30, 36, 53, 55, 56, 65, 84-86, 147, 168, 170, 171, 176, 177,
Afghanistan, 3-9, 11, 15, 21-31, 35, 39, 43, 50, 57, 60, 65, 66, 72-85, 89-101, 109-115, 118, 121-126, 129-140, 148-153, 156, 161, 163-173, 181, 185, 186, 188-190
Afghan Campaign to Ban Landmines (ACBL), 170
Afghan Technical Consultants (ATC), 118, 121, 166
African National Congress (ANC), 64
AFRICOM, *See US Africa Command*
Ahmed Haroun, 160
Aid, 2, 5-7, 11, 17, 23-27, 36, 41, 42, 44, 50-60, 63-65, 73, 76, 79-96, 136, 99, 104, 109, 113, 114, 146, 148, 152, 155, 157, 160, 174, 178, 179, 183
Air-balloons, 5
AKD Mungos, 137
Al Qaeda, 45, 113, 152
Alliant Techsystems, 44, 66

American Civil War, 13
American Friends Service Committee (AFSC), 18, 19, 24
Anderson, Mary, 146
Angola, 8, 22, 29, 39, 72, 74, 75, 83, 86, 87
Animal Holding Facility, 91
Anthropology, 10, 86, 179
Antipersonnel Landmine Ban Treaty, See *Ottawa Convention*.
Antipersonnel landmines, 3, 188
Antivehicle mines, 3
ArmorGroup, 29, 45, 77, 81, 136, 149, 151, 153, 159
Artillery, 13
Australia, 25
Austria, 25, 85
Axworthy, Lloyd, 68
Azerbaijan, 75
Bad Honnef Framework, 24, 179, 181, 183
Banks, Eddie, 44, 49, 117, 118, 125, 130, 181
Bees, 5
Berger Group, 98, 110-111
BH Demining, 102
Bin Laden, Osama, 113, 152
Biokomerc, 158

Bio-politics, 50
Black & Veatch Special Projects Corp, 98
Blackwater, 45
Bosnia and Herzegovina, 1, 5, 7-9, 11, 23, 28, 35, 41, 44, 50, 60, 75, 80, 83, 89, 93, 98, 100-107, 111, 112, 115, 118-137, 140, 141, 144, 145, 148, 150-169, 171, 173, 181, 183, 190-191
Bosnia and Herzegovina Mine Action Center (BHMAC), xiv, 100, 107, 148, 152, 163
Bosniaks, 100-102, 158, 167
Boutros Ghali, Boutros, 52
Boyd, Ronald, 81
Boyer, William W., 1, 189
Branfman, Fred, 19
Brcko District, 106, 163, 191
Breivik, Per, 125
Bull, Hedley, 40
Bureaucracy, 2, 6, 35, 37, 40, 42, 53, 104, 173
Bush, George W., 30, 70, 72, 101, 114, 184, 185
Cambodia, 4, 15-17, 20, 22-24, 39, 66, 72-75, 86. *See also Indochina*
Cambodian Mine Action Center (CMAC), 74
Camp David Accords, 112
Canada, 4, 25, 26, 51, 62, 68, 69, 189
Capacity, 7, 9, 16, 56, 57, 73, 74, 77, 85, 97, 99, 106, 109-111, 118, 123, 147-156, 160, 164-169, 171, 182, 184
Catholic Church, 54
Cengic, Enes, 158
Centers for Disease Control and Prevention, 80
Central Intelligence Agency (CIA), 17, 42, 73, 75, 90, 101, 113
Chandrasekaran, Rajiv, 43
Chechnya, 28, 30, 57
Child Soldiers, 52, 186
China, 27, 31, 43, 71, 87, 185
Churches, , 7, 63, 113, 176

Civil Society, 20, 51-53, 60, 65-69, 72, 83, 88, 89, 100, 116, 166, 170-173, 179, 186
Clausewitz, 38, 40
Clinton, William Jefferson 26, 67-70, 184
Cluster Munition Convention, See *Oslo Convention*
Cluster Munitions, 4, 7, 11, 12, 15, 17, 19-22, 27, 29, 30-34, 38, 39, 43, 44, 47, 49, 52, 55, 56, 62, 65-68, 71, 72, 86-88, 115, 170, 171, 174, 176, 181-186, 189
Cold War, 8, 12-18, 21-28, 32, 33, 44, 45, 51, 72, 113
Collective Security, 10, 36, 39, 40, 52, 60, 64, 89, 112
Commercial Demining, 1, 5-7, 9, 11, 13, 23, 24, 29, 32, 33, 35, 41-49, 53, 55, 57-65, 72, 77-79, 81, 84, 85, 88-92, 96-106, 109-132, 136-156, 159, 161-183
Communism, 16, 22
Competition, 26, 37, 44, 48, 49, 58, 65, 77, 84, 98, 109, 117-124, 129, 137, 141-145, 153, 177, 185
Comprehensive Peace Agreement (CPA), 89, 108, 110-112, 115, 148, 159, 160, 170, 186
Conference on Disarmament, 28, 68, 69
Congo, Democratic Republic of, *See Zaire*
Contracting, 3, 6, 9, 29, 35, 44, 45, 48-53, 56, 58, 63, 65, 77-79, 81, 96, 97, 98, 103, 104, 109-111, 117-119, 123-125, 129-132, 137, 139, 140-142, 151, 156, 166, 178, 180-183
Convention on Conventional Weapons (CCW), 17, 21, 24, 26, 28, 30, 31, 66-72, 86, 176
  Amended Mines Protocol, 67, 68
  Protocol on Explosive Remnants of War, 30, 71
Corruption, 43, 50, 56, 103-105, 140, 148, 157, 158, 169
Cosmopolitanism, 9, 53, 59, 65, 89, 106, 107, 162, 163, 168, 178

Covert Action, 8, 15, 16, 22, 91, 113, 114, 159, 161, 184
Cranfield University, 77
Croats, 100-102, 157, 158, 167
Croatia, 83
'Crosslines' Demining (Sudan), 108, 111, 169, 170
DanChurchAid (DCA), 23, 108, 111, 124, 136, 164, 167, 170, 177
Danish Demining Group (DDG), 23, 99, 110, 167, 169
Darfur, 113, 115, 151, 160, 171
Dayton Peace Accords, 100-102, 105, 112, 148, 167
Decolonization, 4, 13
DECOP, 102, 158
Defense Security Cooperation Agency (DSCA), 79
Defense Systems Limited (DSL), *See ArmorGroup*
Demining Commission (Bosnia and Herzegovina), 100, 103, 104, 157, 158
Demobilization, 1, 24, 29, 148, 149, 151
DENEL, 29, 161
Denmark, 62
Density, *See Task Complexity*
Department of Defense, *See US Department of Defense*
Department of Education, *See US Department of Education*
Department of Peacekeeping Operations, *See UN Department of Peacekeeping Operations*
Deregulation, 44, 98, 124, 139, 141, 156, 159
Diana, Princess of Wales, 25, 171, 189
Disability, 52, 186
Disarmament, 17, 21, 25, 27, 64, 68, 164
Dogs, 5, 16, 73, 80, 90, 91, 97, 150, 156, 164, 190
'Doing no harm', 179, 180
Donais, Timothy, 50, 117, 145, 146, 157, 159, 180

Donors, 6-11, 14, 18, 19, 23, 27, 31, 35, 48, 53, 56, 58, 60, 61, 69, 70, 76, 78, 82, 83, 89, 90, 96, 101, 103, 108, 110, 114-116, 119, 124, 130, 140, 142, 144-148, 151-156, 161, 164, 165, 168, 169, 171-177, 180-183
Duffield, 6, 31, 34, 35, 40, 41, 51, 58, 159, 173, 174
Dyck, Lionel, 159
DynCorp, 29, 44, 45, 77, 81, 97, 98, 149, 150, 154, 170
Dyno-Nobel, 67
Easterly, William, 58
Edelmann, Stephen, 81
Eiffe, Dan, 114
Egeland, Jan, 7, 11, 61-66, 87, 176
Egypt, 20, 72
Eisenhower, Dwight D., 50
Ellis, John, 13
'Embeddedness', 7, 36, 40, 60, 146, 147, 154-156, 159, 162, 165, 171, 178
Enclavization, 31, 32, 35, 42, 43, 46, 53, 58, 59, 183, 187
Energoinvest, 102, 152
'English School' of Internation Relations, 39
Explosive Ordnance Disposal (EOD), 16, 97, 110
EOD Technology, Inc., 137-139
Equatorial Guinea, 50
Erinys, 45
Eritrea, 83
Ethics, 53, 63, 118, 123, 136, 137, 141-147, 177
Ethiopia, 25, 83
Ethnography, 10. *See also Anthropology*
European Union (EU), 101, 105
European Commission, 62, 110, 119
Executive Outcomes, 45
Expatriates, 43, 58, 85, 92, 99, 107, 153, 165, 166
FAFO, *See Institute of Applied Social Science*
Fair Trade, 53

Falklands/Malvinas, 28, 91
Federal Bureau of Investigation (FBI), 19
Federation of Bosnia and Herzegovina, 101, 102, 167
Financial Police (Bosnia), 103, 158
Finland, 85
Fortification, 43, 46, 183, 184, 187
FRELIMO, 159
Friedman, Milton, 44, 58
Friends World Committee for Consultation, 20
FSD, *See Swiss Foundation for Mine Action*
Gallimore, Andrew, 12
Garang, John, 114
Gender, 83, 93, 166, 167, 183. *See also Masculinity; Women Deminers*
General Assembly, *See UN General Assembly*
Geneva, 14, 20, 21, 24, 27, 50, 68, 84, 92, 108, 119, 140, 189
Geneva Accords, 92
Geneva Call, 27, 108
Geneva Conventions, 14, 20, 24
  Additional Protocols, 20
Geneva International Center for Humanitarian Demining (GICHD), 50, 84, 189
Geneva Protocol, 14, 68
Georgia, 28, 181, 185
Germany, 13, 85
Globalization, 5, 25, 10, 39-41, 45, 173
Goodhand, Jonathan, 146
Goose, Steve, 186
Governance, 1, 5, 6, 10, 13, 31-33, 34-60, 148, 149, 173-175, 183-187
Gradasovic, Adnan, 158
Gradasovic, Mirzad, 158
Grants, 7, 9, 11, 44, 53, 56, 65, 77, 78, 84, 96, 98, 102, 110, 115, 117-119, 129, 121, 123, 125, 130, 133, 141, 142, 154, 145, 162, 166, 177, 178, 180, 182, 186
'Great Game', The, 14

Great Powers, 6, 13, 14, 21, 22, 26-28, 30, 31, 33, 35, 38, 39, 41, 51, 57, 61, 64, 71, 87, 93, 174, 176, 185, 186
Griffiths, Hugh, 50, 104, 159
Group 4 Securicor, 81
Guantanamo Bay, 66
Guatemala, 54
Guerilla Warfare, 3, 15, 16, 86
Gulf War, 28, 29, 66
Hagen, Egil, 114
Hague Regulations, 14
Halliburton, 44
HALO Trust, 8, 23, 24, 57, 77, 83, 84, 91-93, 99, 100, 110, 111, 120, 124, 125, 136, 150-153, 160, 163, 164-169, 177, 190, 193
Handicap International, 18, 91, 120, 164
'Hard Power,' 59, 186
Hardin, Garrett, 46
Hayek, Friedrich, 44
HELP, 23
'Herzeg-Bosna', 8, 100, 157
Hezbollah, 28, 32
HIV/AIDS, 40, 123
'Hollow State', The, 44
Homayoun, Farid, 91, 100
Honduras, 22, 67, 73, 80
Honeywell, 19, 66
Human Rights, 7, 25, 26, 33, 39, 50, 51, 52, 57, 60, 61, 63, 64, 66, 81, 101, 113, 114, 168
Human Rights Watch, 24, 29, 67, 161
Human Security, 6, 11, 24, 32, 33, 35, 36, 42, 50, 51, 52, 53, 54, 55, 56, 57, 58, 59, 60, 62, 63, 87, 88, 116, 142, 145, 147, 148, 152, 162, 163, 164, 166, 172-175, 186
Human Security-Civil Society Complexes, 6, 11, 32, 33, 35, 36, 42, 50-63, 88, 116, 142, 145-148, 162-166, 172-175, 186
Humanitarian Demining Research and Development Program, 80

INDEX

Humanitarian Demining Training Center, 79
Humanitarian Intervention, 51, 52, 60, 74
Humanitarian Law, 20, 21, 24, 39, 68, 71, 72
Humanitarianism, 4, 6-9, 11, 14-29, 32, 33, 35, 36, 39, 41, 42, 44-55, 59, 62, 64, 65, 68, 70-92, 96, 99, 100, 105, 106, 109, 111, 113-119, 124, 125, 140, 142, 145, 147, 151, 152, 160-166, 170-177, 180, 181, 185
Hutson, Dean, 96, 98
'Ideal Types', 6, 35, 41, 59, 60, 62, 80,
Idealism, 2, 36-39, 50, 53, 54, 57, 60, 84, 176
Ideals, 6, 7, 35, 41, 45, 53, 59, 60-62, 64, 80, 83, 88, 113, 162, 172, 175-178
IED, *See Improvised Explosive Device*
Ilic, Radoslav, 157, 158, 162
Impartiality, 91, 149
Improvised Explosive Device, 3, 22, 27, 28, 30
Incentives, 17, 26, 48, 56, 59, 69, 105, 119, 137, 141, 142, 152, 153, 156, 172, 176, 181-183
Inclusivity, 7, 147, 166-168, 172, 174, 183, 184, 185
India, 4, 27, 31, 43, 185
Indochina, 10, 13-19, 66, 72, 97. *See also Cambodia; Laos; Vietnam*
Infrastructure, 80, 90, 98, 102, 110, 124, 148, 151, 163, 183
Institute of Applied Social Science (FAFO), 84
Intelligence, 28, 38, 42, 45, 64, 105, 114, 161
Intelligent Munitions System (IMS), 71
Interests, 2, 6-9, 11, 13-17, 23, 24, 26, 30, 32, 35-39, 42, 47, 48, 50-53, 55, 58-66, 72-76, 83, 87-93, 96, 97, 99, 103-105, 112-116, 146, 147, 152-155, 157, 161-172, 174-177, 187

International Campaign to Ban Landmines (ICBL), x, 18, 24-28, 30, 53, 68-70, 84, 85, 91
International Committee of the Red Cross (ICRC), x, 15, 20, 21, 25, 26, 84, 91, 185
International Criminal Court, 52, 53, 68, 160, 186
International Criminal Tribunal for the Former Yugoslavia, 157, 158, 164
International Mine Action Standards (IMAS), 108
International Security Assistance Force (ISAF), 99
International Trust Fund for Demining and Mine Victim Assistance (ITF), 75, 78, 81, 102, 103, 119, 121, 130, 152, 156, 158
International War Crimes Tribunal (for Vietnam), 19
Inter-Service Intelligence (ISI) (Pakistan), 90
Iran, 27, 31, 87
Iraq, 3, 8, 9, 24, 27-30, 43, 45, 50, 60, 65, 66, 75-79, 81, 83, 86, 93, 98, 102, 115, 161, 181, 186
Israel, 27, 28, 30-35, 43, 51, 71, 185
James Madison University, 77
JASMAR, 108, 110, 111, 148, 169, 170
Jihad, 93, 168
Joint Integrated Demining Units (JIDU), 109
Joint Military Commission (JMC), 108, 110, 115, 148
Journal of Mine Action, 77, 179
Juba, 10, 110, 111, 193
Kabul, 10, 91, 92, 99, 118, 121, 129, 137, 138, 149, 150, 166, 168, 189
  Kabul International Airport, 118, 121, 137, 138, 188, 189
Kaldor, Mary, 51
Kandahar, 97, 150, 162
Kant, Immanuel, 39
Karadzic, Radovan, 103, 104, 157, 159, 162

Kellogg, Brown and Root (KBR), 44
Kenya, 184
Khartoum, 10
Kidd, Richard, 77
Kojic, Radomir, 104, 157-159, 162
Korea, 15, 26, 66, 67, 69, 70. *See also North Korea, South Korea*
Kosovo, 23, 24, 27, 28, 29, 30, 60, 66, 82, 86, 93, 98, 109, 185
Kozul, Sandi, 158
Kuwait, 28, 29, 161
Landmine Impact Surveys (LIS), 107, 111, 125, 163, 164
Landmine Monitor, 4, 28
Landmine Survivors Network, 23, 84, 171
Laos, 15-24, 66, 75, 79. *See also Indochina*
Lawlessness, 34, 150, 182. *See also Rule of Law*
Leahy, Patrick J., 67-71, 80, 87. *See also Patrick J. Leahy War Victims Fund*
Lebanon, 8, 28, 30, 32, 71, 75, 77, 78, 83, 86, 112, 181
Liberalization, 44, 50, 98, 102, 156
'Lifeboat Ethics', 46
Lisica, Darvin, 107, 118, 148
Litvinenko, Alexander, 185
Local Ownership, 85, 99, 154
Louis Berger Group, Inc., *See Berger Group.*
Lord Curzon, 34
Lucerne, 20, 22
Lugano, 20
Lutheran Church, 63
Machiavelli, Niccolo, 37
Malvinas, *See Falklands/Malvinas*
Management, 2, 13, 28, 43, 44, 48, 58, 65, 96, 102, 109, 110, 129, 132, 139, 140, 165, 179
Mandela, Nelson, 25
Masculinity, 169, 183. *See also Gender*
Maslen, Stuart, 79
Markets, 24, 43, 45, 46, 81, 117, 124, 136, 137, 141, 142, 149, 151, 153, 163, 178
Mazar-i-Sharif, 99

McGrath, Rae, 91
Mechanical Demining, 4-5, 105, 121, 137
MECHEM, 29, 120, 159, 161
MEDECOM, 104, 157, 158
Mennonite Central Committee (MCC), 18, 19, 23, 24
Menschen gegen Minen, 23
Metal Detectors, 5, 17, 18, 90, 189
Mexico, 20
Middle Powers, 6, 13, 18, 19, 21, 22, 25, 26, 28, 30, 32, 33, 35, 41, 51, 54, 55, 57, 59, 60, 69, 174, 176, 185, 186
Militarism, 29, 32, 33, 173, 185, 186
Militarization, 6, 35, 36, 81, 83, 97, 112, 115, 116, 162, 167, 184
Military and Professional Resources, Inc. (MPRI), 45
Milosevic, Slobadan, 81
Mine Action Information Center, 77
Mine Action Program for Afghanistan, (MAPA), 90-93, 98, 100
Mine Awareness, *See Mine Risk Education.*
Mine Ban Treaty, *See Ottawa Treaty*
Mine Clearance and Planning Agency, (MCPA), 152, 162
Mine Dog Center (MDC) (Afghanistan), 91, 92, 100, 162, 190
Mine Protection and Removal Agency (MPRA), 100, 158
Mine Risk Education, 27, 73, 74, 79, 91, 93, 107, 110, 123
Mine Tech, 102, 120, 124, 136, 151
Mines Advisory Group (MAG), 23, 77, 83, 84, 91, 110, 111, 119, 164, 167, 169, 170
Mitchell, Colin, 91
Molam, Paul, 124
Money Laundering, 157
Monin, Lydia, 12
Mostar, 106, 163
Mozambique, 23, 29, 39, 73, 74, 83, 86, 120, 121, 153, 159, 164
Mustard Seeds, 5
Myrdal, Alva, 21

Myanmar, *See Burma*
Nairobi, 184
National Mine Action Authority (NMAA) (Sudan), 160
National Security, 10, 17, 22, 24, 32, 36, 38, 52, 57, 60, 75, 87, 89, 116, 147, 152, 179, 185
National Security Council, 75
Natsios, Andrew, 65
'Neo-Medievalism', 35, 54
'Neo-Trusteeship', 74
Netherlands, 62
Neutrality, 8, 39, 52, 86, 91-93, 100, 114, 149, 150, 156, 162, 166, 186
'New Wars,' 4, 10, 13, 22, 23, 28, 36, 40, 41, 45, 53
NGOs, 1, 6-9, 11-14, 18-25, 28, 29, 32, 33, 35, 39, 41, 53-65, 69, 73, 74, 77, 78, 83-102, 106-126, 129-132, 136-179, 182, 185-187
Nicaragua, 22, 67, 72, 73, 80, 112
No Man's Land, 1
Nobel Peace Prize, 28, 63
Non-Aligned Movement, 20
Nonproliferation, Antiterrorism, Demining and Related Projects (NADR), 75, 77, 78
Nonviolence, 146, 162, 168, 171
Norms, 2, 6, 7, 12, 20, 21, 22, 24, 26, 27, 30, 33, 35-39, 45, 49-52, 61-65, 68, 71, 82, 106, 142, 174, 176, 178, 181, 185, 186
North Atlantic Treaty Organization (NATO), 6, 8, 20, 28, 30, 67, 68, 72, 73, 79, 98, 99, 101, 105, 112, 115, 137, 148, 150, 156, 157, 162, 168, 185
North Korea, 66. *See also Korea*
Northern Alliance, 93, 168
Norway, 7-9, 11, 18, 20, 22, 23, 26, 31, 51, 54, 60, 61-74, 77, 82-96, 99, 100, 101, 105-115, 119, 147, 156, 162, 164-168, 171, 175-178, 182
Norwegian Labor Party, 85

Norwegian Ministry of Foreign Affairs (MFA), 65, 71, 82, 83, 84, 106
'Norwegian Model' of Foreign Affairs, 51, 65, 88
Norwegian People's Aid (NPA), 9, 23, 74, 77, 83-88, 105-107, 111, 114, 119-125, 130, 132, 136, 140, 148, 156, 163-165, 167, 169, 171, 177, 190, 191
Norwegian Red Cross, 84
Nuba Mountains, 9, 108, 110, 111, 115, 124, 148, 160, 164, 169, 170
Night Vision and Electronic Sensors Directorate (NVESD) (USA), 80
Obama, Barack, 187
OCHA, *See UN Office for the Coordination of Humanitarian Affairs*
Office of the High Representative (Bosnia), 103, 104, 158
OKTOL, *See SI/OKTOL*
'Old War', 36-40, 60
OMAR Mine Museum, 189
Operation Enduring Freedom, 81, 99
Operation Lifeline Sudan (OLS), 114
Operation Restore Hope, 73
Operation Salam, 23, 92
Operation Save Innocent Lives (OSIL), 107, 108, 110, 111, 148, 170
Opium. *See Poppy Eradication*
Organization for Economic Cooperation and Development, (OECD), 64, 65
Organization for Mine Clearance and Afghan Rehabilitation (OMAR), 118, 121
Organization of American States (OAS), 77, 84
Oslo Convention, 4, 31, 72.
Oslo Process, 31, 66, 71.
Ottawa Convention, 1, 4, 22, 25, 27, 30, 29, 66, 69, 70, 71, 76, 82, 83, 85, 106-108, 170, 186.
Ottawa Process, 26, 30, 68, 69, 71, 176.
Ottawa Treaty, *See Ottawa Convention*
Oxfam, 18

Pakistan, 22, 27, 30, 31, 43, 73, 90, 91, 166, 185
Pale, 157
Palestine, 32, 34, 35
Palestinian Liberation Organization (PLO), 86
Participation, 85, 93, 107, 148, 179, 181
Pashtuns, 166
Patrick J. Leahy War Victims Fund, 80, 87
Patronage, 157, 159, 169. *See also Corruption*
Peace Research Institute, Oslo (PRIO), 84
Peacebuilding, 6, 39, 54, 57, 60, 73, 85, 86, 106, 110, 124, 146, 167, 169, 171, 175, 185
Peacekeeping, 1, 9, 29, 39, 44, 54, 73, 79, 101, 105, 109-112, 115, 123, 151, 152, 162
Pentagon, *See US Department of Defense*
'Peripheries', 14, 16, 18, 23, 32, 51-54, 57, 59, 186, 187
Personal Protective Equipment (PPE), 137, 138
Petrisch, Wolfgang, 104, 158
Physicians for Human Rights, 67
Political Economy of War, 3, 6, 7, 57, 146, 147, 154, 155, 159, 161, 162, 165-168, 174, 175, 179, 180, 185
Poppy Eradication, 97, 156, 161.
Post-Conflict Reconstruction, 1, 178. *See also Reconstruction*
Power, Samantha, 105
Post-Modern Warfare, 13, 28, 29
Price of Demining, 117-120, 124, 125, 129, 130, 137, 140, 141, 175, 182
'Principal-Agent Model', 36, 48, 55, 59, 65, 77, 118, 142, 181
'Principal-Steward Model', 55, 56, 65, 77, 84, 118, 141, 142, 144, 179, 181, 182
Prisoners of War (POWs), 14
Private Security, 6, 29, 33, 34, 42-46, 50, 77, 81, 87, 116, 149-151, 156, 161, 169, 174, 175, 179, 180, 183, 185
Privatization, 3, 6, 11, 29, 39, 41, 43, 44, 48-50, 53, 58, 63-65, 87, 98, 102, 124, 129, 130, 146-150, 156-159, 161,172, 174, 175, 183
Productivity of Demining, 44, 111, 119, 122, 140. *See also Speed of Demining*
Profit, 24, 32, 36, 48, 53, 83-85, 119, 121, 123, 124, 136, 137, 142, 150, 152, 158, 174, 175, 177
PROVITA, 102, 158
Proxy Wars, 14, 22, 23, 72
Psychological Operations (PYSOPS), 79
Public-Private Partnerships, 5, 47
Pusic, Berislav, 157, 158
Quality of Demining, 6, 7, 11, 36, 42, 44, 48, 49, 56, 100, 108, 109, 117, 129, 130, 132, 136, 137, 139, 140-142, 144, 145, 151, 152, 157, 174, 175, 177, 178, 180, 181. *See also 'Race to the Bottom'*
Quick Reaction Demining Force (QRDF), 110
'Race to the Bottom', 27, 69, 118, 137, 139, 141, 142, 144, 180, 181. *See also Quality of Demining*
Radar, 5, 8
Rats, 5
Razac, Olivier, 13
Reagan, Ronald, 22, 72
'Reagan Doctrine', 22, 72
Realism, 2, 36-39, 42, 47, 60, 185
Reconstruction, 6, 9, 11, 49, 50, 74, 85, 96, 97, 102, 104, 106, 110, 124, 150, 152, 155, 166, 172, 183, 185. *See also Post-Conflict Reconstruction*
Refugee Convention, 106
Rent-Seeking, 43, 58, 170
Republika Srpska, 8, 100, 101, 102, 157, 167
Rice, Condoleezza, 38
Rimli, Lisa, 150
Robots, 5
RONCO Consulting Corporation, 29, 31, 73, 77, 79, 81, 85, 88, 90, 91, 97, 100-

104, 108, 110, 111, 115, 120, 121, 124, 125, 136, 137, 148, 150, 153, 155, 156, 158, 162, 170
Roshan, 98, 151
Rowe, David, 119, 159
Royal Ordnance, 29
Rubin, Barnett 113
Rule of Law, 38, 49, 53, 59, 61, 139, 182, 183, 187. *See also Lawlessness*
Russell, Bertrand, 19
Russia, 13, 27, 28, 31, 37,43, 57, 71, 87, 185, 188. *See also Union of Soviet Socialist Republics*
Ruru, Kerei, 152
S3Ag, 118, 121
Saddam Hussein, 28, 161
Safety of Demining, 5, 6, 11, 49, 53, 61, 93, 106, 111, 117, 129, 130, 132, 135-138, 140-142, 144, 145, 157, 163, 169, 174, 175, 177, 180, 181
Sandinistas, 86
Sandline International, 45
Sarajevo, 5, 10, 64, 106, 112, 123, 157, 158, 163, 164, 191
Sartre, Jean-Paul, 19
Saudi Arabia, 27
Security, 1-3, 6-9, 11, 14, 17, 20, 24, 26, 28-60, 63, 66, 74, 81, 93, 98-101, 104, 105, 108, 112-116, 123, 139, 145-156, 161, 162, 164, 167, 171-175, 181, 184-187
Schmeidl, Susanne, 150
School of the Americas, 73
Shafer, D. Michael, 154
Securitization, 9, 63, 65, 83, 115, 185
Security, 3, 28, 35, 37, 51, 52, 54, 60, 93, 147, 161, 184
Security Council, *See UN Security Council*
'Security Dilemma', 37
Sediq, Mohammed, 149, 168
September 11, *See 9/11*
Serbs, 100, 101, 102, 103, 157
Stabilization Force (SFOR), 102
SI/OKTOL, 102, 103, 150, 158

Sierra Leone, 28
Slovakia, 62
'Smart Mines', 66
Sociology, 86, 179
'Soft Power', 51, 59, 64, 186
Somalia, 27, 69, 73
South Africa, 25, 29, 43, 44, 64, 66, 159
South Korea, 66, 154
South Sudan, 109-111, 115, 149, 152, 160, 161, 166, 167. *See also Sudan*
South Sudan Demining Commission (SSDC), 160, 167
Soviet Union, *See Union of Soviet Socialist Republics*
Special Forces, 73, 79, 99, 101, 156
Speed of Demining, 5, 49, 117, 121-125, 132, 136, 137, 141, 144, 145, 159, 174, 175. *See also Productivity of Demining*
Spider Networked Munitions System, 70
Sri Lanka, 54, 78, 83, 86
Standing Operating Procedures (SOPs), 131, 136
State Department, *See US Department of State*
Stewardship, *See 'Principal-Steward Model'*
Stigler, George, 44
Stockholm International Peace Research Institute (SIPRI), 20
STOP Mines, 102
Støre, Jonas Gahr, 61
Strategic Provincial Roads, 98
Strategic-Commercial Complexes, 6, 11, 32, 35, 36, 41-55, 59, 62, 78, 80, 87, 116, 142, 145-147, 149, 151, 153, 154, 160-163, 171, 174, 175, 177
Subcontracting, 97, 98, 110, 158, 159
Sudan, 7-9, 11, 20, 35, 54, 75, 86, 89, 107-119, 122-125, 128, 130, 132, 137, 140, 146, 148, 150-153, 159-161, 164-173, 178, 186, 193
Sudan Campaign to Ban Landmines (SCBL), 107, 171

Sudan Integrated Mine Action Service (SIMAS), 107
Sudan Landmine Information and Response Initiative (SLIRI), 108
Sudan Landmine Response (SLR), 151, 160, 169, 193
Sudan People's Liberation Army (SPLA), 27, 86, 107-109, 111, 113, 114, 115, 148, 160, 164, 167, 169, 170, 171
Sudanese Association for Combating Landmines, *See JASMAR*
Sudanese Red Crescent Society, 107
Suez Canal, 72
Sun Tzu, 37
Sweden, 20, 21, 68
Swiss Foundation for Mine Action (FSD), 23, 109, 110, 119
Switzerland, 20, 108
Taliban, 57, 93, 98, 113, 162, 168, 186
Tamil Tigers, 86
Tanovic, Danis, 1
Task Complexity of Demining, 106, 117, 126-128, 141-144
Technocracy, 2, 49, 58, 82, 154, 160, 179, 180, 183
Tendering, 7, 9, 11, 29, 44, 56, 48-50, 65, 77, 78, 84, 89, 102, 103, 105, 109, 116-119, 121, 123, 130-133, 142, 144, 149, 153, 177, 178, 181, 182. *See also* Contracting
Terrorism, 20, 40, 42, 45, 51, 57, 63, 79, 93, 97, 113, 162. *See also 'War on Terror'*
Thailand, 16, 22, 73, 90
The Development Initiative, 152
Triple Canopy, 45
Tromsø Trauma Care Foundation, 83
Tuzla, 106
United Kingdom of Great Britain and Northern Ireland (UK), 25, 45, 51, 151, 152
United Nations, 1, 5, 6, 8, 9, 16, 17, 20, 21, 23, 26, 29, 30, 32, 33, 35, 39, 52, 53, 54, 65, 67, 68, 69, 73, 74, 79, 83-86, 89-93, 96, 97, 99, 100, 105, 108-110, 113-118, 125, 140, 146, 148, 151, 152, 159-161, 166, 168, 170, 174, 175, 178, 181, 185. *See also World Food Program; World Health Organization*
UN Children's Fund (UNICEF), 85, 163, 178
UN Department of Peacekeeping Operations, 23, 29, 33, 83, 96, 109, 115
UN Development Programme (UNDP), 29, 83, 84, 109, 148, 152
UN Environment Programme (UNEP), 17
UN General Assembly, 20, 23, 52, 67
UN High Commissioner for Refugees (UNHCR), 74, 110, 167, 169
UN Mine Action Center (UNMAC) (Bosnia), 100, 106, 125, 136, 157
UN Mine Action Center for Afghanistan (UNMACA), 92, 96, 99, 100, 110, 129, 136, 139, 149, 152, 162, 163, 168
UN Mine Action Office (UNMAO) (Sudan), 108, 109, 110, 123, 140, 152, 153, 160
UN Mine Action Service (UNMAS), 23, 29, 79, 96, 108, 109, 159, 178, 181
UN Mission in Sudan (UNMIS), 108, 109, 111
UN Office for Project Services (UNOPS), 29, 96, 109, 160, 181
UN Office for the Coordination of Humanitarian Affairs, (OCHA), 29, 83, 92, 93, 96, 109
UN Office in Mozambique (UNOMOZ), 74
UN Security Council, 28, 51, 161
UN Transitional Administration in Cambodia (UNTAC), 74
Unexploded Ordnance (UXO), 1, 3-6, 8-19, 21, 23-25, 28, 29, 31, 32, 36, 41, 72,

74, 87, 88, 91, 97, 100, 107, 108, 110, 118, 125, 174, 175, 178, 181, 186

Unilateralism, 67, 69-72, 107, 184, 185

Union of Soviet Socialist Republics (USSR), 4, 14, 16, 17, 21, 22, 24, 44, 67, 73, 90, 92, 97, 102, 168, 185. *See also Russia*

UNIPAK Demining, 102, 137, 157, 159, 191

United Arab Emirates, 62

United States of America, 3-5, 7-9, 11, 13, 15-32, 37, 38, 42-45, 50, 51, 60-120, 124, 137, 139, 146, 148, 150-153, 156, 158, 159, 161, 162, 164, 167, 170, 171, 175-178, 181, 184-186

   Civil War, 4, 13

   US Africa Command (AFRICOM), 115

   US Agency for International Development (USAID), 9, 29, 44, 64, 73, 74, 76, 80, 81, 90, 96-98, 102, 109-111, 114, 115, 148, 150, 152, 153, 161. *See also Patrick J. Leahy War Victims Fund*

      Cross-Border Humanitarian Assistance Program for Afghanistan, 90-92, 161

   US Air Force, 16, 118, 137

   US Army, 16, 31, 73, 80, 97, 124, 148, 152

   US Commerce Department, 63

   US Department of Agriculture, 80

   US Department of Defense, 19, 24, 66, 69, 73-75, 79, 80, 87, 90

   US Department of Education, 80

   US Marine Corps, 16

   US Department of State, 9, 63, 64, 69, 72, 73, 76-79, 81, 90, 97, 98, 102-105, 109-113, 152, 153, 162, 164, 167, 170

      Office for Weapons Removal and Abatement (WRA), 76-78, 81, 87, 170

UXB International, 81, 97, 102, 118, 121, 137, 152, 156

van Zyl, Johannes 'Sakkie', 159

Victim Assistance, 1, 78, 80, 103, 107

Vientiane Peace Agreement, 17

Viet Cong, 15, 16, 159

Vietnam, 4, 15-20, 23, 42, 66, 67, 75, 79. *See also Indochina*

Vietnam Veterans of America Foundation (VVAF), 23, 67

'Village Deminers,' 16

Wackenhut Services Incorporated, 81

War Crimes, 19, 52, 103-105, 156, 157, 161, 164, 180

'War on Terror', 13, 28-30, 33, 45, 63, 65, 96, 114, 170, 179. *See also Terrorism*

War Victims Fund, *See Patrick J. Leahy War Victims Fund*

Wareham, Mary, 186

Warlordism, 40, 93, 97, 155, 161, 173

Weapons Removal and Abatement (WRA), *See US Department of State Office for Weapons Removal and Abatement*

Weber, Max, 6, 42, 60, 87, 88, 173

Weizman, Eyal, 34, 35

Willoughby, Guy, 91, 125

Williams, Jody, 25, 28, 173

Women, *See Gender, Women Deminers*

Women Deminers, 167, 183

World Bank, 50, 99, 102-105, 152, 156, 157, 158, 178

World Food Program (WFP), 108-110, 119, 160

World Health Organization (WHO), 17

World Vision, 91

World War I, 13, 14, 38

World War II, 4-6, 13-16, 18, 24, 28, 31, 32, 37, 38, 40, 66

Yemen, 79

Yugoslavia, 20

Zaire, 22

Zambia, 154

Zimbabwe, 124, 152, 153, 159